Physical Methods in Modern Chemical Analysis

Volume 1

PHYSICAL METHODS IN MODERN CHEMICAL ANALYSIS

Edited by

THEODORE KUWANA

Department of Chemistry
The Ohio State University
Columbus, Ohio

Volume 1

ACADEMIC PRESS New York San Francisco London 1978

A Subsidiary of Harcourt Brace Jovanovich, Publishers

ACADEMIC PRESS, INC.
111 Fifth Avenue, New York, New York 10003

United Kingdom Edition published by
ACADEMIC PRESS, INC. (LONDON) LTD.
24/28 Oval Road, London NW1 7DX

Library of Congress Cataloging in Publication Data

Main entry under title:

Physical methods in modern chemical analysis.

Includes bibliographies.
1. Chemistry, Analytic. I. Kuwana, Theodore.
QD75.2.P49 543 77-92242
ISBN 0-12-430801-5 (v. 1)

PRINTED IN THE UNITED STATES OF AMERICA

82 83 9 8 7 6 5 4 3 2

Contents

List of Contributors

Numbers in parentheses indicate the pages on which the authors' contributions begin.

Bruce N. Colby (57), Systems, Science and Software, P.O. Box 1620, La Jolla, California 92038

Catherine Fenselau (103), Department of Pharmacology and Experimental Therapeutics, The Johns Hopkins University, School of Medicine, Baltimore, Maryland 21205

Peter N. Keliher (255), Chemistry Department, Villanova University, Villanova, Pennsylvania 19085

Judy P. Okamura (1), Department of Chemistry, San Bernardino Valley College, San Bernardino, California 92403

Donald T. Sawyer (1), Department of Chemistry, University of California, Riverside, California 92521

Thomas J. Vickers (189), Department of Chemistry, Florida State University, Tallahassee, Florida 32306

Preface

The practitioners of chemistry today are faced with a multitude of increasingly complex problems concerned with *chemical analysis*. They are, for example, requested to find and identify trace amounts of materials in complex mixtures. Moreover, trace quantities, rather than being in the microgram range as thought of several years ago, are now being extended below the nanogram level to femtograms. The problem of identification is also nontrivial, extending from organic and inorganic compounds in various matrices to complex biological macromolecules. New tools often associated with sophisticated instrumentation are also constantly being introduced. Surface analysis is a good example of an area for which recent years have seen the advent of many new methods, and the abbreviations ESCA, SIMS, XPS, LEEDS, etc. are now common in the literature. These methods have made it possible to analyze and characterize less than monolayers on solid surfaces. Thus the demand upon a practicing chemist is to have a working knowledge of a wide variety of physical methods of chemical analysis, both old and new: the new ones as they are developed and applied, and the old ones as they are better understood and extended. It is the aim of "Physical Methods in Modern Chemical Analysis" to present a description of selected methodologies at a level appropriate to those who wish to expand their working knowledge of today's methods and for those who wish to update their background. It should also be useful to graduate students in obtaining a basic overview of a wide variety of techniques at a greater depth than that available from textbooks on instrumental methods.

"Physical Methods in Modern Chemical Analysis" will contain chapters written by outstanding specialists who have an intimate working knowledge of their subject. The chapters will contain descriptions of the fundamental principles, the instrumentation or necessary equipment, and applications that demonstrate the scope of the methodology. The chapters are not written as a review in which the description and evaluation of the method are incomplete.

It is hoped that these volumes continue the standard exemplified by the earlier volumes, "Physical Methods in Chemical Analysis," edited by Walter Berl in the 1950s and 1960s.

Volume 1 contains chapters on gas chromatography (Okamura and Sawyer); principles and instrumentation of mass spectrometry (Colby); its applications, scope, and structural problems (Fenselau); fluorescence and atomic absorption spectroscopy (Vickers); and flame and plasma emission methods of analysis (Keliher). That mass spectroscopy (MS) chapters are preceded by a discussion of gas chromatography (GC) seems appropriate in view of the importance of GC to MS. The methods of analytical atomic spectrometry are also treated in two complementary chapters. Future volumes will contain chapters on x-ray analysis, x-ray photoelectron spectroscopy, Auger spectroscopy, high performance liquid chromatography, photoacoustical spectroscopy, refractive index measurements, laser resonance spectroscopy, magnetic circular dichroism, ion cyclotron resonance, and other methods.

The patience and assistance of my wife Jane during the editing process are gratefully acknowledged.

Gas Chromatography

Judy P. Okamura

Department of Chemistry
San Bernardino Valley College
San Bernardino, California

Donald T. Sawyer

Department of Chemistry
University of California
Riverside, California

ISBN 0-12-430801-5

I. Introduction to Gas Chromatography

A. Descriptions

1. *Gas Chromatography*

Gas chromatography is a technique that provides the means to accomplish the efficient separation and sensitive analytical determination of the components in complex mixtures. The relative residence time of a given component in a column system is an identification characteristic, while the magnitude (area) of the detector response for the component as it emerges from the column is proportional to its concentration. Even when gas chromatography is done under much less than optimum conditions, results are achieved that would require a great amount of work and finesse by other methods. Still better results, however, can be obtained when the theory is applied in the laboratory to the wide variety of practical problems that must be met. Because the theory can be complex, the chemist in the laboratory is often required to use an empirical approach. The goal of this chapter is to outline practical methods to achieve desired separations, which have been developed from the basic theory of gas chromatography.

2. *The Gas Chromatograph*

The term "gas chromatography" refers in reality to only a small portion of what we think of as a gas chromatograph, i.e., the process of partitioning the sample between the mobile gas phase and another material, liquid or solid, which has some attraction for the sample. However, when we speak of a gas chromatograph we envision an entire apparatus. Figure 1 illustrates a simple gas chromatograph.

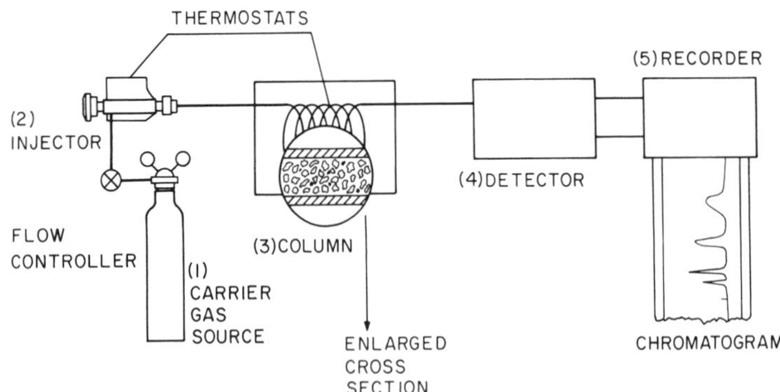

Fig. 1 Components of a gas chromatographic system.

A fundamental element of the gas chromatographic system is a source of carrier gas (1). This normally is H_2, He, or N_2, and is brought from the pressurized gas cylinder through a pressure regulator to an injector (2). The injector is the means of introducing the sample into the carrier gas stream as a gas. The simplest system is a "T" in the line with one "arm" closed with a rubber septum. The sample is injected with a syringe through the septum into the moving gas stream. If the sample is gaseous itself, there is no problem. However, if the sample is a liquid, the injector (and the rest of the system) must be heated so that the sample is volatilized rapidly and remains gaseous until leaving the detector. There are more sophisticated injector systems for samples which present special problems, such as being a solid.

The sample is transported by the carrier gas to the column (3), which is where the actual separation takes place. The various components of the sample are retained by the column for differing lengths of time and therefore are separated from one another. A component of the sample which is not retained by the column will travel through as fast as the carrier gas and will emerge first, while others will emerge at later times. Because temperature control of the column normally is necessary we have illustrated (Fig. 1) the column in a simple bath which could contain a boiling liquid or it could be a Dewar flask containing ice or liquid nitrogen slushes to maintain a subambient temperature. Most commercial instruments achieve temperature control by placing the column in an oven with precise control of its temperature.

The problem, after the components of the sample have been separated, is to detect them as they emerge from the column still mixed with carrier gas. Ideally, the detection system should yield a quantitative response for the sample components. The gas stream from the column normally goes directly to the detector (4). The variety of detection methods is enormous and ranges from the analyst smelling the outlet gases to determine when an odoriferous component emerges to the use of a mass spectrometer to provide almost immediate qualitative identification. However, there are three types of detector which have widespread practical application and are easy to use. These are the thermal conductivity detector, the flame ionization detector, and the electron capture detector, and they will be discussed in detail in a later section. When a gas chromatograph is purchased, one of these detectors along with its associated electronics normally is a part of the system together with the injector, the oven compartment for the column, and the heaters for the detector and injector; most instruments are packaged as a single unit. The electronic signals from the detector usually are connected to a strip-chart potentiometric recorder (5). The position of the recorder pen fluctuates with the amount of sample in the detector and yields a tracing like that illustrated in Fig. 2.

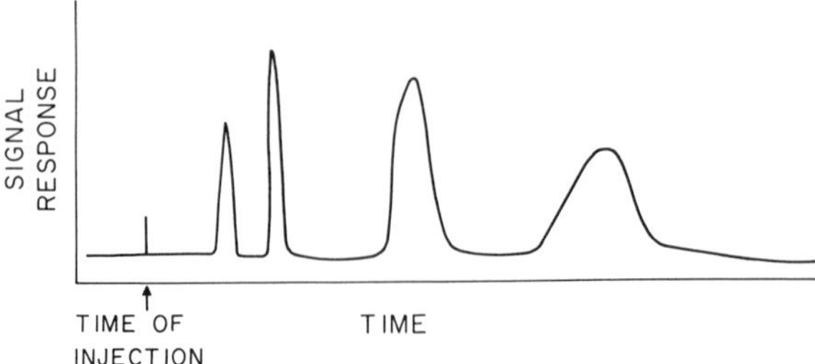

Fig. 2 Recorder trace of four separated sample components that have been eluted from a gas chromatographic column.

The separation that occurs in the column makes gas chromatography a powerful technique. However, the availability of sensitive, quantitative detection systems for almost all types of molecules has been a vital factor to its widespread use.

3. *The Column*

Let us return to that mysterious part which actually does all the work, the column. Commercial gas chromatographic instruments cost from \$1500 to \$10,000, yet the essential work is done in something that can be fabricated for about \$5, or obtained commercially for about \$35. The column usually consists of a piece of metal tubing that is about 6 ft long with an internal diameter from $\frac{1}{8}$ to $\frac{1}{4}$ in. (2 to 4 mm), which contains either small particles of solid adsorbent (gas–solid chromatography) or a viscous liquid coated on small solid particles, which are then termed the support (gas–liquid chromatography). Although gas chromatographic separations were first achieved by gas–solid chromatography (and it remains the only method by which some types of separations can be made), the majority of present-day analyses are done by gas–liquid chromatography.

A column is simple to prepare, but the preparation of a highly efficient one is an art. For instance, to prepare a column of 10% Carbowax 20M on 100/120 mesh Chromosorb W, 1 g of the liquid phase (the Carbowax 20M) is dissolved in a volatile solvent (in this case, chloroform or methylene chloride). Next, 9 g of 100/120-mesh Chromosorb W are added and mixed with the solution of liquid phase plus solvent. The solvent is then removed by evaporation, using a rotary evaporator, to give Chromosorb W with a uniform coating of the liquid. One end of a metal column is then plugged with glass wool and the coated support is added slowly, with tapping, to the column. When the column is full, another plug of glass wool is added to hold in the packing. The column normally is "conditioned" by flowing carrier gas through it with the temperature at 25°C above the maximum temperature to be used in the analytical separations. This is done with the exit end *disconnected* from the detector and serves the purpose of removing solvent or any other material which might contaminate the detector during use of the column.

B. Theory

The fact that a column will retain one compound longer than another must be investigated in more depth to provide an understanding of the controlling factors in the separation process. This understanding is necessary in order to optimize these factors for a particular analysis. In Fig. 3, the symbols △ and ● represent molecules of two different components of a sample. The △ molecules have a partition coefficient K, which is high and favors absorption by the liquid phase. The ● molecules have a low partition coefficient which favors the gas phase.

$$K = \frac{\text{amount of solute per unit volume of liquid phase}}{\text{amount of solute per unit volume of gas phase}} \tag{1}$$

Figure 3 has the carrier gas flowing from left to right across a liquid phase (in the lower half of each segment) of a four-segmented column. The consequences of moving the carrier gas in four increments across the segments are illustrated by (a)–(d). In Fig. 3a the compounds have each reached equilibrium in the first stage of the column. Movement of the carrier gas causes those molecules in the gas phase (of the first segment) to be carried to the second segment where they equilibrate with the liquid phase. At the same time, fresh carrier gas has been introduced above the liquid phase in the first segment and the molecules which remained behind in the liquid phase reequilibrate. These processes are illustrated by Fig. 3b. In Fig. 3c the consequences of a further movement of carrier gas are illustrated. Partitioning occurs once again with fresh carrier gas in segment 1 and sample

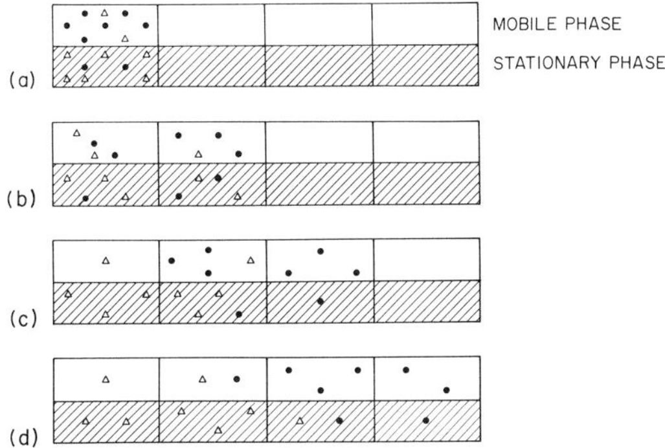

Fig. 3 Separation of molecules △ and ● as the mobile phase flows across the stationary phase in four steps.

molecules carried to segment 3. Also, additional sample molecules have been carried from the gas phase in segment 1 to segment 2, where some molecules have remained in the liquid phase. These then must also reequilibrate to establish their partition coefficients. The same steps occur in Fig. 3d when the carrier gas is moved another increment. At this point some separation already has been achieved in that the first and fourth segments contain pure △ and ● molecules, respectively.

As the process continues, the compounds become more and more separated until the ● molecules emerge from the column first, with the △ molecules eluted later. During the process some spreading of the molecular band occurs. Hence, operating conditions must be adjusted to minimize this spreading or it will be difficult to obtain complete separation. Consideration of Fig. 3 and of the partition process confirms that the greater the difference in partition coefficients, the greater the ease of separation. In real columns this process can be viewed in two ways:

(a) the separation stages are such small segments of the column that the system approaches a continuous process; and

(b) the number of equilibrations is equivalent to the number of theoretical plates (a term we will discuss later).

The height of a theoretical plate (often 1 mm or less in a high performance column) is considered to be equivalent to one of the illustrated segments in Fig. 3.

While chemists often speak of the retention time of a compound, this can be used only when all of the variables such as carrier gas flow rate, column size, and amount of liquid phase are known. The important factor in discussing the retention of a compound is the amount of carrier gas required to elute a compound from 1 ml of liquid phase. The specific retention volume per gram of liquid phase (corrected to 0°C), V_g, is equal to the partition coefficient times the quantity $(273/T\ P_l)$, where T is the column temperature and P_l is the density of the liquid phase. For a thorough discussion of the relationship between retention volume and partition coefficients see Purnell (1962, Chapter 2).

II. Instrumentation

A. Detectors

1. *Introduction*

When one is considering an analysis, a choice of instrumental conditions and of detectors must be made. The selection of a detection system can be extremely important because a component can be missed entirely if the detector is not sufficiently sensitive or does not respond to that particular type of compound. In addition, a detector which is selectively sensitive to the compound of interest can ease the problem of separating it from other classes of compounds. Another desirable feature of a detector is a signal response that is linear to the amount of compound in the sample.

The general characteristics of the three most commonly used detectors will be discussed, as well as some of their limitations for specific analyses, before each detector is treated in detail. The most common detection system is the thermal conductivity detector (TCD). It is fairly inexpensive, stable, and easy to use. Its most important characteristic is that it is a universal detector, which will detect all types of compounds. However, it is only moderately sensitive and cannot easily be used to detect trace quantities. The flame ionization detector (FID) is much more sensitive to organic compounds. It will not, however, detect inorganic compounds. It is a great deal more expensive, because it requires more extensive electronics. Also, it requires in addition to carrier gas a supply of hydrogen and of air, and its special sensitivity requires greater care in the selection of septums and column packings. The electron capture detector (ECD) requires the same electronics as the flame ionization detector but no additional gases. This detector is relatively insensitive to most compounds, but highly sensitive to compounds which contain elements that are strongly electron attracting, e.g., the halogens.

Obviously, attempts to analyze trace amounts of halogenated pesticides by thermal conductivity would be difficult and require a large amount of sample and several concentration steps. The use of the flame ionization detector would provide the necessary sensitivity, but would require the removal or separation of many interfering compounds in the sample system. Because of the selective response of the electron capture detector, the unresolved interfering organic compounds would not be a problem. This makes the EC detector the one of choice for this type of problem. However, an EC detector is not useful for the determination of most organic components and the FID detector does not respond to most noncarbon containing compounds. The TCD is ideal for the determination of compounds when they are at moderate levels of concentration. When, however, water is an interference in the determination of an organic compound, such as butanol, the analysis is greatly simplified by use of an FID because of the absence of a response for water.

2. *Thermal Conductivity Detector* (*Katharometer*)

(*a*) *Basis of Operation* A thermal conductivity detector cell is illustrated in Fig. 4. A heated wire is placed in the path of the gas and the signal is measured as the resistance of the wire. The passage of carrier gas over the

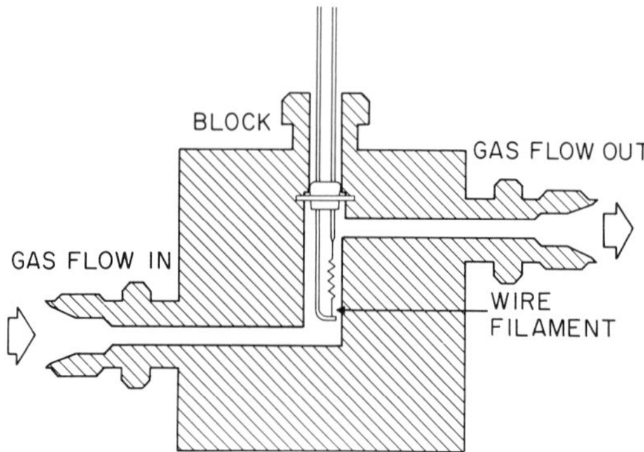

Fig. 4 Thermal conductivity detector cell.

wire removes heat at a rate controlled by the flow rate and thermal conductivity of the gas. As the sample passes through with the carrier gas (and if it has a different thermal conductivity than the carrier gas), the rate of heat removal from the wire will change. This will cause the temperature of the wire and, in turn, the resistance of the wire to change. Reference to Table I indicates that the thermal conductivities of H_2 and He are uniquely large, and therefore the best choices as carrier gases. The large difference in thermal conductivity between them and typical sample compounds yields a significant positive signal for all components of a sample. If one is interested, however, in the measurement of either of these two gases, then N_2 is an effective gas (the sign of the signal response will be negative). Helium is the normal choice for the carrier gas for reasons of safety, but in some countries its cost is prohibitive and H_2 is commonly used.

TABLE I

Thermal Conductivities for Several Compounds in the Gas Phase

H_2	434	H_2O	40.5	C_8H_{18}	33
He	352	CH_4	78	CH_3OH	48
N_2	60	C_2H_6	48	CH_2Cl_2	29
O_2	62	C_3H_8	40	C_6H_6	38

Because even small changes in carrier gas flow rate produce a signal, the TC detector is used in at least a two-cell configuration. Both cells are machined into the same block with a bypass system of carrier gas run into one cell. The column effluent passes through the other cell of the detector. The reference cell (pure carrier gas) and the sensing cell are two resistance arms of a Wheatstone bridge. The signal then is the imbalance of the bridge caused by the presence of sample in the sensing cell. Often four cells, two references, and two sensing cells are used to increase the sensitivity.

(*b*) *Detection Characteristics* The thermal conductivity detector is a universal detector of medium sensitivity. Because detection depends on the difference in thermal conductivity between the carrier gas and the sample, the sensitivity will vary from compound to compound. This means that each constituent must be compared to a standard of the same compound. The standard must be injected under the same conditions as the sample, including flow rate.

The sensitivity of a detector often is given in terms of its detectability, which is commonly defined as a signal which is twice the level of the noise. For most compounds and thermal conductivity systems, at least 10.0 ng of sample must be injected in order to observe a peak. Good column performance dictates that the size of the sample be less than 10 μl for analytical systems. With these limits the TC detector yields a response for 1 ng/μl or a minimum of 1 ppm concentration. For reasonable quantitative precision, five to ten times this concentration is needed. The response of the TC detector is linear from the lower detectable limit up to about 1 mg. Beyond this upper limit, standards must be close to the sample concentrations if useful analyses are to be achieved.

(*c*) *Problems* 1. Oxygen must be excluded from the system to prevent oxidation of the detector filaments. This also means that injection of air with the sample should be minimized.

2. Because it is a nonspecific detector, any sample that contains large amounts of water (which usually gives a broad unsymmetrical elution peak) must have the component of interest eluted before the water or well past the water peak. This increases the demands on the column for selective, efficient separations.

(*d*) *Advantages* The thermal conductivity detector is an inexpensive, universal, and fairly sturdy detection system. It can be used for field applications much easier than other detectors, and is the only common detector which is nondestructive. Therefore, it is almost always used for preparative separations when the components are to be collected after detection.

3. *Flame Ionization Detector* (*FID*)

(*a*) *Basis of Operation* The principle of the FID is the combustion of the organic molecules in the sample to produce ionized fragments. The ions and electrons are collected to give an ion current, which is the measured quantity. The noble gases, N_2, O_2, and compounds such as SO_2, CO, CO_2, and H_2O give little or no signal. As is shown in Fig. 5 the carrier gas and

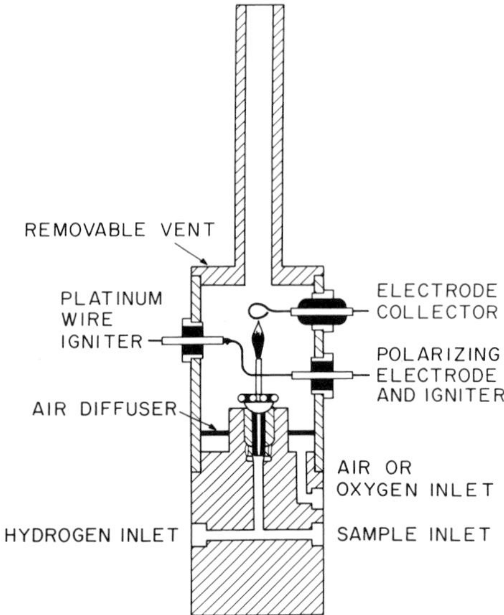

Fig. 5 Flame ionization detector.

sample leave the column and are mixed with H_2. The resulting mixture is burned in a compartment which has a separate supply of air or O_2; N_2 normally is used as the carrier gas. The sample molecule is ionized in the flame and subjected to a polarization voltage of 300-V dc. The burner jet itself is made the negative electrode in combination with a positive collector electrode. Because the detector has a high impedance and a small ion current, an electrometer is required to give a useful signal for a millivolt recorder. This normally is an integral part of commercial instruments.

(*b*) *Detector Characteristics* The detector is sensitive to almost all organic compounds with the exception of highly oxygenated molecules such as HCOOH. The presence of other elements in the molecule besides carbon and hydrogen decrease the sensitivity. While all pure hydrocarbons have the same sensitivity (on a per carbon basis), and therefore do not have to be calibrated separately, any compound that contains other elements must be calibrated with a standard of that compound. The FID detects the total mass of compound, but changes in the flow rate of greater than 25% will frequently alter the sensitivity of the detector itself and therefore recalibration will be required. Because current is the measured quantity (coulombs per second), the detector actually is a rate meter that responds to mass per unit time through the detector. The FID detector is extremely sensitive with a lower limit of about 20 pg or 2 ppb. The range for linear response is approximately 10^7; hence, samples with components of more than 10 μg give a nonlinear response. The detector can be readjusted to be linear for larger amounts, but this results in a reduction in sensitivity.

(*c*) *Problems* 1. The FID obviously cannot be used to detect those compounds which do not burn and produce ions in the flame.

2. To achieve the low-level detectability of the detector requires special care beyond that for the thermal conductivity detector. Septums and columns can bleed organic material to the detector to create excess noise and false peaks. If there is a choice available between a liquid phase and a solid phase column, the latter is preferable because it precludes any problems of stationary phase bleeding to the detector. The gases that are used with the FID detector (N_2, H_2, and air) must be free of organic contaminants or they will contribute to detector noise.

3. The FID does not make a good field instrument because of the necessity of protecting the flame from drafts and the additional gases which are required. Furthermore, the presence of a flame in a plant environment requires special explosion-proof housings for the FID detector.

(*d*) *Advantages* The flame ionization detector owes most of its popularity to its great sensitivity. Its selectivity can also be used to advantage. Not only can samples be run which contain water, but more intriguing is the use of water-saturated carrier gas to deactivate columns and the use of water as the stationary liquid phase.

4. *Electron Capture Detector* (*ECD*)

(*a*) *Basis of Operation* An electron capture detector is illustrated in Fig. 6 and is an extremely simple device. It consists of a radioactive β source which ionizes the carrier gas (usually N_2 or argon with methane) in the presence of a 300-V dc ion collection system with an anode and a cathode. The source is most frequently ^3H, but ^{63}Ni also is used. The ionization of the carrier gas produces an ion current due to cations and electrons. When a compound with a strong affinity for electrons (high capture cross section) comes through the detector, the ion current is decreased in a manner analogous to the decrease of light in absorption spectroscopy; that is, the decrease is logarithmic and follows Beer's law. The reason for the decrease in current is that the rate of recombination of cations and anions is much faster than is the recombination of cations and electrons. The latter condition prevails with just carrier gas going through the detector and results in a standing current, I_0. When a compound passes through with a high electron capture coefficient ε, the current is decreased to I. Hence, the detector gives a response according to the relation $\varepsilon Cl = \log(I_0/I)$, where C is the concentration of

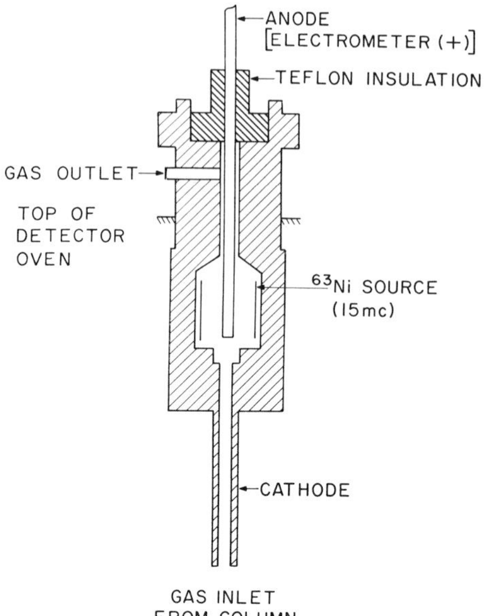

Fig. 6 Electron capture detector with ^{63}Ni source.

the sample species and l is the length of the ionization path. Table II summarizes values for the electron capture coefficients for various types of compounds.

The ECD is most sensitive to compounds which are halogenated, or contain nitrates or conjugated carbonyls. However, it does respond in varying degrees to other types of compounds. The most frequent use of the ECD is in the trace analysis of pesticides, where its selectivity reduces the cleanup and separation necessary to isolate the pesticide from the great multitude of other organic compounds that are present in plants or in animal tissue.

(*b*) *Detection Characteristics* Because each compound has a different electron capture coefficient, each must be calibrated. Also, the detector is concentration dependent, which requires that the flow rate be constant. The sensitivity of the ECD varies greatly, but can be as low as 0.1 pg of highly halogenated substances. Because of its logarithmic response, the ECD has a limited linear range; about 500 for the ^{3}H detector.

(*c*) *Problems* 1. The ^{3}H form of the ECD is easily contaminated because it cannot be heated above 220°C (limit of stability of metal tritide). When this is a serious problem, the ^{63}Ni detector can be used, but it is slightly less sensitive and has an even smaller linear range.

TABLE II

Electron Capture Coefficients of Various Compounds and of Classes of Compounds for Thermal Electrons

ε, Electron Capture Coefficient[a]	Compounds and Classes	Electrophores
0.01	Aliphatic saturated, ethenoid, ethinoid, and diene hydrocarbons, benzene, and cyclopentadiene	None
0.01–0.1	Aliphatic ethers and esters, and naphthalene	None
0.1–1.0	Aliphatic alcohols, ketones, aldehydes, amines, nitriles, monofluoro, and chloro compounds	OH, NH_2, CO, CN, halogens
1.0–10	Enols, oxalate esters, stilbene, azobenzene, acetophenone, dichloro, hexafluoro, and monobromo compounds	CH:C, OH, CO, halogens
10–100	Anthracene, anhydrides, benzaldehyde, trichloro compounds, acyl chlorides	CO·O·CO, phenyl·CO, halogens
100–1000	Azulene, cyclooctatretrene, cinnamaldehyde, benzophenone, monoidodo, dibromo, tri, and tetrachloro compounds, mononitro compounds	Halogens, NO_2, phenyl·CH:CH·CO
1000–10^4	Quinones, 1,2-diketones, fumarate esters, pyruvate esters, diiodo, tribromo, polychloro, and polyfluoro compounds, dinitro compounds	CO·CO, CO·CH:CH·CO, quinone structure, halogens, NO_2

[a] Values are relative to the absorption coefficient of chlorobenzene, which is arbitrarily taken to be unity.

2. The detector is merely selective and not specific. A peak may indicate a large amount of some other chemical rather than a small amount of pesticide.

3. Great care must be taken to keep O_2 out of the carrier gas and system because the detector responds to O_2 down to 1 ppm. For this reason, great care is necessary to a avoid leaks in the system. The septums in the injector are another source of spurious peaks, because the septum material often contains large amounts of highly halogenated material.

4. Solvents must be free of contaminants which will give peaks; "Pesticide Grade" solvents (interfering contaminants removed) often are used, especially with the EC detector. Note that this is a different purity requirement than for the "Spectrograde" solvents; the two are not interchangeable in their uses.

(*d*) *Advantages* The great advantages of this detector are its sensitivity, selectivity, and ease of use.

5. *Other Detectors*

Several other selective detectors are commercially available, including the alkali flame ionization detector and the flame photometric detector (FPD). The alkali flame is selective toward phosphorus-containing compounds. It has a configuration similar to the FID with the addition of the presence of a source of alkali salt. The vaporization of the salt in the flame apparently selectively increases the formation ions from organophosphorus compounds. With the alteration of the alkali salt and other parameters, this detector can also permit selective detection of compounds that contain halogens or nitrogen.

The FPD is selective for sulfur and phosphorous-containing compounds. When such compounds are burned in a flame, they emit light at 526 and 394 nm, respectively. A filter is used to determine which light will reach the photomultiplier tube. While selectivity compared to alkanes is good, there is poor discrimination between phosphorus and sulfur.

There are several other detectors designed specifically for gas chromatography. In addition, almost any type of analytical instrumentation, such as nmr, ir, uv, and mass spectroscopy, can be used as a detector. Of these, mass spectroscopy has become the most important due to the additional identification information it provides (see Section V.B.4). For further reading on detectors, see Hartmann (1971), David (1974), and Adlard (1975).

B. Other Important Features

1. *Injectors*

Figure 7 illustrates a typical type of injection system for liquid or gaseous samples to be injected by syringe. In order to prevent spreading of the peak

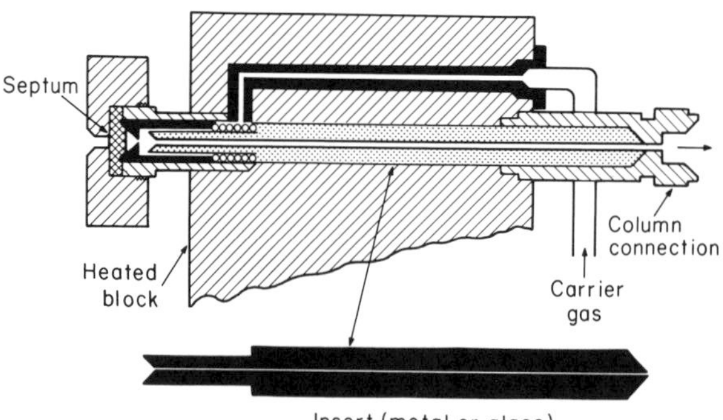

Fig. 7 Injection system.

and thereby increase the separation requirements, the sample should be introduced as close to the column as possible and it should be vaporized as quickly as possible. The injection port temperature should be at (or above) the boiling point of the compound.

To assist in rapid vaporization (and also to avoid overloading the liquid phase on the column), small sample sizes should be used. Ten microliters (μl) is considered an upper limit for optimum analytical work. Gas samples can be as large as 5 ml, and usually are injected with a gas-tight syringe. Because the holdup in the needle is small compared to the overall volume, corrections are not required. Quantitative injection of small liquid samples requires practice and the development of a reproducible technique. The syringe should be filled with more liquid than will be used and bubbles removed by pointing the needle upward while tapping or gently pumping the syringe plunger. The plunger is then depressed to have the amount of liquid desired. To avoid the injection of air, a smooth flowing technique is required; the syringe needle is pushed through the septum and the sample immediately injected by pushing in the plunger, followed by the immediate withdrawal of the syringe. A more accurate method for the measurement of injection volumes is to pull the liquid back into the barrel of the syringe after adjusting the contained volume in order to determine the total volume, inject, and then read the volume of liquid left in the syringe. Subtraction of the latter from the total volume yields an accurate reading of the amount which was actually injected. This is termed a needle correction. If the pressure of the carrier gas is fairly high, the plunger can be blown back out of the syringe. Therefore, a wise precaution is to keep your thumb behind the plunger. A good injection technique is to avoid use of more than 60% of the capacity of a syringe.

Solids usually are injected after dissolution in a suitable solvent. When this is not practical, special equipment must be used. The simplest form is a modified syringe which will hold the solid and inject it through a septum to an injection post that provides rapid volatilization. The sample also can be enclosed in a sealed glass ampoule and placed into a special chamber equipped with the means to crush it. Because of the low sample capacity of open tubular columns, there has been renewed interest in the development of solid injection systems to eliminate the need for solvent. These must, however, meet the special requirements imposed on the injector by the use of open tubular columns. One method is illustrated in Fig. 8.

A simple way to solve some special injection problems (such as the need for pyrolized samples and for desorption of the sample from an adsorbent) is to use a valve which normally is switched to route the carrier gas straight through the column, but which can be switched to route it through a sample loop (see Fig. 9). For an extensive discussion of the use of valves to solve

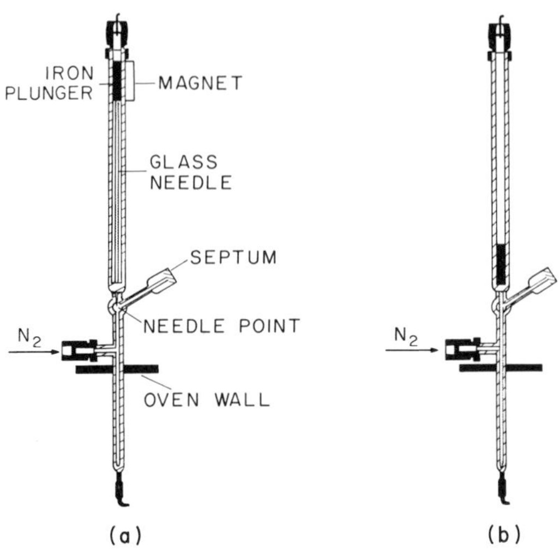

Fig. 8 An injection system for open tubular columns. (a) Load position; (b) inject.

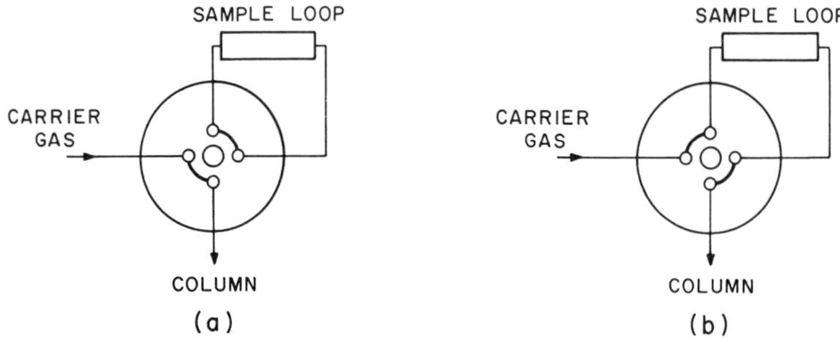

Fig. 9 Valve injection system with a sample loop. (a) Isolated sample; (b) injected sample.

specialized problems, see "Imagineering with Carle Micro Volume Valves," Carle Instruments, Inc., Fullerton, California.

2. *Temperature Programming*

The technique of changing the column temperature (according to a timed program) during a chromatographic run can be of great assistance, but it has several drawbacks. Temperature programming frequently is used for the analysis of mixtures that include components with a wide range of boiling points. If the early peaks of such a mixture are adequately separated in isothermal operation, then the higher boiling compounds will come out much later and be so spread out as to be difficult to detect or determine quantitatively. The solution to this is illustrated in Fig. 10. The sample is injected onto the column at a low temperature, which will provide adequate resolution of early peaks, and then the temperature of the column is increased so as to decrease the residence time of the later peaks. This decreases the broadening of the peak and makes it easier to obtain accurate peak areas. Such an approach also decreases the analysis time required per sample.

For effective use of temperature programming in gas chromatography, the programmer must be highly reproducible. In addition, a flow control device must be used (under constant pressure the viscosity of a gas increases with temperature, which causes the flow rate to decrease). Most liquid phases give some background signal by slowly eluting small amounts of the liquid ("bleeding"). This "bleeding" of the liquid phase increases with increasing temperature and results in a steadily increasing baseline. The use of dual columns with differential detection can partially correct these problems. Contaminants from septums and regulators do not reach a steady state condition with temperature programming, but may concentrate at the lower temperatures and then be given off as extraneous peaks as the temperature is raised. Also, this technique cannot be used if the higher boiling components are difficult to separate; resolution normally decreases with

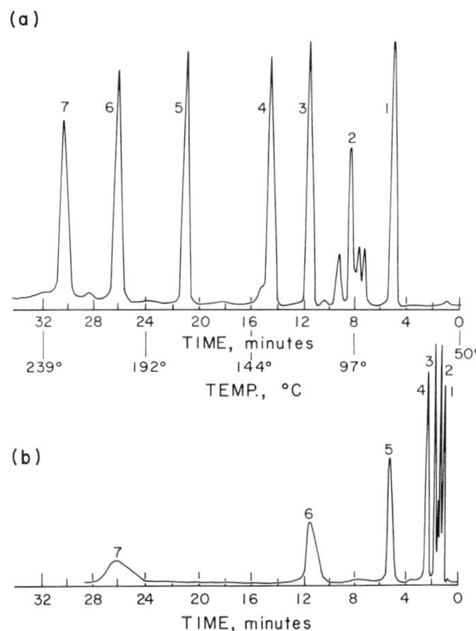

Fig. 10 Sample mixture separated by (a) temperature programming and (b) isothermally so as to give about the same analysis time.

increasing temperature. An excellent chapter on temperature programming can be found in Del Nogre and Juvet (1962).

3. *Collection of Samples*

For most analytical separations the sample components are allowed to pass into the air without being recovered. Indeed, most of the detectors other than thermal conductivity are destructive; they can be used for preparative separations only if part of the sample is split from the main stream and passed through the detector. To collect a sizable amount of a particular component it must be well separated from other components. For such a system, large columns ($\frac{1}{4}$–$\frac{1}{2}$ in.) can be used which in turn permit large sample sizes to be injected. The sample size is limited by the ability of the injection port to achieve rapid and complete volatilization.

A practical design for a collector is shown in Fig. 11. Most sample collection is done to obtain enough pure material to do qualitative analyses (by IR, NMR, and mass spectrometry, or by chemical test) or to obtain pure compounds. Another application is the collection of a group of unresolved peaks (frequently on an absorbent) for subsequent separation on another column. A valve similar to the one illustrated in Fig. 9 is a convenient means for such collection and transfer.

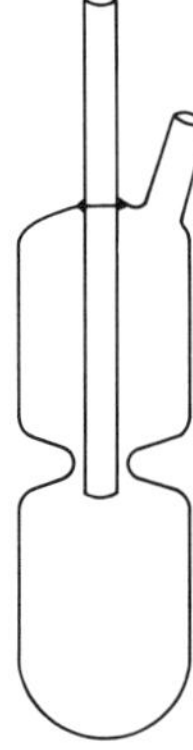

Fig. 11 One type of sample collector.

III. Choosing a Column

A. Introduction

Once the proper instrumentation has been selected to attempt to solve a problem, the choice of a column which will perform the desired separation becomes the important consideration. If the separation of interest has been described in the literature, the chemist needs only to duplicate the column and operating conditions. (One often discovers, however, that the necessary information is missing in published procedures.) Factors that affect the separation (the separation of the peak maxima) and the resolution (the separation of the area of one peak from the area of another) of a column include the liquid phase, the support, the mesh size of the support, the percent loading of the liquid phase on the support, the column length and diameter, the carrier gas flow rate, and the column temperature. Different commercial brands of a given support or liquid phase usually are equivalent. Use of the information that is given in this section often makes it possible to substitute an equivalent support or liquid phase for the one cited in the literature and still achieve comparable analytical results. If a relevant separation procedure cannot be found in the literature, another useful resource is the Gas and Liquid Chromatography Abstracts, which have been put out by various publishers under the auspices of the Gas Chromatography Discussion Group of the Institute of Petroleum (Great Britain) since 1958, or the abstracts (in the form of bound issues or microfilm) that are published by Preston Technical Abstracts Company, 909 Pitner Avenue, Evanston, Illinois.

Another approach is to use one of the compilations of gas chromatographic data, e.g., Lewis (1963), Schupp and Lewis (1967), and Schupp (1971a). If the compounds which one desires to separate are listed in the compilation for the same column and temperature conditions, the degree of separation of the compounds can be estimated quickly. If the sample mixture includes components that are not listed in the literature or components with unknown chemical identities, then selection of the best column system and conditions becomes a matter of experience as well as "trial and error."

B. Adsorption Columns

1. *Introduction*

For analytical separations both solid and liquid stationary phases are useful. Solid phase adsorbents provide high efficiencies, selective separation of isomeric mixtures, and the absence of column bleeding. Although gas chromatography actually began with the use of solid adsorbents, they fell into disfavor because

(a) higher temperatures are required to elute a given sample mixture than are required with a liquid phase column,

(b) sample sizes must be kept small to prevent overloading of the column,

(c) the heterogeneous surfaces of most adsorbents cause the elution peaks to tail, and

(d) most adsorbents can act as catalysts and potentially promote the decomposition of the sample components.

The development of sensitive detectors, especially FID has made the use of solid columns more desirable because of their attendant sensitivity to liquid phase bleeding. At the same time such detectors allow the use of the small sample sizes that are necessary to prevent overloading. Modifications of adsorbent surfaces have reduced tailing and catalytic activity. In addition, an entire class of solid materials of a polymeric nature have been devised which probably retain compounds by a combination of adsorption and absorption mechanisms. This section also includes liquid-modified adsorbents with the liquids placed on the adsorbent both by traditional techniques and by chemical bonding to the support. Adsorption rather than absorption appears to be the dominant retention mechanism with these materials.

Because gas–solid chromatographic (GSC) columns retain molecules more strongly than gas–liquid chromatographic (GLC) columns, they most commonly are used for the separation of fixed gases and of low boiling stable organic molecules. Modified GSC columns also can accomplish unique separations for mixtures of much larger molecules.

2. *Graphitized Thermal Carbon Black*

Graphitized thermal carbon blacks are excellent nonspecific adsorbents which retain compounds in relation to their polarizability and geometric configuration. Graphitized carbons are especially effective for the separation of isomers or stereoisomers. Numerous examples of their application to specific chromatographic separations are presented in the book by Kiselev and Yashin (1969). If some additional selectivity is desired, it can be induced by the addition of a modifier, for instance, a salt such as lanthanum chloride which increases the retention of aromatic rings by surface complexation.

3. *Alumina and Silica Gel*

Unlike thermal carbon blacks, alumina and silica gel are polar adsorbents. Because most compounds adsorb on the acidic sites of alumina (Snyder, 1968) and on the hydroxyls of silica gel, the general order of retention is similar for both adsorbents. The exact activity of each adsorbent varies from batch to batch and is extremely dependent upon the water content and pretreatment of the adsorbents. Given similar activities, however, silica gel usually is less catalytic. It is frequently used for the analysis of permanent gases. Alumina has better selectivity for alkenes and groups such as halogens, which have excess electron density, and frequently is used for the separation of low boiling hydrocarbons and the isotopic separation of molecular hydrogen and deuterium. An improved material whose surface resembles silica gel is porous silica with controlled pore structures (Porasil®). Its adsorption characteristics are like those for silica gel, but high efficiency columns are much easier to prepare and the batch-to-batch reproducibility appears to be better.

In addition to the control of adsorbent activity by adjustment of the water content, the surfaces of alumina and silica adsorbents can be modified and stabilized by coating with inorganic salts (Okamura and Sawyer, 1973a). Not only are the peak shapes improved by reduction of the heterogeneity, but the use of different salts provides separation selectivity.

4. *Porous Polymer Beads*

Porous beads made from polymeric materials were first used for gas chromatography by Hollis (1966) and are available commercially as Porapaks® and Chromosorbs®. While these materials are solid, it is believed the dominant mode of retention is absorption in the polymeric material rather than adsorption on the surface. Some of these are nonpolar sorbents, which resemble graphitized carbon in their selectivity. Because water is eluted in a symmetrical peak before organic solvents, these are excellent materials for trace water determinations. By altering the monomer the selectivity of the resulting polymeric beads can be changed. The porous polymer beads, which resemble other solid materials in their uses and advantages, usually are applied to the separation of lower boiling materials. They have good mechanical resistance to breakage, give highly efficient separations, and do not bleed or change retention characteristics with time when used below their decomposition temperature.

5. *Liquid-Modified Gas–Solid Chromatography*

Low loadings of water and of organic liquid phases on solid adsorbents have been used to deactivate adsorbents for some time. However, this approach has not been pursued in detail, with the exception of the use of water and of a few of the more common liquid phases. The number of possible partition equilibria that are involved in the separation process increases dramatically with the addition of a liquid phase to an adsorbent. At submonolayer levels there is still adsorption by the solid surface as well as retention by the liquid phase. With a high surface-area adsorbent, the liquid phase also will have a high surface area at low loadings, and adsorption on the liquid surface can become a dominant retention mechanism. Even if the solute partitions into the liquid phase, the retention characteristics of that liquid phase may be altered by the solid adsorbent orienting the molecules or by occupying a significant portion of the hydrogen-bonding sites of the surface. Despite the lack of theoretical understanding, liquid-modified gas–solid chromatography is useful because

(a) it has many of the advantages of GSC,
(b) polar compounds can be eluted with good peak shape, and
(c) wide variations of selectivity can be made by simply altering the concentration of the liquid phase [e.g., see Okamura and Sawyer (1973b)].

Even when selectivities are similar to those for the bulk liquid phase, the liquid phase as a modifier normally can be used at higher temperatures than when in bulk.

6. *Chemically Bonded Stationary Phases*

These materials have been developed to extend the useful temperature range of liquid phases and to decrease bleeding in gas chromatography. Research in this area has been encouraged by the development of bonded stationary phases for high pressure liquid chromatography, which requires organic stationary phases that do not bleed when organic mobile phases are used. Two types of bonded stationary phase exist:

(a) the brush type in which the reactive end of a long-chain organic molecule is reacted with the Si–O–H group of the support material to give a long chain that extends like the hairs on a brush, and
(b) material which is polymerized in situ on the support.

Exhaustive extraction of some of the latter materials indicates that many of them are not bonded directly to the support but only are attached better mechanically. Some of these materials are available commercially and can be quite useful, but their retention properties are not well understood because of multiple retention mechanisms. For example, when *n*-octane is bonded to Porasil C it is much more polar than bonded oxypropionitrile (OPN) on Porasil C. This is the opposite of the behavior that is observed for bulk liquid phases, and is attributed to residual SiOH groups on the Porasil after reaction with *n*-octanol (Little *et al.*, 1974). A good review of bonded stationary phases has been published by Rehak and Smolkova (1976).

C. Columns for Gas–Liquid Chromatography

1. Preferred Liquid Phases

The selection of a liquid phase for a separation can be a bewildering task because of the fantastic number of available materials. At least 700 different materials have been used, and there are well over 100 materials that are used regularly. Only recently has some sort of order emerged to replace the chaos that resulted from the absence of a standard method for the characterization of liquid phases. Two similar methods, Rohrschneider constants and later McReynolds constants, have gained fairly wide acceptance and have been used to characterize over 200 liquid phases (Supina and Rose, 1970; McReynolds, 1970). Many liquid phases have similar or identical chromatographic characteristics, and a vast majority of liquid phases increase in polarity in a regular manner, i.e., the increase in relative retention of any polar compound is related as a first approximation to the increase in the retention of benzene (see Table III).

This has led to the suggestion that certain "standard liquid phases" or "preferred liquid phases" can be selected and used as much as possible. By this approach, one could have a "library" of six to 25 columns which would permit duplication of most literature separations. For the development of a new separation, one of these columns should be capable of achieving the separation.

Table III lists 21 liquid phases (Haken, 1975) that one or more groups believes to be uniquely useful, or that were included in the list of Mann and Preston (1970) as one of the most popular liquid phases in 1968 and 1969. They are listed by chemical name, but some examples of commercial names are given. Also given are the upper temperature limits and the McReynolds constants for the compounds. The numbers to the side indicate the six preferred phases listed by Haken. A set of twelve preferred phases has been proposed by Leary *et al.* (1973).

Given a list of liquid phases, how does one decide which one to use for a particular separation? To separate a group of compounds that have a particular functional group, e.g., alcohols, or to separate an alcohol from other compounds whose structure may or may not be known, a reasonable selection from the list in Table III would be a polyethylene glycol. Because "like dissolves like," a liquid phase which has functional groups similar to those of the compound or compounds to be separated will retain that compound longer, and hopefully retain it until after other classes of compounds have eluted. By increasing their retention, the separation for members of the group will be improved. As another example, for the separation of hydrocarbons, squalane would be a logical choice because it is a long-chain

TABLE III

	Phase	Temperature Limit (°C)	X	Y	Z	U	S
	Squalane	120	0	0	0	0	0
(1)	Dimethylpolysiloxane (OV 101, SE 30, SF 96)	350	15	53	44	64	41
	Apiezon L and M	250	32	22	15	32	42
	10% Phenyl substituted siloxane	350	44	86	81	124	88
	Dexsil 300	500	47	80	103	148	96
	20% Phenylsiloxane	350	69	113	111	171	128
	20% Phenylsiloxane	350	74	116	117	178	135
	Dinonylphthalate	150	83	183	147	231	159
(2)	50% Phenylsiloxane (OV 17, SP 2250)	300	119	158	162	243	202
	65% Phenylsiloxane	350	160	188	191	283	253
(3)	Trifluoropropylsiloxane (OV 210, QF 1, SP 2401)	250	144	233	355	463	305
	Tricresyl phosphate	125	176	321	250	374	299
	Polyphenylethers	200	182	233	228	313	293
	25% Cyanoethyl substituted siloxane	275	204	381	340	493	367
(4)	Polyethylene glycols (Carbowax 20M, etc.)	250	322	536	368	572	510
	Ethylene glycol adipate	225	371	576	454	655	617
	Diethylene glycol adipate	200	378	603	460	665	658
(5)	Diethylene glycol succinate (DEGS)	225	496	746	590	837	835
(6)	di-3-Cyanopropylsiloxane (Silar 10 CP, SP 2340)	250	523	757	659	942	801
	TCEP	175	593	782	677	920	837
	Cyanoethyl formamide	125	690	991	853	1110	1000

hydrocarbon. If its low maximum temperature (120°C) presents a problem, the nonpolar dimethylpolysiloxanes offer an alternative. This approach has led to lists of recommended liquid phases for particular classes of compounds; an example of such a list is given in Table IV.

2. *Development and Use of McReynolds' Constants*

When a liquid phase from Table IV fails to give a satisfactory separation, what is the alternative to random "trial and error" of other liquid phases? For example, the recommended liquid phases in Table IV for the determination of pesticides are DC 11, QF 1, SE 30, OV 1, and OV 17. If the initial attempt to separate the pesticide from some unknown biological material is not satisfactory on OV 1, then which liquid phase should be tried next? The use of column characterization by McReynolds' constants can help supply that answer. An understanding of how these numbers are obtained is helpful.

(*a*) *Kovat's Indices* In gas chromatography, because retention times vary with so many factors and specific retention volumes are difficult to measure accurately, many retentions are reported for a specific temperature as relative to that for some standard compound. The normal alkanes represent another system to relate retention to a set of standards. Such an approach provides greater utility for the correlation of retention to structure, the qualitative identification of compounds (see Kovats, 1965) and the characterization of liquid phases. When a graph of the logarithms of adjusted retention times (y axis) versus carbon numbers of the alkanes (x axis) is prepared, a straight line results. If the logarithm of the retention time of, for instance, butanol is measured and evaluated in terms of the line for the normal alkanes, the equivalent carbon number is determined. For example, if the logarithm of the retention time for butanol falls exactly midway between that for pentane (5 carbons) and for hexane (6 carbons), it has an equivalent carbon number of 5.5. In the Kovat's indices a normal alkane is given an index value equal to 100 times the number of carbon atoms; hence, pentane is 500, hexane is 600, and the Kovat's index I for the butanol example would be 550.

(*b*) *Rohrschneider and McReynolds' Constants* Rohrschneider (1966) has suggested that column packings be characterized by measuring the I values for five compounds at 120°C. The five compounds are benzene, ethanol, methylethyl ketone, nitromethane, and pyridine (frequently symbolized as x, y, z, u, and s). The resultant I values are related to those for the least polar liquid phase, squalane. For example, if benzene has an index value of 797 on QF 1 and of 653 on squalane, the Rohrschneider constant

TABLE IV

Recommended Liquid Phases by Sample Type

Classification of Compounds	Stationary Phase
Acids	
C_1–C_{18} (free)	FFAP
Bile and urinary	SE 30
Fatty acid–methyl esters	DEGS, FFAP, Apiezon L, TCEPE, EGSS-X
Alcohols	
C_1–C_5	Hallcomid M-18 OL, Carbowax 600 or 1540
C_1–C_{18}	FFAP, Carbowax 20M
Di-poly	FFAP, QF 1
Aldehydes	
C_1–C_5	Ethofat
C_5–C_{18}	Carbowax 20M
Alkaloids	QF 1, SE 30
Amino acid derivatives	
N-Butyl trifluoroacetyl esters	DEGS/EGSS-X
Amines	
See Nitrogen compounds	
Boranes	Apiezon L
Essential oils	
General	FFAP, Carbowax 20M
Esters	
Mixed	Dinonyl phthalate, Porapak Q
Ethers	Carbowax 20M
Glycols	Porapak Q
Halogen compounds	Carbowax 20M, QF 1 (FS 1265), FFAP
Freons	Dibutyl tetrachlorophthalate, UCON Polar 2000
Hydrocarbons	
Aliphatic	
C_1–C_5	Propylene carbonate, Carbowax 400, tributyl phosphate
C_5–C_{10}	Didecylphthalate, SE 30
Aromatics	Tetracyanoethylated pentaerythritol, dibutyl tetrachlorophthalate
Hydroxy	2,4-xylenyl phosphate
Olefins	
C_1–C_6	$AgNO_3$/benzyl cyanide, dimethylsulfolane, propylene carbonate
C_6–up	Carbowax 20M
Polynuclear	SE 30 on DMCS-treated support, FFAP on DMCS-treated support, PMPE (5 ring)
Ketones	Lexan, FFAP
Nitrogen compounds	
Amines	Dowfax 9N9/KOH
Amides	Versamid 900
Ammonia	Ethofat or Carbowax 600 on Chromosorb T

(*Continued*)

TABLE IV

(Continued)

Classification of Compounds	Stationary Phase
Nitriles	Tetracyanoethylated pentaerythritol, FFAP, XF 1150
Organometallic	FFAP, SE 30
Pesticides	DC 11, QF 1 (FS 1265), SE 30, OV 1, OV 17
Phosphorus	SE 30, STAP
Silanes	SF 96, FFAP
Steroids	STAP, XE 60, QF 1 (FS 1265), SE 30, OV 1, OV 17
Sugar derivatives	
Trimethylsilylethers	QF 1, SE 52
Sulfur compounds	Carbowax 20M, FFAP, dinonylphthalate, Porapak Q
Water	Porapak Q
Gases	
Ar and O_2	6-ft activated molecular sieve 5A at −72°C
H_2, O_2, He, N_2, CO, CH_4	20-ft molecular sieve 13X
CO_2, H_2S, CS_2, COS	Silica gel or Porapak Q
H_2 isotopes	6-ft activated molecular sieve 13X
CO_2, N_2, O_2, CH_4, CO	$2\frac{1}{2}$-ft silica gel internal column, 20-ft molecular sieve 13X external column
N_2O, CO_2, NO	Porapak Q

would be

$$(I_{(QF\ 1)}) - I_{(squalane)})/100 = (797 - 653)/100 = 1.44 \tag{2}$$

These constants change somewhat with temperature. The polarities of 30 compounds in relation to the five test compounds have been determined, which in turn makes it possible to predict their index on characterized liquid phases with good accuracy. This method of characterization has increasingly become popular; at least 70 liquid phases (Supina and Rose, 1970) have been evaluated.

Because of the short retention times for three of the Rohrschneider compounds, McReynolds (1970) has suggested that butanol replace ethanol, 2-pentanone replace methylethyl ketone, and nitropropane replace nitromethane. These suggestions have resulted in a set of terms known as McReynolds' constants. However, by convention the latter are not divided by 100. Thus, the example of benzene on QF 1 would give an index value 144 rather than 1.44. McReynolds also suggested the addition of five other compounds for characterization, but these have not been widely adopted. His paper includes the constants for 226 liquid phases. The listing is too extensive to be repeated here, but should be obtained as a standard reference.

(*c*) *Use of McReynolds' Constants* Returning to the earlier question of which liquid phase should be tried for a pesticide analysis that does not work on OV 1, reference to Table V indicates that SE 30 and OV 1 have almost identical McReynolds' constants. Therefore, it would be useless to try one when the other has failed. Also, the constants for DC 11 are similar and it would not be expected to provide a significantly better separation. The other two suggested phases, OV 17 and QF 1, have substantially different constants from the other phases and from each other, and warrant an attempt at the separation. If neither proved satisfactory, then other liquid phases which have different sets of constants should be tried.

TABLE V

McReynolds' Constants for Liquid Phases That Are Recommended for Pesticide Analysis

Phase	X	Y	Z	U	S
SE 30	15	53	44	64	41
OV 1	15	56	44	65	42
DC 11	17	86	48	69	56
OV 17	119	158	162	243	202
QF 1	144	233	355	463	305

Another use of the constants in Table V is the prediction of a satisfactory substitute for a liquid phase. For instance, the DC 11 has a maximum temperature of 300°C, while SE 30 and OV 1 can be used to 350°. Hence, if the temperature limit or column bleed for DC 11 is a problem in an otherwise good separation, a change to SE 30 or OV 1 is a worthwhile consideration. To reproduce a separation from the literature which uses a liquid phase that is unavailable in the laboratory, a liquid phase with similar McReynolds' constants should be used.

Consideration of the large list of McReynolds' constants indicates that the majority of liquid phases can be subdivided into groups which are chromatographically similar. According to Wold (1975), only about 28 of McReynolds' 226 phases do not follow general polarity trends. While there probably will never be any official regulation of liquid phases, and Haken (1975) outlines some important objections to preferred liquid phases, use of a limited number of phases for initial attempts at general separation problems will ease the burden on the chromatographer a great deal. A thorough review of McReynolds' constants can be found in Ettre (1974). Also, additional values for McReynolds' constants are summarized in the 1975 edition of the Supelco Company Catalog.

D. *Specific Stationary Phases*

Some stationary phases are unusually retentive for certain types of compounds and these are normally termed specific stationary phases. A preferable term might be superselective because few are really specific, but they are more selective than those phases which follow general polarity trends. The selectivity may be based on compound size, shape, or its chemical nature.

1. *Specific Adsorbents*

Molecular sieves are the most well known of the specific solid stationary phases. These are aluminosilicates with cavities whose size can be varied by ion exchange or other chemical treatment. Compounds which are small enough to pass through the access holes into the cavities are exposed to a much greater surface area (700–800 m^2/g versus 1–3 m^2/g for the external surface area) and are therefore retained much longer. For instance, molecular sieve 5 Å can be used to separate butane, (which can pass into the cavities) from isobutance which cannot. Molecular sieves are the stationary phases most often used for the separation of permanent gases.

A clay adsorbent, Bentone 34, can be used to separate isomers based on shape. It is used to separate the *o*, *m*, and *p* isomers of xylene and related isomeric mixtures. The adsorbent is frequently combined with a liquid phase to produce more efficient separations. Many adsorbents have also been impregnated with transition metal compounds or complexes, and frequently exhibit increased selectivity for unsaturated compounds.

2. *Specific Liquid Phases*

Transition metal compounds or complexes dissolved in typical liquid adsorbents also are frequently selective for unsaturated compounds. The classical example is the addition of silver nitrate to polar liquid phases. Aqueous solutions of silver and mercury(II) salts also have been used as liquid phases. In addition, some metal chelates are liquids at typical gas chromatographic temperatures (such as the *n*-nonyl-β-diketonates) and can be specific for other complexing compounds such as alcohols.

Two types of specific liquid phases that are used for separations based on molecular geometry are becoming popular. Liquid crystals, which are used for the separation of geometric isomers, exhibit a transition from a normal solid to a "liquid crystal" (a highly ordered liquid phase). When used at their transition temperature, they have a selective affinity for linear molecules. Enantiomers (most frequently, amino acids) can be separated by use of optically active liquid phases. The mechanism is thought to be the formation of diastereoisomeric complexes between the liquid phase and the solutes, which are held together by hydrogen bonding. A good review of specific stationary phases can be found in Baiulescu and Ilie (1975, pp. 201–224 and 312–332).

E. Column Preparation

After a liquid phase has been selected, decisions concerning the other column parameters (the support material, the percent loading on the support, the type of tubing to be used, and its length and width) must be made. Support materials are discussed in terms of the Chromosorbs® but equivalent supports can be purchased from other manufacturers. (See the Supelco Company Catalog, or write Phase Separations, Ltd., Deeside Industrial Estate, Queensferry Flintshire, England, for lists of equivalent supports.)

1. *Support Materials*

The most commonly used support materials for analytical separations are Chromosorb G and Chromosorb W. They both are fairly inert forms of flux calcined diatomaceous earth, but the Chromosorb G is much harder than the friable Chromosorb W. The Chromosorb G, however, has a lower surface area which requires lower loadings of the liquid phase. Chromosorb W normally is used when a support is needed that is as inert as possible.

Chromosorb A is designed to hold a large amount of liquid phase and is used for preparative gas chromatography. Chromosorb T, which is a Teflon material, is difficult to pack and gives low column efficiencies. Therefore, it is used only when necessary for extremely polar or reactive compounds. Chromosorb P is a hard, but relatively adsorptive, material which is used primarily for nonpolar sample systems. Also available as supports are porous glass beads with a wide choice of surface areas. These beads are somewhat adsorptive but have the advantage that their spherical shape makes it much easier to pack an efficient column, and they are not friable.

When an inactive support material is needed, it should be acid-washed and then silanized to mask the surface hydroxyl groups. This can be done with DMCS (dimethyldichlorosilane) or HMDS (hexamethyldisilazane). Such surface treatment is indicated by the term Chromosorb W–AW DMCS. When a polar liquid phase is used, silinization of the support material may not be necessary because the polar groups of the liquid phase often mask the surface hydroxyl groups. However, with nonpolar liquid phases, such as squalane, surface deactivation is imperative. When acidic compounds are to be separated, a small percentage of an acid (usually phosphoric) is added to the liquid phase to deactivate any remaining basic surface sites. Similarly, when bases are to be chromatographed, a small percentage of KOH is added to the liquid phase to block any acidic sites.

The optimal size of the support depends upon the diameter of the tubing and the chromatographic application. Preparative separations with $\frac{1}{2}$-in.-o.d. columns often make use of 20/30-mesh material. For analytical columns, $\frac{1}{4}$-in.-o.d. tubing with 60/80- or 80/100-mesh support material is common, while high efficiency systems consist of $\frac{1}{8}$-in.-o.d. tubing with 100/120-mesh (149–125 microns) support. Larger particle sizes (lower mesh size) give less efficient columns, while particles smaller than 125 microns cause excessive pressure-drops for the column.

2. *Percent Loading of the Liquid Phase*

The appropriate amount of liquid phase to add to support material depends on the surface area of the support. If too much liquid is used, puddles will result rather than a uniform thin layer; this disrupts the gas flow and causes slow partition equilibration. If too little liquid phase is used, uncoated sites result or the partition process is a combination of gas–liquid and gas–solid equilibria. A loading of 20% wt/wt, on most supports, eliminates surface effects. However, for separations of mixtures with low volatility, a lower loading (3–10%) is needed to achieve separations within reasonable analysis times. As an example of the tradeoffs, the analysis time can be halved (without changing the carrier-gas flow rate) by

(a) increasing the column temperature about 30°C (this may lead to problems of thermal stability with either the compound or the liquid phase),
(b) halving the liquid loading, or
(c) halving the column length.

As the loading is increased, the liquid phase usually exhibits somewhat more polar character.

To prepare the column packing material, the liquid phase is dissolved in an appropriate solvent (see supplier's catalog), to which is added the solid support. The solvent then must be gently vaporized to allow the liquid phase to form a uniform layer on the support. This normally is done with a rotary evaporator, with heat added when necessary by means of a steam bath or a heat lamp.

3. *Packing Procedures*

Columns are packed by use of straight tubing (normally 3–6 ft of $\frac{1}{8}$-in.-o.d. stainless steel for high performance analytical columns), which has been plugged at one end with glass wool. The packing is added via a funnel that is connected to the column opening by a piece of rubber tubing. To pack a column uniformly without crushing the support is an art that requires practice (a selling point that is used by suppliers of prepacked columns). The method which we have found most successful is to alternate between bouncing the bottom of the column on the floor and tapping it on the side with a small rod. When the column is full, some of the packing is removed to make space for a plug of glass wool to close the column. Finally, the column is coiled to fit the oven and inlet–outlet fittings of the instrument.

If a compound is extremely unstable, a glass column may be required to avoid decomposition. Such columns either are purchased or prepared coiled to fit the instrument before they are packed. For the latter case, the column is packed by applying a vacuum to its plugged end and a mechanical vibrator often is used simultaneously on the walls of the column.

4. *Column Conditioning*

After filling, GLC columns need to be conditioned to remove the last traces of solvent and other volatile materials. This is accomplished by placing the new column in the instrument with the outlet end disconnected. Carrier gas (at a low flow rate) is passed through the column while it is heated to 25°C above the anticipated maximum temperature of usage. However, the column must not be heated to a temperature above the limit set by the supplier of the liquid phase. Columns usually are conditioned for 12 to 24 hr.

For trace analysis (especially pesticides), another type of conditioning also may be necessary. Because small amounts of many sample compounds are adsorbed irreversibly by a new column, large quantities must be injected to condition the column and thereby achieve reproducible results at trace levels. For some systems such conditioning must be repeated once a day.

F. Capillary Columns

Much of the previous discussion has been directed to the use of packed columns. However, capillary columns are used for an increasing number of analyses, especially for mixtures with 20 or more components. Such columns (also called open tubular columns) are prepared by use of 5- to 50-m lengths of tubing with an internal diameter of 0.2 to 0.5 mm and can be divided into three types. Classical open-tubular columns result when a coating of the liquid phase is applied to the internal wall of the capillary tubing (WCOT). If the internal wall of the tubing is treated to make it more porous, a larger amount of liquid phase can be applied; such columns are referred to as PLOT columns. The third type of capillary column actually has a support material coated on the internal wall to increase the capacity for liquid phase (SCOT). The latter two types of columns also have been used successfully for gas solid chromatography.

The great advantage of open-tubular columns is their high resolving power; separation efficiencies that are 5–30 times better than for packed columns are common. However, the traditional measure of column efficiency (see Section IV) does not provide an accurate comparison of the two types of column. Several other methods of comparison have been proposed, including the time required to make a specific separation (Ettre and March, 1974). For example, the time required for a comparable resolution of methyl stearate and methyl oleate on the liquid phase diethyleneglycol succinate (with columns of similar length) is 35 min for a packed column, 2.5 min for a WCOT column, and 3.2 min for a SCOT column. Two reasons account for the limited use of capillary columns in spite of their superior ability to separate complex mixtures. One is that most commercial gas chromatographs must be modified because of the slow gas flow rates needed for capillary columns. Low dead-volume injectors and detectors also are required. Another is the low capacity of capillary columns, which requires that only a small amount of sample be used. This is much smaller than the amount that can be injected conveniently, and usually requires the use of a sample splitter. Small sample sizes require the use of a highly sensitive detector, and for trace analysis the problem is even greater. The use of solid injection methods (where the solvent is evaporated from the sample components before injection) avoids the need to use a splitter which discards 99% of the sample. For the analysis of complex multicomponent mixtures, the initial effort to obtain an instrumental setup for effective capillary chromatography will be repaid many times over.

A useful general discussion of capillary columns, as well as specific instructions for their preparation, are presented in the excellent book by Ettre (1965) and in a chapter by Prevot (1969).

IV. Operating Conditions

After an instrument has been selected and a column has been prepared, those operational factors which affect separation and/or resolution of the components of a sample must be considered. Of these factors, temperature and carrier gas flow rate normally are the most important.

A. Temperature

In general, because the heats of absorption of many compounds are similar, separation can be improved by a decrease of the column temperature.

(To a first approximation, the retention volume of a compound doubles with a 30°C drop in column temperature.) However, if two compounds have different heats of absorption (as the result of different functional groups), the degree of separation actually may be decreased by lowering the temperature and increased by raising it. Another factor is the minimum operating temperature for the liquid phase. Many liquid phases can be operated down to 0°C, but some become viscous when operated at ambient temperature and the efficiency of the column deteriorates. For instance, the lower temperature limit for Apiezon L and SE 30 is 50°C. Also, the upper temperature limit must be respected or there will be excessive column bleeding which will create noise, contaminate the detector, and alter the column loading.

B. *Gas Flow Rate and the van Deemter Equation*

1. *Resolution*

Although a change of the carrier gas flow rate does not affect the separation of the two peaks, it does change their resolution. Reference to Fig. 12 indicates that quantitative analysis by use of the peaks of trace 1a will be much easier than with those of trace 1b. For the latter, accurate determination of the position of the second peak maximum is difficult because it is on a sloping base line that results from the tail of the first peak. The resolution R which represents the separations of the bodies of the peaks, is defined as

$$R_{12} = 2(t_{R_2} - t_{R_1})/(y_{t_2} + y_{t_1}) \tag{3}$$

where t_{R_2} and t_{R_1} are the retention times of peaks 1 and 2, and y_{t_2} and y_{t_1} are the widths of the peak bases (in units of time). Two peaks are considered to be completely resolved when $R = 1.5$. (See Fig. 14 for an illustration of terms.)

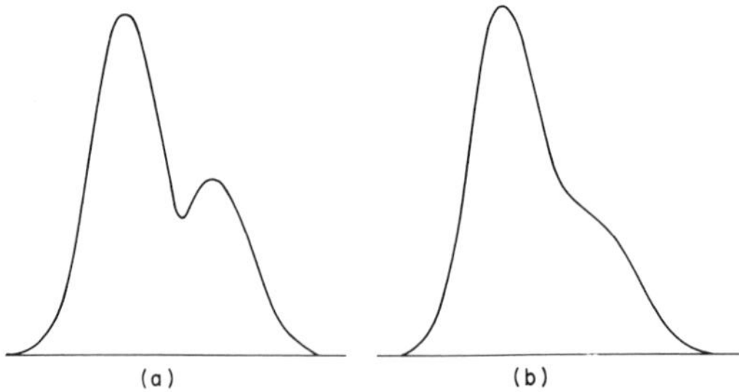

Fig. 12 A two component sample run under conditions of (a) better and (b) poorer resolution.

2. *Predicting the Efficiency Needed for a Separation*

If the ratio of the adjusted retention times for two compounds is known for a particular liquid phase, the efficiency required to achieve a resolution of 1.5 can be calculated. For packed columns, where the compounds are well retained by the stationary phase, the efficiency is calculated as n, the number of theoretical plates, by use of the relation

$$n = 16R^2\{(t'_{R_2}/t'_{R_1})/[(t'_{R_2}/t'_{R_1}) - 1]\} \tag{4}$$

The number of theoretical plates can be viewed as the number of equilibrations that a compound undergoes as it passes through the column. Conversely, the height of a theoretical plate H is the length of column that is required for one equilibration, i.e.,

$$H = L/n \tag{5}$$

where L is the length of the column. The height of a theoretical plate for a particular column is affected by the column preparation procedure and by the rate of gas flow through the column. These factors have been given quantitative consideration by van Deemter.

3. *van Deemter Equation*

For moderate gas flow rates, a relation for column efficiencies has been developed by van Deemter *et al.* (1956):

$$H = 2\lambda d_p + \frac{2\gamma D_g}{\overline{u}} + \frac{2}{3}\frac{k' d_f{}^2 \overline{u}}{(1 + k')^2 D_l} \tag{6}$$

where H is the height of a theoretical plate, λ a constant which is a function of the uniformity with which the column is packed, d_p the particle diameter, γ the tortuosity factor (how tortuous a path the gas must take), D_g the diffusion coefficient of the sample component in the carrier gas, $\overline{u}$ the average linear gas velocity (cm/sec), k' the partition ratio (capacity factor) of the compound,

$$k' = K\frac{[\text{volume liquid phase}]}{[\text{volume gas phase}]}$$

d_f the thickness of the stationary phase, and D_l the diffusion coefficient of the sample component in the liquid phase. Many of these terms have been mentioned in Section III.E and influence the practices that have been recommended for the preparation of columns. For instance, because of the d_p and γ terms, small-diameter particles appear to be desirable, but they are difficult to pack well. This often causes λ to increase as the particle diameter decreases. Another problem is that a high pressure drop causes an undesirable variation in the flow rate of the gas through the column; smaller particles cause higher pressure drops.

The first term of the van Deemter equation is absent for wall-coated open-tubular columns, and results in smaller H values. However, the most important difference is the small pressure drops that exist for capillary columns. As a result, WCOT and SCOT columns can be much longer than packed columns and made possible the attainment of many more theoretical plates.

The diffusion coefficient of the sample in the carrier gas, D_g, varies inversely with the molecular weight of the carrier gas, but the choice of carrier gases often is dictated by expense and detector requirements. The effect of d_f, the thickness of the liquid phase, is one of the reasons that lower loaded columns have been used in the past few years. Once again, however, the situation is not as simple as implied by the van Deemter equation. Thin films may result in the support interacting with the solute to cause peak broadening.

The diffusion coefficient of the compound in the liquid phase, D_l, has not been studied extensively. However, the liquid phase should be used at a temperature which ensures that it is not too viscous. Otherwise, this term will decrease by a large factor to cause a significant increase in H.

To study only the effect of the gas flow rate on the height of a theoretical plate, the terms which are constant for a given column are grouped into a set of constants to give the simplified expression

$$H = A + (B/\bar{u}) + C_l \cdot \bar{u} \tag{7}$$

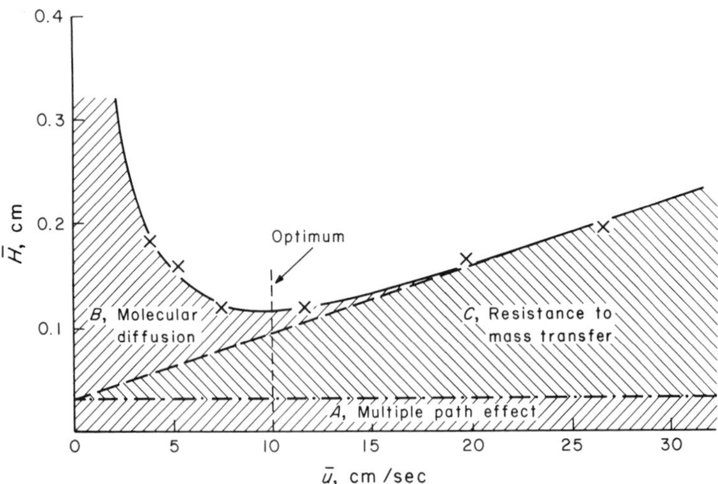

Fig. 13 A van Deemter plot which illustrates the various contributions to $\bar{H}$.

Although the optimum value of u to minimize H can be calculated from the relation

$$\bar{u}_{\min H} = (B/C_l)^{1/2} \tag{8}$$

the constants frequently are not known. As a result, the optimum flow rate normally is determined graphically. Reference to Fig. 13 indicates that the C_l term affects the optimum $\bar{u}$ value for minimum H and directly controls the rate of increase of H with increases in flow rate beyond the optimum value. For columns with thick liquid layers the change is quite large, but for small values of d_f the minimum value of H is smaller and it does not increase a great deal with increased carrier gas velocity. This is the reason why low-liquid loadings are particularly desirable for fast analyses. A detailed discussion of the derivation of the van Deemter equation and of its experimental effects is presented by Purnell (1962).

4. *Finding $\bar{u}$ for Minimum H*

The value for $\bar{u}_{\text{opt}}$ is dependent on the column temperature and the nature of the sample species because these factors affect D_g, D_l, and k'. Therefore, the sample molecule or a related species should be used at the desired column temperature to optimize the flow rate. As will be discussed later, the value of H also varies with sample size. In practice, $\bar{u}$, the average linear gas velocity, is not determined (nor is u_o, the linear gas velocity at the outlet). Instead, F_o, the gas flow rate at the exit is measured as millimeters per minute at 1 atm.

After the sample is injected for a wide variety of flow rates, the number of theoretical plates n is calculated from the chromatogram at each flow rate as well as the value of H [from Eq. (5)]. The resulting data are used to plot a curve analogous to that of Fig. 13. Optimum flow rates often are in the region of 20 ml/min. Hence, flow rates from about 5 to about 60 ml/min should initially be tried. Figure 14 illustrates a chromatogram and some of the points of measurement that are used to calculate n. Of the several methods,

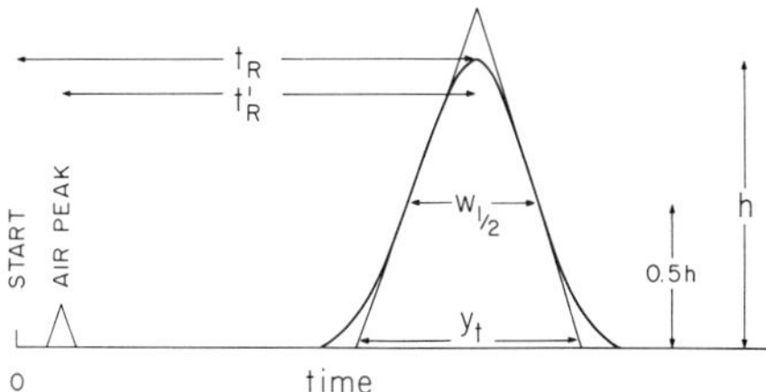

Fig. 14 Points of measurement on a chromatogram which can be used to calculate theoretical plate numbers.

the most common is

$$n = 16(t_R/y_t)^2 \tag{9}$$

where t_R is the time from the point of injection and y_t is the width of the peak base in units of time as determined by drawing tangents through the points of inflection and extending them to the baseline. A simpler method is to measure the width of the peak at exactly one-half of its height, $w_{1/2}$; the equation then becomes

$$N = 5.545(t_R/w_{1/2})^2 \tag{10}$$

This may also be done in units of distance on the chart paper.

An H versus $\bar{u}$ (or F_o) curve should be made for a column when it is first prepared, and rechecked on a periodic basis. This will provide an indication of any column deterioration that results from the crushing or settling of packing material.

When a separation is first attempted, the initial chromatograms should be made at $\bar{u}_{opt}$. However, when there is more than adequate resolution, the time for subsequent analyses can be decreased by increasing the gas flow rate. The H versus $\bar{u}$ graph can be used in conjunction with Eq. (4) to determine the largest value of $\bar{u}$ that is allowable while maintaining adequate resolution.

C. *Other Factors Which Affect Efficiency*

Many factors not included in the van Deemter equation can affect the efficiency of a column. The importance of a small inlet-to-outlet pressure ratio has been noted. In addition, sample introduction and detection effects are important.

1. *Sample Introduction*

If the sample, as it is introduced into the column, is already in the form of a broad band rather than a narrow plug, then the efficiency of the column will be adversely affected. Such extra column effects manifest themselves as substantial increases in the A and B terms of the van Deemter equation. The most important aspect of the sample introduction is the development of an efficient, reproducible injection technique. Poor injection technique, with the sample only slowly introduced rather than by a rapid injection, will contribute to peak broadening and will make quantitative reproducibility difficult. The volume of the injector and of the tubing that connects it to the column must be small so that the sample band does not have time to spread. Another cause of peak broadening in the injector is the slow vaporization of the sample that results from too low a temperature. However, to avoid thermal decomposition as low an injector temperature as possible should be used (not much above the column temperature). To test that the temperature is not too low, it can be raised to see if the peak becomes less broad. If this does not occur, the lower temperature can be used with confidence. Because larger samples require more heat and time to be vaporized, better column efficiencies usually are obtained with smaller samples; with high resolution analytical columns sample sizes of 10 μl or less are commonly used. The large sample sizes that are associated with preparative separations can cause large losses in efficiency unless special precautions are taken.

2. *Detector Effects*

The volume of the detector and of the connecting tubing also can adversely affect column efficiencies. This normally is not a problem with commercial instrumentation and packed columns, but it can cause significant loss of resolution with capillary columns. In addition, capillary columns may elute peaks which are so sharp that the response time of the detector and the recorder can become a limiting factor.

A dirty detector and other instrumental malfunctions can effect the real or apparent efficiency of a column. A very good set of symptoms, possible causes, and remedies can be found in McNair and Bonelli (1968).

V. Qualitative and Quantitative Analysis

A. Determination of the Retention of a Compound

The problem of quantitative and qualitative analysis are intertwined. A quantitative analysis requires that the identification of the compound or compounds under analysis be known. While absolute identification usually requires an ancillary technique, such as mass spectroscopy, most identifications are made as the result of some knowledge of the sample in combination with the measurement of the retention of the compound.

1. *Absolute Retention Information*

When a compound is run in a particular laboratory on a specific column and instrument at a specified temperature and gas flow rate, it can be characterized by t_R, the retention time of the compound. This method, however, does not allow any changes in operating conditions and is not a reliable method to identify a compound. The specific retention volume V_g is the only measurement which can yield the information that is necessary to calculate solubilities, partition coefficients, and other thermodynamic parameters. The retention time t_R can be calculated by measuring d_R, the actual distance on the chromatogram, and dividing it by the chart speed (see Fig. 14). The gas hold-up time t_M is calculated in the same way by use of an unretained solute. This measures the length of time that is required for a molecule which is not retained by the liquid phase or an adsorbent to go from the injector through the column to the detector; normally air (for a TC or EC detector) or methane (for a FID detector) is used. If, however, the temperature is low enough, these substances can be significantly retained and the problem becomes more complicated. If the specific retention volume V_g is to be calculated, the column temperature can be raised until the species that is used to measure the dead volume no longer is retained; its V_g can be calculated and subtracted from the V_g of the compound or compounds of interest. Otherwise, the gas holdup is calculated on the basis that for *n*-alkanes above butane the logarithms of the retention volumes increase linearly with carbon number. The value for the column dead-volume which results in a linear curve should be close to the true void space of the column.

The first calculation of importance is that of the adjusted retention time $t_R{}'$:

$$t_R{}' = t_R - t_M \tag{11}$$

This in turn allows the calculation of the adjusted retention volume, $V_R{}'$;

$$V_R{}' = t_R{}' \cdot F_o \tag{12}$$

where F_o is the flow rate of the carrier gas at the outlet of the column. Although rotameters often are provided to measure the gas flow rate, a much more accurate method is to use a modified pipet or buret with soap solution. The carrier gas needs to be equilibrated to room temperature by running it through some copper tubing for heat exchange before the measurement. The outlet flow rate F_o does not represent the true average flow rate of the carrier gas inside the column because the head of the column is at greater pressure, which results in compression of the carrier gas. Therefore, to calculate the net retention volume, V_N, the relation

$$V_N = j \cdot V_R{}' \tag{13}$$

must be used, where

$$j = \frac{3}{2}\frac{[(p_i/p_o)^2 - 1]}{[(p_i/p_o)^3 - 1]} \tag{14}$$

and p_i is the inlet pressure and p_o the pressure at the outlet (corrected for water vapor saturation in the soap bubble meter). The specific retention volume V_g is calculated by adjusting V_N to what it would be at 0°C and dividing it by the weight of liquid phase in the column, w_l,

$$V_g = (V_N/w_l)(273/T) \tag{15}$$

where T is the temperature at which the carrier gas flow rate is measured.

When V_g is calculated in one step from the necessary information, the expression has the form

$$V_g = (t_R - t_M)F_o \cdot \frac{3}{2}\frac{[(p_i/p_o)^2 - 1]}{[(p_i/p_o)^3 - 1]}\frac{273}{Tw_l} \tag{16}$$

Values of j for various p_i/p_o ratios are tabulated in Table VI.

TABLE VI

Tabulated Values of $j = \frac{3}{2}\left[\frac{(p_i/p_o)^2 - 1}{(p_i/p_o)^3 - 1}\right]$ for Values of (p_i/p_o) between 1 and 2.6[a]

p_i/p_o	0.0000	0.0100	0.0200	0.0300	0.0400	0.0500	0.0600	0.0700	0.0800	0.0900
1.00	1.0000	0.9950	0.9900	0.9851	0.9803	0.9754	0.9706	0.9658	0.9611	0.9563
1.10	0.9517	0.9470	0.9424	0.9378	0.9333	0.9287	0.9242	0.9198	0.9154	0.9110
1.20	0.9066	0.9023	0.8980	0.8937	0.8895	0.8853	0.8811	0.8769	0.8728	0.8687
1.30	0.8646	0.8606	0.8566	0.8527	0.8487	0.8447	0.8408	0.8371	0.8333	0.8295
1.40	0.8257	0.8219	0.8182	0.8145	0.8109	0.8073	0.8037	0.8001	0.7965	0.7930
1.50	0.7895	0.7860	0.7825	0.7791	0.7757	0.7723	0.7690	0.7657	0.7624	0.7591
1.60	0.7558	0.7526	0.7494	0.7462	0.7430	0.7399	0.7368	0.7337	0.7306	0.7275
1.70	0.7245	0.7215	0.7185	0.7155	0.7126	0.7097	0.7068	0.7039	0.7010	0.6982
1.80	0.6954	0.6926	0.6898	0.6870	0.6843	0.6815	0.6788	0.6762	0.6735	0.6708
1.90	0.6682	0.6656	0.6630	0.6604	0.6579	0.6553	0.6528	0.6503	0.6478	0.6453
2.00	0.6429	0.6404	0.6380	0.6356	0.6332	0.6308	0.6285	0.6261	0.6238	0.6215
2.10	0.6192	0.6169	0.6146	0.6124	0.6101	0.6079	0.6057	0.6035	0.6013	0.5992
2.20	0.5970	0.5949	0.5928	0.5906	0.5885	0.5865	0.5844	0.5823	0.5803	0.5783
2.30	0.5763	0.5742	0.5723	0.5703	0.5683	0.5664	0.5644	0.5625	0.5606	0.5587
2.40	0.5568	0.5549	0.5530	0.5512	0.5493	0.5475	0.5456	0.5438	0.5420	0.5402
2.50	0.5385	0.5367	0.5349	0.5332	0.5314	0.5297	0.5280	0.5263	0.5246	0.5229

[a] From Purnell (1962, p. 69).

Because all of the variable factors are included in the equation for V_g, the quantity represents a useful identification parameter for qualitative analyses. However, the potential for errors in its evaluation is extremely high and a great deal of care must be taken in the measurement of every factor. Actually, V_g values are most widely used for the determination of thermodynamic parameters for the partition equilibria.

2. *Relative Retention Ratios*

The relative retention ratio r_{12} is defined as

$$r_{12} = t'_{R_1}/t'_{R_2} = d'_{R_1}/d'_{R_2} \tag{17}$$

Once again it is extremely importantthat only the adjusted retention times be used. Many of the possible experimental errors are compensated for by use of this ratio, but Kaiser (1970a) lists many factors which are important to the achievement of reproducible retention ratios. Another difficulty in the use of relative retention ratios from the literature is the diversity reference standards. Adlard *et al.* (1965) have suggested that normal alkanes be used as standards, that the slope for the curve of the retention volume logarithms of the alkanes versus their carbon numbers be given, and that the relative retention ratio always be in relation to *n*-nonane.

3. *Retention Index Values*

The index values most often used for qualitative analyses are those proposed by Kovats (1958). The basis for their use, the linearity of the logarithm of the adjusted retention times of the alkanes when plotted versus carbon number, has already been described in Section III.C.2a. Adlard *et al.* (1965) found in a cooperative test that retention index values were reproduced better than relative retention data. Kaiser (1970a,b) also found that the Kovats indices are much more accurate when calculated than when determined graphically. The simplest calculation of a Kovats index is provided by

$$I_x = 100z + \left(\frac{\log t'_{R_x} - \log t'_{R_z}}{\log t'_{R_{(z+1)}} - \log t_z{}'} \right) \tag{18}$$

where I_x is the retention index of compound x, t'_{R_x} the adjusted retention time of compound x, t'_{R_z} the adjusted retention time of the *n*-alkane standard with z carbon atoms, and $t'_{R_{(z+1)}}$ the adjusted retention time of the *n*-alkane standard with $z + 1$ carbon atoms. Kovats indices, in addition to providing a method for identification of a compound, can be broken down to components indices for various functional groups to predict retention. See Kovats (1965) for a detailed review of both the Kovats indices and their use in qualitative analysis.

B. Qualitative Analysis

1. *Use of Retention Data*

In a survey by the Gas Chromatography Discussion Group in 1964, it was found that most chromatographers used some sort of absolute retention information in comparison with a known sample to determine the identity of a compound. The next most popular approach was relative retention ratios; the use of retention indices was found to be the third most popular method at that time. Since then the use of retention indices undoubtedly has become much more common. The survey also indicated that only 22% of the chromatographers used literature values. The use of literature values requires exact reproduction of the experimental conditions as well as of the liquid phase. Many liquid phases, such as squalane, tend to oxidize with use to give different users different values for retention indices and different values for the same column at different times.

The use of any of these methods as a means of identification usually relies on a fair knowledge of what the sample contains. Even though retention data now can be determined with extremely high precision [see, e.g., Rijks and Cramers (1974)], many compounds still appear to have exactly the same retention. The reliability of an identification increases greatly when the sample is run on two separate columns with different polarities.

What is the procedure if one has an unknown peak, there is no reference standard to compare it with, and the literature retention information does not appear to be applicable? Some types of samples contain compounds that are members of homologous series, such as petroleum samples and the saturated fatty acids in lipid samples. When that is the case, a plot of the logarithm of the retention time of several members of that series versus the number of carbon atoms can be used to predict the retention times of members of that series which are not available for direct comparison. Once again the identification is much more certain when two columns of differing polarity are used. If this approach is not feasible, other methods of qualitative analysis must be used.

2. *Spectroscopic Methods*

When a sample contains several unknown peaks, the components can be trapped (see Section II.B.3) for subsequent spectroscopic investigation by ultraviolet, infrared, and mass spectrometry. (Chemical testing can also be done on the trapped sample.) However, such techniques have been supplanted to a great extent by the development of integrated GC–MS instruments. These have resulted through the development of

(a) separators for the coupling of packed columns to mass spectroscopy units,
(b) direct mass spectrometer coupling to capillary columns, and
(c) commercial GC–mass spectrometer systems.

More than 400 papers per year on the use of GC–MS instruments now are being published. Beyond the use of simple retention information, this is by far the single most important technique in qualitative analysis. The advantages of direct GC–MS include the use of submicrogram quantities of material, high speed mass spectral scans (in less than 1 sec), and individual mass fragmentation patterns for the eluted components of the sample mixture. A good discussion of GC–MS is presented by Leathard and Shurlock (1970) and in an excellent book on the subject by McFadden (1973).

3. *Other Detection Methods*

The use of selective gas chromatographic detectors can aid in the identification of compounds, especially in conjunction with retention data. Element selective detectors have been discussed by Natush and Thorpe (1973). After suitable postcolumn reactions, the electrolytic conductivity detectors and the more specific coulometric detectors can be used. Piezoelectric detectors can be altered to provide varying selectivities through use of a specific adsorbent on the detector, and gas density detectors provide approximate molecular weights. In the developing field of plasma chromatography, mobility measurements can give specific compound identification for samples as small as 10^{-10} g.

Another method for the qualitative analysis of compounds is the use of two parallel detectors with different selectivities. A good example is the identification of sulfur compounds in oil spills by use of a flame ionization detector in parallel with a flame photometric detector; the details of the method are presented by Bentz (1976) in a review on oil spill identification.

4. *Sample Modification*

(*a*) *Derivitization* Derivitization can contribute to qualitative analysis in two ways. If the sample chromatographs well, derivitization can provide a second retention time which greatly improves the chance for an accurate identification. In addition, many compounds which contain polar functional groups tend to chromatograph poorly because of low volatility or adsorption on the chromatographic support. This problem can frequently be solved by derivatizing the compound, which will allow accurate retention data to be obtained. A thorough review of chemical derivitization has been written by Drozd (1975).

(*b*) *Other Chemical Modifications* In addition to traditional derivitization techniques, components of samples may be modified to identify functional groups or alter retention times to assist in identification. The most drastic modification is the complete removal of certain classes of compound, e.g., alcohols by use of boric acid. Usually the abstracting agent is distributed on a support material and placed as a precolumn reactor in the gas chromatographic system. Other approaches are the pyrolytic decarboxylation of carboxylic acids and hydrogenation of multiple bonds; these are known as carbon–skeleton gas chromatography. Postcolumn reactions also may be of use, but their main utility is the enhancement of detector sensitivity. An excellent monograph on functional group identification by gas chromatography has been written by Ma and Ladas (1976).

(*c*) *Pyrolysis* Gas chromatography also may be used for the qualitative analysis of samples which are impossible or difficult to chromatograph in their original forms. In pyrolysis gas chromatography, a sample, frequently a polymer, is heated to a high temperature and the pyrolysis products are then chromatographed to provide a "fingerprint" or pyrogram of the compound. This is especially useful when the resultant pattern is compared to pyrograms that have been obtained for standard materials in one's own laboratory. Correlation trials (Coupe *et al.*, 1973) indicate that interlaboratory comparisons of pyrograms are not reliable and that more standardization of methods is required.

Two books on qualitative analysis have been prepared by Leathard and Shurlock (1970) and by Crippen (1973). Also helpful is a monograph on "Analytical Reaction Gas Chromatography," by Berazkin (1968), and a short review by Leathard (1975).

C. *Quantitative Analysis*

Achievement of accurate quantitative analysis by gas chromatography is frustrated by three factors:

(a) detectors have different sensitivities for almost every compound,
(b) syringes are used for most injections and are not extremely accurate, and
(c) unresolved peaks require the use of approximations.

The degree of difficulty depends on whether the goal is the concentration or weight of one component or the determination of all components.

1. *Single Component*

(*a*) *Absolute Method* When the degree of precision afforded by syringe injection is sufficient (this will vary from operator to operator and can be tested by a plot of peak area or peak height versus amount of sample), then development of a quantitative procedure for one component is straightforward; peak heights or areas are compared with standards of similar concentrations. (See Section II.B.1 for quantitative syringe injection techniques.) Peak heights may be used only if the peak is symmetrical and there is no change in the retention time of the compound between the running of the standard and the running of the sample. Peak areas are less sensitive to changes in conditions, but altering conditions, such as flow rate, also can alter the sensitivity of the detector (see Section II.A).

(*b*) *Internal Standard* When syringe accuracy is not sufficient for a single component analysis, an internal standard is required. A known quantity of a standard is added to a known amount of the sample and the ratio of the peak areas (or heights) is determined. This same internal standard is then added to the unknown sample. The amount of compound is determined by measuring its peak area relative to that for the internal standard. In this way it is not necessary to know exactly how much is being injected, although similar sample sizes should be used. The internal standard should be completely resolved from other peaks, but fairly close to the peak of the sample component. It also should have a similar peak area and be of a similar chemical nature to the component that is to be determined. Some workers believe that an *n*-alkane should be used whenever possible.

2. *All Components*

(*a*) *Absolute and Internal Standard* If the goal is the determination of all components in a mixture, an absolute calibration can be made for each of the components. Results normally are calculated in terms of the weight of compound injected to obtain the weight percent of any compound:

$$\mathrm{wt}\%_A = (\mathrm{wt}_A / \textstyle\sum \mathrm{wt}) \times 100 \tag{19}$$

An internal standard also may be used, but if peaks have greatly different retention times, peak area, or the sample components have different chemical natures and/or concentration levels, more than one internal standard may be required.

(*b*) *External Standard* When an analysis is done routinely, a set of mixtures that have concentrations close to those of the sample components can be prepared for comparison and standardization.

(*c*) *Relative Response Factors* This approach can require considerable effort, but unlike the absolute calibration method it is not limited by the accuracy of the syringe. Quantitative mixtures of compounds are injected and their peak areas compared to find the relative response of the detector. Only in the case of pure hydrocarbons detected by the FID is an almost constant response per gram observed. Tables of relative response factors have been prepared by Dietz (1967). However, because relative response factors can vary with sample size, column temperature, and gas flow rate, improved accuracy is obtained when values are used that have been determined for the analyst's particular type of sample and specific experimental conditions. The percent composition of a sample can be determined by use of the relation

$$\%A = \frac{\text{area}_A / F_A}{\sum \text{area}_i / \text{factor}_i} \times 100 \tag{20}$$

where F_A is the relative response factor of compound A.

3. *Peak Area Determination*

(*a*) *Direct Methods* (1) *Electronic integrator* The electronic integrator (or data processing system) provides by far the greatest accuracy for the evaluation of elution peak areas. This will be discussed in greater depth in Section V.D.

(2) *Disk integrator* The disk integrator (illustrated in Fig. 15) is a much less expensive mechanical system and provides fairly accurate data. The integrator pen runs parallel to the recorder pen on a small section of the graph paper. By a mechanical arrangement with a ball-and-disk assembly, it oscillates; the number of oscillations is proportional to the area of the peak. If the peak is on a sloping base line, the area of the trapezoid that represents the base line can be determined and subtracted from the total integrated area.

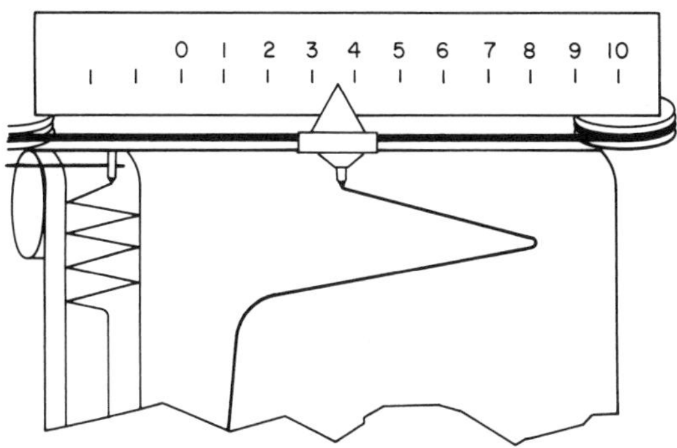

Fig. 15 Disk integrator tracing.

(3) *Cut and weigh* This is the next most accurate method and may be the one of necessity for an asymmetrical peak when other methods are not available. However, it is a method of last resort which requires a great deal of time and effort, and results in the destruction of the chromatographic record unless a photographic copy is made.

(4) *Planimeter* This is a mechanical device that is designed to measure the areas of irregular peaks. It is not accurate for small peaks and is not any more precise than geometrical approximations for symmetrical peaks. Its effective use requires practice and considerable patience.

(*b*) *Geometrical Approximations* (1) *Triangulation* Consideration of Fig. 14 indicates that the area of the peak, A, can be determined by the relation

$$A = \tfrac{1}{2} h y_t \tag{21}$$

where h is the height of the triangle and y_t is the peak width at the base as determined by tangents through the points of inflection. This would appear to be the most accurate procedure, but considerable error can be introduced in drawing the tangents.

(2) *Height times width-at-half-height* This method also is illustrated in Fig. 14 and is based on the relation

$$A = h w_{1/2} \tag{22}$$

where $w_{1/2}$ is the width at half of the height of the peak and h is the height of the peak.

(3) *Height times retention time* It frequently happens that the width of peaks varies almost linearly with the retention time; however, this must be established for each system. For such response characteristics the product of the peak height times the retention time is proportional to the peak area. This method is especially valuable for chromatograms with many narrow peaks (which make direct width measurements difficult)—often the case for capillary columns. The method gives values which are proportional to the area of the peak, but if it is used with any of the other area-determination methods the relative proportionality constant must be determined.

(*c*) *Unresolved Peaks* Unresolved peaks can cause a great deal of error in quantitative analysis. Several methods for determining the areas of unresolved peaks have been discussed by Pecsok (1959). The two cases illustrated in Figs. 16a and 16b represent the extremes. When peak sizes are similar and well-shaped (Fig. 16a), a simple method is to draw a dividing line through the minimum between the peaks. For a sample with a trace on the tail of a much larger peak (Fig. 16b), a hypothetical base line is drawn to conform to the trailing edge of the peak. When peaks are poorly resolved, of similar size, and not gaussian (Fig. 16c), the other estimation methods that are given by Pecsok can be tried, or an attempt can be made to trace out the

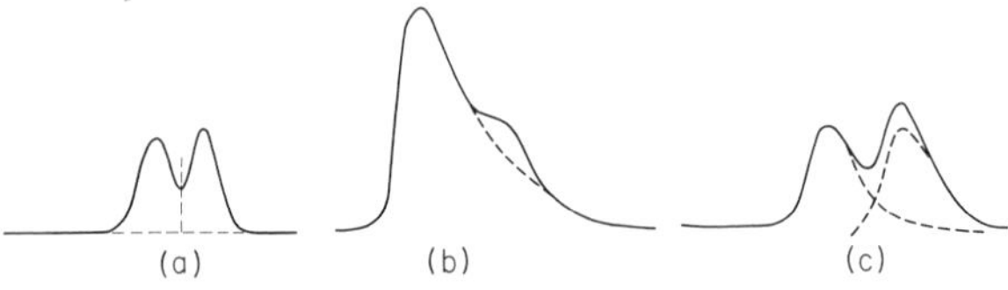

Fig. 16 Methods for estimating the areas of unresolved peaks.

individual curves. However, a better approach is to obtain better resolution before any accurate quantitative analysis is attempted.

D. Electronic Integrators and Computers

Automation in gas chromatography has become increasingly important in the last decade. Automatic quantitative analysis by means of electronic integrators or computer-based data-reduction systems has grown from 23% in 1966 to an estimated 50% in 1976 (Gill, 1976). In addition, the operation of many gas chromatographs is being automated. Advances in microelectronics, which have so drastically altered the electronic calculator market in the last five years, are having a comparable impact in gas chromatography.

In addition to becoming acquainted with the hardware, one must also be aware of the decision-making processes that are programmed into an integrator. An important point to consider in a system is the method used to sense a peak (whether an absolute threshold or a change in slope is used as criterion). There also are various ways that an electronic device may be used to treat a drifting base line. Two good reviews in this area are by Johnson (1968) and Derge (1972b,c). The use of computers in chromatography was the subject of symposia at Mainz, Germany [*Chromatographia* **5**, 63–165, 166–211 (1972] and at Washington, D.C. (*J. Chrom. Sci.* **9**, No. 12 (1971); **10**, No. 1 (1972)]. The papers from the symposium entitled "Impact of Microelectronics in Chromatography Automation" at the April 1976 ACS Meeting have been published [*J. Chrom. Sci.* **14**, 165–200 (1976)].

Anyone considering electronic data systems should talk with manufacturers about newly available equipment and should read the latest literature to learn of new levels of electronic sophistication. While automation can be an economical, labor-saving addition to a gas chromatographic system, it does not assure improved quantitative results. All too often the other steps in the analysis prior to the measurement of the peak areas present the limiting factors in experimental accuracy. Some of these are discussed by Derge (1972a). An example is in pesticide residue analysis, where it is rare that 100% of the pesticide is extracted from the plant material.

VI. Applications of Gas Chromatography

A. General Analytical Methods

Gas chromatography has been used for the analysis of most types of compounds, including inorganic mixtures. Because of its speed, simplicity, and sensitivity, the method of gas chromatography (only 25 years old) probably is the world's most important analytical technique. While there are many compounds which cannot be chromatographed directly, many can be analyzed after transformations by pyrolysis or derivatization.

Most books on gas chromatography include example applications, but the possibilities are so diverse that reference should be made to primary literature, including the abstracts that are discussed in Section III.A. Another place to find recent references is the biannual reviews that appear in the April issue of *Analytical Chemistry*.

Ettre (1974) estimates that 95% of gas chromatography papers can be found in six journals. These are *Analytical Chemistry, Bulletin de la Societe Chimique de France, Chromatographia, Journal of Chromatographic Science, Journal of Chromatography*, and *Zeitschrift für analytische Chemie*. Reference to any of these journals will yield many, many applications. For specific areas of applications such as food chemistry or oil analysis, one may find many gas chromatographic methods in journals devoted to these types of analysis. In addition, books are available that are devoted entirely to gas chromatographic methods for specific areas of analysis. In addition, *J. Chrom. Sci.* devotes special issues to state-of-the-art reviews of specific areas; reviews also may be found in the Chromatographic Reviews section of *J. Chrom.* and in *Advances in Chromatography*. The problem is not a lack of information on applications, but rather the challenge of sifting through the great quantity available.

B. Preparative Gas Chromatography

In addition to its use in analysis, gas chromatography enjoys wide application as a method for the separation of chemical mixtures. This is termed preparative gas chromatography and has not received much attention in this chapter because its methodology usually is quite different. The goal of preparative separations also is high efficiency *but* with as large a sample size as possible. This requires

(a) injectors that are capable of vaporizing large quantities of sample,
(b) large diameter columns (which are difficult to pack),
(c) usually automatic sample collecting devices, and
(d) often automatic sample injection systems.

For the occasional preparative separation, an analytical instrument may prove adequate. However, routine preparative chromatography should be done with instruments that are designed for that purpose. Chemists who are interested in preparative separations should consult Zlatkis and Pretorius (1971), Sawyer and Hargrove (1968), and Rijnders (1966). In addition, many books on gas chromatography contain sections on preparative gas chromatography.

C. Elemental Analysis

An example of the use of gas chromatography for other than direct quantitative or qualitative analysis is its use in elemental analysis. Organic samples that contain C, H, and N are converted in precolumn reactions to CO_2, H_2O, and N_2. These are then chromatographed and sensed with a thermal conductivity detector. The peak areas are compared to those for standards to determine the weight of C, H, and N. Several commercial models for elemental analysis are available. A review of methods and instruments has been written by Rezl and Janak (1973).

D. Surface Area Measurements

The measurement of surface areas by the Brunauer–Emmet–Teller (BET) method has been modified by Nelsen and Eggersten (1958) to provide a simpler and more rapid gas chromatographic procedure. Three different concentrations of nitrogen in helium are used. For each concentration the nitrogen is adsorbed at $-196°C$ and then desorbed at room temperature; the peak areas are measured by a thermal conductivity detector. The surface area is then calculated by the BET equation. A commercial gas chromatograph is available for such surface area measurements.

E. Physicochemical Constants

One of the most interesting areas of application is the use of gas chromatography to obtain physicochemical constants. It has been used extensively to monitor the progress of reactions and thereby determine reaction rate constants or to sample phases to determine equilibrium constants. However, gas chromatography also may be used directly to measure these and other physicochemical constants. The retention of a compound is determined by its equilibrium constant, and the spreading of the chromatographic peak is due to rate processes. A general review of physical measurements by means of gas chromatography has been prepared by Conder (1968).

1. *Information from Retention Data*

In gas–liquid chromatography the gas–liquid partition coefficient K_L can be determined directly from the specific retention volume:

$$V_g = K_L/P \tag{23}$$

where P is the density of the liquid phase. In gas–solid chromatography

$$V_g = K_A \tag{24}$$

where V_g is the specific retention volume per meter squared of adsorbent and

$$K_A = \frac{\text{conc/m}^2 \text{ adsorbent}}{\text{conc/ml of gas phase}} \tag{25}$$

In addition, for some systems a solute may have a significant contribution to its retention by adsorption on the liquid surface. For retention volumes adjusted to STP, the appropriate relation is

$$V_R = K_L V_L + K_A A_L \tag{26}$$

where V_L is the volume of the liquid phase, and A_L the surface area of the liquid phase (usually determined by the BET method with the liquid in the frozen state).

In addition, because changes in temperature are easy to make with a gas chromatographic system, the data for plots of $\log K$ versus $1/T$ are easy to obtain and yield the value for heat of adsorption or absorption, ΔH. The slope is equal to $-\Delta H/2.3R$, and the intercept is equal to $-\Delta S/2.3R$.

The equilibrium constant can be used to estimate the solubility of a solute in the liquid phase if the saturation vapor pressure is known. The relation is

$$K = c_{\text{liquid}}/c_{\text{gas}} \qquad \text{(at saturation for each phase)} \tag{27}$$

Also, the activity coefficient α can be evaluated by means of the relation

$$\alpha = (1.7 \times 10^7)/V_g p_o M_l \tag{28}$$

where p_o is the saturation vapor pressure of the solute and M_l is the molecular weight of the liquid phase. Because of the nonideality of the carrier gas, more complicated equations often must be used. This area has been reviewed by Clever and Battino (1975).

The determination of complexation constants from partition coefficients in gas chromatography has been treated by Purnell (1967). Additional work in this area recently has appeared (Laub and Pecsok, 1974). Several comparisons of data obtained by gas chromatography with that obtained by spectroscopy or nuclear magnetic resonance indicate that the gas chromatographic data may be superior.

Solutes also may be used as probes to obtain information on the nature of the liquid phase. One method of studying complexation between two materials is to prepare several columns with various ratios of the materials as the liquid phase. Juvet (in Purnell, 1967) found that when the materials became tied up in complexed form, the retention time of a solute molecule changed drastically. As a result, the ratios at which complexation occurs can be determined as well as an estimate of the relative strengths of the complexation constants from the degree of change of the retention time. This method has been used by Bogoslovsky *et al.* (1972).

Another method which involves the use of a solute as a probe is the determination of the phase-transition temperature for a liquid crystal. A solute is injected at various temperatures on the liquid crystal column; at the temperature of phase transition a sharp break usually occurs or there is a change in the slope of the retention-time versus temperature curve.

2. *Reactions on Columns*

Gas chromatography has long been used as an analytical technique to investigate reactions. More than 20 years ago Kokes *et al.* (1955) recognized the advantages of a precolumn microreactor to facilitate the study of reactions on catalysts. Only recently Phillips (1971) has treated in depth the study of reactions that are catalyzed by or react with the stationary phase of a chromatographic column. Although different methods are discussed for the various type of reactions, the technique of stopped-flow chromatography is used most often. The sample is injected onto the column, the flow stopped to allow the reaction to take place, and then the flow is resumed in order to chromatograph the products and remaining reactant. Elution curves also have been used by Yanovskii and Berman (1972) to study cumene cracking, oxidative dehydrogenation, and catalyst poisoning.

Reaction kinetics also may be studied in a continuous flow situation by the distortion of the chromatographic elution peaks. This area has been reviewed by Van Swaay (1969). Li *et al.* (1974) have investigated the use of peak profiles to determine the rates of formation and dissociation of complexes that are formed on gas chromatography columns. The use of moment analysis to characterize reacting systems has been discussed in detail by Suzuki and Smith (1975).

3. *Peak Broadening*

In addition to the use of peak shape to study reactions on columns, Suzuki and Smith (1975) have used moments to analyze axial dispersion in packed beds, intraparticle diffusion in porous particles, and rates of adsorption. Diffusion coefficients also may be determined by gas chromatographic peak broadening. This has been reviewed by Maynard and Grushka (1975) and by Choudhary (1974).

References

Adlard, E. R. (1975). *Critical Rev. Anal. Chem.* **5**, 1–36.

Adlard, E. R. *et al.* (1965). *In* "Gas Chromatography–1964" (A. Goldup, ed.), pp. 238–265. Institute of Petroleum, London.

Baiulescu, G. E., and Ilie, V. A. (1975). "Stationary Phases in Gas Chromatography." Oxford Univ. Press (Pergamon), London and New York.

Bogoslovsky, Yu. N., Sakharov, V. M., and Shevchuk, I. M. (1972). *J. Chromatogr.* **69**, 17–24.
Choudhary, V. R. (1974). *J. Chromatogr.* **98**, 491–510.
Clever, H. L., and Battino, R. (1975). *In* "Solutions and Solubilities—Part I" (M. R. J. Dack, ed.), pp. 380–441. Wiley (Interscience), New York.
Conder, J. R. (1968). *In* "Progress in Gas Chromatography" (J. H. Purnell, ed.), pp. 209–270. Wiley (Interscience), New York.
Coupe, N. B., Jones, C. E. R., and Stockwell, P. B. (1973). *Chromatographia* **6**, 483–488.
Crippen, R. C. (1973). "Identification of Organic Compounds with the Aid of Gas Chromatography." McGraw-Hill, New York.
David, D. J. (1974). "Gas Chromatographic Detectors." Wiley, New York.
Del Nogre, S., and Juvet, R. S. (1962). "Gas–Liquid Chromatography." Wiley (Interscience), New York.
Derge, K. (1972a). *Chromatographia* **5**, 415–421.
Derge, K. (1972b). *Chromatographia* **5**, 284–290.
Derge, K. (1972c). *Chromatographia* **5**, 335–338.
Dietz, W. A. (1967). *J. Gas Chrom.* **5**, 68–71.
Drozd, J. (1975). *J. Chrom.* **113**, 303–356.
Ettre, L. S. (1965). "Open Tubular Columns in Gas Chromatography." Plenum Press, New York.
Ettre, L. S. (1974). *Chromatographia* **7**, 261–268.
Ettre, L. S., and March, J. (1974). *J. Chromatogr.* **91**, 5–24.
Gill, J. M. (1976). *J. Chrom. Sci.* **14**, 165.
Haken, J. K. (1975). *J. Chrom. Sci.* **13**, 430–439.
Hartmann, C. H. (1971). *Anal. Chem.* **43**, 113A (No. 2).
Hawkes, S., Grossman, D., Hartkopf, A., Isenhour, T., Leary, J., Parcher, J., Wold, S., Yancey, J. (1975). *J. Chrom. Sci.*, **13**, 115–117.
Hollis, O. L. (1966). *Anal. Chem.* **38**, 309–316.
Johnson, H. W. (1968). *Adv. Chrom.* **5**, 175–228.
Kaiser, R. (1970a). *Chromatographia* **3**, 127–133.
Kaiser, R. (1970b). *Chromatographia* **3**, 383–387.
Kiselev, A. V., and Yashin, Ya. I. (1969). "Gas Adsorption Chromatography." Plenum Press, New York.
Kokes, R. J., Tobin, H., Jr., and Emmett, P. H. (1955). *J. Am. Chem. Soc.* **77**, 5860–5862.
Kovats, E. (1958). *Helv. Chim. Acta* **41**, 1915.
Kovats, E. (1965). *Adv. Chrom.* **113**, 303–356.
Laub, R. J., and Pecsok, R. L. (1974). *J. Chromatogr.* **69**, 3–15.
Leary, J. J., Justice, J. B., Tsuge, S., Lowry, S. R., and Isenhour, T. L. (1973). *J. Chrom. Sci.* **11**, 201–206.
Leathard, D. (1975). *Adv. Chrom.* **13**, 265–303.
Leathard, D. A., and Shurlock, B. C. (1970). "Identification Techniques in Gas Chromatography." Wiley (Interscience), New York.
Lewis, J. S. (1963). "Compilation of Gas Chromatographic Data." ASTM STP 343, Philadelphia, Pennsylvania.
Li, K. P., Duewer, D. L., and Juvet, R. S. (1974). *Anal. Chem.* **46**, 1209–1214.
Little, J. N., Horgan, D. F., Cotter, R. L., and Vivilecchia, R. V. (1974). *In* "Bonded Stationary Phases in Chromatography" (E. Grushka, ed.), pp. 39–58. Ann Arbor Science Publ., Ann Arbor, Michigan.
Ma, T. S., and Ladas, A. S. (1976). "Organic Functional Group Analysis by Gas Chromatography." Academic Press, New York.
Mann, J. R., and Preston, S. T., Jr. (1973). *J. Chrom. Sci.* **11**, 216.

Maynard, V. R., and Grushka, E. (1975). *Adv. Chrom.* **12**, 99–140.
McFadden, W. H. (1973). "Techniques of Combined Gas Chromatography/Mass Spectrometry." Wiley, New York.
McNair, H. M., and Bonelli, E. J. (1968). "Basic Gas Chromatography." Varian Instruments, Palo Alto, California.
McReynolds, W. O. (1970). *J. Chrom. Sci.* **8**, 685–691.
Natush, D., and Thorpe, T. (1973). *Anal. Chem.* **45**, 1184.
Nelsen, F. M., and Eggertson, F. T. (1958). *Anal. Chem.* **30**, 1387.
Okamura, J. P., and Sawyer, D. T. (1973a). *In* "Separation and Purification Methods" (E. S. Perry and J. Van Oss, eds.), Vol. 1, pp. 409–475. Dekker, New York.
Okamura, J. P., and Sawyer, D. T. (1973b). *Anal. Chem.* **45**, 80–84.
Pecsok, R. L. (1959) "Principles and Practice of Gas Chromatography." Wiley, New York.
Phillips, C. S. G. (1971). *In* "Gas Chromatography–1970" (R. Stock, ed.), pp. 1–17. Institute of Petroleum, London.
Prevot, A. F. (1969). *In* "Practical Manual of Gas Chromatography" (J. Trauchant, ed.), pp. 137–158. Elsevier, Amsterdam.
Purnell, H. (1962). "Gas Chromatography." Wiley, New York.
Purnell, J. H. (1967), *In* "Gas Chromatography 1966" (A. B. Littlewood, ed.), p. 3. Institute of Petroleum, London.
Rehak, V., and Smolkova, E. (1976). *Chromatographia* **9**, 219.
Rezl, V., and Janák, J. (1973). *J. Chromatogr.* **81**, 233–260.
Rijks, J. A., and Cramers, C. A. (1974). *Chromatographia* **7**, 99–106.
Rijnders, G. W. A. (1966). *Adv. Chrom.* **3**, 315–258.
Rohrschneider, L. (1966). *J. Chromatogr.* **22**, 6–22.
Sawyer, D. T., and Hargrove, G. L. (1968). *In* "Progress in Gas Chromatography" (J. H. Purnell, ed.) pp. 325–359. Wiley, New York.
Schupp, O. E. (1971a). "Gas Chromatographic Data Compilation," 2nd ed., Suppl 1. ASTM AMF 25A S-1, Philadelphia, Pennsylvania.
Schupp, O. E. (1971b), *J. Chrom. Sci.* **9(7)**, 12A.
Schupp, O. E., and Lewis, J. S. (eds.) (1967). "Gas Chromatographic Data Compilation." ASTM DS-25, Philadelphia, Pennsylvania.
Snyder, L. R. (1968). "Principles of Adsorption Chromatography," pp. 155–189. Dekker, New York.
Supina, W. R., and Rose, L. P. (1970). *J. Chrom. Sci.* **8**, 214–217.
Suzuki, M., and Smith, J. M. (1975). *Adv. Chrom.* **13**, 213–263.
Van Deemter, J. J., Zuiderweg, F. J., and Klinkenberg, A. (1956). *Chem. Eng. Sci.* **5**, 271.
Van Swaay, M. (1969). *Adv. Chrom.* **8**, 363–385.
Wold, S. (1975). *J. Chrom. Sci.* **13**, 525–532.
Yonovskii, M. I., and Berman, A. D. (1972). *J. Chromatogr.* **69**, 3–15.
Zlatkis, A., and Pretorius, P. (eds.) (1971). "Preparative Gas Chromatography." Wiley (Interscience), New York.

Mass Spectrometry: Instrumentation

Bruce N. Colby

Chemistry and Chemical Engineering
Systems, Science and Software
La Jolla, California

I. Introduction

Mass spectrometry is an analytical technique wherein a material is ionized and the mass and relative abundance of ions analyzed to produce a mass spectrum. The sample may be ionized in a variety of ways depending on its

ISBN 0-12-430801-5

physical state and on the type of information desired. The mass and relative abundance of the ions detected are both qualitative and quantitative measures of the species present in the sample.

A. History

The mass spectrometer was first introduced by J. J. Thomson (1911) in the early 1900s. An electrical discharge was used to produce vapor phase ions which were electrically accelerated into a region of parallel electric and magnetic fields; here the ions were separated according to their kinetic energy and momentum. The ions struck a photographic plate forming a latent image which, on development, indicated mass and relative abundance by the location and extent of exposure, respectively (Fig. 1). Using this apparatus it was shown that all the atoms of an element have one of a limited number of well-defined masses and that the atomic weight of an element is the weighted average of these well-defined masses called isotopes.

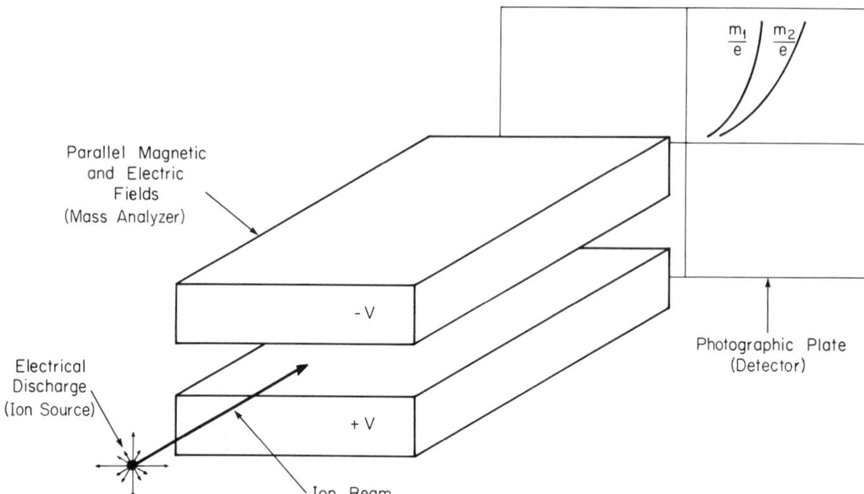

Fig. 1 J. J. Thomson's "positive ray analyzer" in which ions, accelerated into parallel magnetic and electric fields, are separated according to their mass-to-charge (m/e) ratio.

The first analytical mass spectrometer was built in 1920 by F. W. Aston (1919, 1927) and for the next 20 years mass spectrometers were used to analyze the isotopes of all the elements to determine their exact masses and their relative abundances. This information was instrumental in establishing the mass equivalent of nuclear binding energy ($E = mc^2$).

Just after the Second World War, mass spectrometers became commercially available, and possessed all the essential features of a modern mass spectrometer (sample inlet, ion source, mass analyzer, and ion detector). These instruments were intended to be used in the search for oil by analyzing the atmosphere in the vicinity of a prospective well. While they were a failure at oil prospecting, precise and accurate determinations could be made of volatile components in the organic mixtures encountered in the petroleum and chemical industries. As more and more compounds were analyzed, mass spectrometry developed into a powerful tool for the identification of molecular structures. Currently, mass spectrometric analysis spans an incredible range of applications. It has been used to establish metabolic pathways, identify insect sex attractants, measure diffusion rates in solids, analyze trace elements in lunar soil, and find leaks in refrigeration systems; one has even been sent to Mars in an attempt to find life.

B. General Description

A mass spectrometer system may be divided into four sections: sample inlet, ion source, mass analyzer, and ion detector (Fig. 2).

Fig. 2 Mass spectrometer system components.

The function of the sample inlet is to provide a means for transporting the sample from the atmosphere into the vacuum of the ion source. A variety of inlet systems are currently available to handle gaseous and solid samples. Liquid samples are generally vaporized and introduced as gases but are occasionally frozen and treated as solids. The most common gas inlets are the batch inlet and the gas chromatograph. Nonvolatile or thermally labile organic samples are usually introduced into the ion source on the tip of a direct probe; heat is applied to the probe tip causing the sample to vaporize directly into the ionizing region. The introduction of solid, nonvolatile materials, such as metals or semiconductors, into an ion source is accomplished either by insertion through a vacuum lock or by venting the ion source vacuum housing to atmospheric pressure.

Ion sources can be divided into two basic categories: those which ionize gases and those which ionize solids. Typically, gas sources are used to analyze organic compounds, while solid sources are used to characterize the elemental composition of materials, such as alloys or semiconductors. Gas sources involve mild ionizing conditions which partially fracture molecules so as to produce structural information. The most commonly used gas sources are the electron ionization source and the chemical ionization source. Ion sources for solids involve large amounts of energy which reduce the sample to elemental species during ionization.

Once the ions have been formed, they are accelerated by an electrical field into the mass analyzer which separates the ions according to their masses. As with sample inlets and ion sources, there are a variety of ways to analyze or separate ions. In a magnetic mass analyzer, they are separated as they pass through a magnetic field perpendicular to their direction of travel; the lighter ions follow a path of smaller radius than do the heavier ones and a location-based mass spectrum is produced. In a quadrupole mass analyzer, an oscillating electronic field is created along the ion flight path such that only one particular mass/charge (m/e) ratio will resonate with it. All other m/e's will have unstable trajectories and fly out of the analyzer before reaching the detector.

There are two basic types of ion detectors: photographic and electronic. The photographic detector consists of a glass photographic plate inserted into the vacuum system of the magnetic analyzer which is exposed by the impact of mass-analyzed ions. The location and extent of the exposed areas indicate the mass and abundance of the ions, respectively. Two electronic detectors are in general use: the Faraday collector and the electron multiplier. The Faraday collector is a metallic cup which captures the desired ions generating an electrical current which is measured directly. The electron multiplier also produces an electrical current, but the ion current is amplified by a factor of up to 10^8.

C. *Performance Specifications*

There are several ways of describing the performance characteristics of a mass spectrometer. Two of the most important are the mass resolving capabilities and the sensitivity. Other important considerations are both long- and short-term mass stability and the ability to accurately reproduce data from day to day.

1. *Resolving Power and Resolution*

Resolving power (RP) is a measure of the ability to separate two adjacent mass peaks. Several definitions of RP exist, the most common one is given by

$$\mathrm{RP} = m_1/(m_2 - m_1) \tag{1}$$

where m_1 and m_2 are the masses involved, m_1 being the lower mass. When known mass values are inserted into Eq. (1), a measure of the instrument performance required to achieve a separation is generated. For example, if a mass spectrum is to be obtained up through mass 600, masses 600 and 601 must be distinguishable and a RP of 600 is required. To evaluate instrument performance in terms of RP, it is necessary to define the extent of mass separation. This is generally done by describing the "percent valley" between the two peaks (Fig. 3). Ideally, the RP of a mass spectrometer is given for symmetrical peaks of equal intensity which overlap to produce a 10% valley between them. In practice, this situation rarely exists. Still, it is desirable to estimate the RP from a real situation. One way to do this is to "synthesize" an adjacent peak of equal intensity with a piece of tracing paper and move it across the mass scale until the tailing edge of one peak crosses the leading edge of the other peak at 5% of the maximum peak height. At this point the combined contribution of the two peaks to the valley would be 10%. Inserting the mass value for the peak used for m_1 and the mass scale equivalent for the distance the synthesized peak was moved as $m_2 - m_1$, a resolving power may be calculated. Resolving power is a variable which depends on several parameters including the method of calculation. Consequently, any RP value should be considered as an approximation of absolute instrument performance. Also, it should be kept in mind that it is relatively easy to distinguish the presence of two peaks even at a 50 or 60% valley.

Resolution is simply the reciprocal value of the resolving power. The term is employed primarily by those doing "high resolution" mass spectrometry where RPs in the range of 10,000 to 100,000 are employed. Here the desire is not just to determine the presence or absence of a peak at some nominal mass,

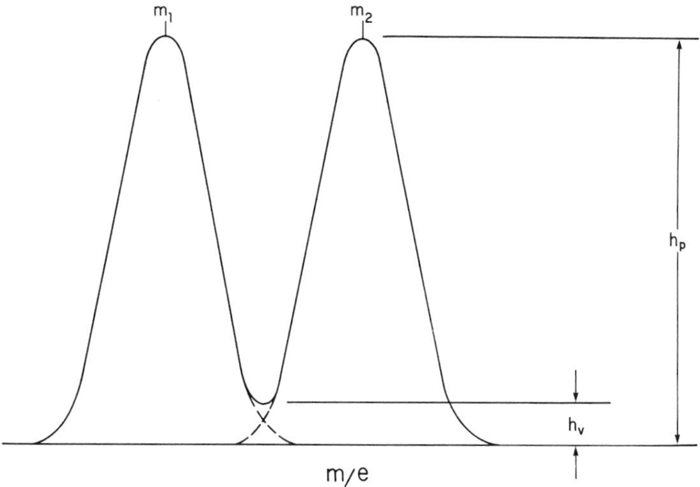

Fig. 3 Illustration of two symmetrical mass peaks resolved with a valley (h_v) 10% of the peak height (h_p); a 10% valley.

but to accurately establish the exact mass of the peak. At a RP of 10,000 it is readily possible to distinguish a difference of 1 part in 10,000 or 100 parts per million (ppm). The peak centroid location, however, can be established with a considerably greater degree of accuracy, say, up to about ± 5 ppm, with the help of sophisticated peak maximum detection circuits or computer techniques. With this kind of resolution, it is possible to determine the empirical formula of an ion based on its exact mass. The exact masses of $C_{14}H_{32}N_2O$ and $C_{15}H_{32}O_2$ calculated from the data in Table I are 244.2514 and 244.2402, respectively. They can be distinguished from each other based on exact mass measurement if the measurement can be made to slightly better than $+25$ ppm. The numerical difference between exact mass and nominal mass is often referred to as the mass defect of an ion species. It should be pointed out that the terms "resolution" and "resolving power" are used almost interchangeably in the literature. Consequently, it is expedient to establish which meaning is intended prior to drawing conclusions.

TABLE I

Exact Mass of Some Common Isotopes

^{1}H	1.007825	^{19}F	18.998405
^{12}C	12.000000	^{28}Si	27.976927
^{14}N	14.003074	^{32}S	31.972074
^{16}O	15.004015		

2. *Sensitivity*

There are two important types of sensitivity to be considered, basic sensitivity and system sensitivity. Basic sensitivity describes the efficiency of the ion production-to-collection process, while system sensitivity includes the efficiency of the inlet system.

Basic sensitivity indicates the number of ions collected at the detector for a given sample flow rate into the ion source and can be expressed in coulombs per mole. Presently there is no uniformly accepted compound or mass peak used for this measurement; however, a volatile, nonlabile compound measured at its most intense (base) peak is a typical choice. Normal values for the basic sensitivity range from about 10^{-2} to 10^{1} C/mole, and for a given instrument, it is approximately inversely proportional to resolving power. These values vary from instrument to instrument as a function of original design, manufacturing quality, and present operating conditions.

Although basic sensitivity describes the efficiency of the ion source, mass analyzer, and detector, it is not a particularly useful parameter for describing the ability of the total system to perform an analysis. A good way to describe system performance is in terms of the ability to produce a recognizable mass spectrum for a given mass scan rate and sample load (moles/sec) through a specified inlet system. The type of compound to use for this evaluation is one subject to thermal decomposition or adsorption at "active sites" in the inlet system and ion source. The spectral quality should be judged in terms of the number of ions collected and the quality of the fragmentation encountered. Due to the complexity of these requirements, it is no wonder that no universally accepted procedure exists, each manufacturer and mass spectrometry expert tending toward some different specific set of criteria. An example of one set of criteria would be as follows: for a gas chromatographic peak 4 sec wide at its half height, an interpretable mass spectrum (about 300 ions in the base peak and peaks 2% of the base peak just detectable above the noise level) at a resolving power of 600 (10% valley) is produced for 10 pmole of sample (methyl stearate) injected onto the column.

D. Vacuum System

In all present forms of mass spectrometers, ions are physically separated by some combination of electrostatic and/or magnetic fields in a vacuum environment. The vacuum must be of sufficient quality to ensure that the accelerated ions do not collide with residual gas molecules and become deflected from their intended paths. This collisional defocusing of the ion beam leads to losses in both sensitivity and resolution but is not a significant factor at 10^{-6} Torr or less except when flight paths greater than 1m are encountered. Other possible problems that can result at high pressures are electrical discharges, shortened filament life in electron ionization sources, high background signal, and rapid contamination of the ion optical surfaces from the deposition of gaseous species that form electrically insulating films.

The pressure in a vacuum system is given by

$$P = Q/C \tag{2}$$

where Q is the quantity of gas loading the system (Torr liters/sec) and C is the conductance of the system through the pumps (liters/sec). Conductance is sometimes referred to as pumping speed. Perhaps a more familiar way to express Q is in terms of the gas law,

$$Q = PV/t = nRT/t \tag{3}$$

where V is volume, T the absolute temperature, R the gas constant, and t the time interval. The gas load of a system has many sources, the most obvious being the sample itself and any leaks in the system hardware. Conductance is a function of the hardware employed along with its configuration. The conductance of items in parallel is the sum of the conductances of the

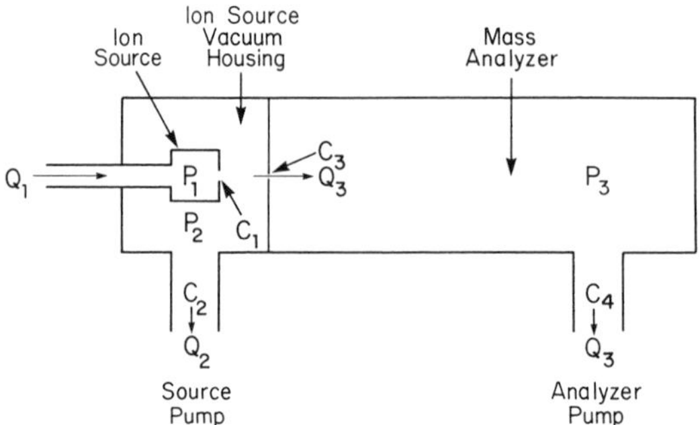

Fig. 4 Differentially pumped mass spectrometer vacuum system.

individual items,

$$C_{\text{parallel}} = \sum C_{\text{i}} \tag{4}$$

For items in series the conductance is given by

$$C_{\text{series}}^{-1} = \sum C_{\text{i}}^{-1} \tag{5}$$

where C_{i}'s are conductances of individual component sections. In a typical mass spectrometer operating at 10^{-6} Torr or less and with a gas load of 2×10^{-5} Torr liters/sec, a conductance of at least 20 liters/sec must exist at the mass analyzer. When the conductances of all the pumping arms, valves, traps, and baffles are taken into account using Eqs. (4) and (5), the pump will have a conductance or pumping speed of about 200 liters/sec. For a mass spectrometer coupled with a gas chromatograph (GCMS) the gas load may be considerably higher, about 2.5×10^{-2} Torr liters/sec. While it is possible to increase pumping speed to handle the increased gas load, a technique called differential pumping is generally employed. This is achieved by providing separate pumping systems for the ion source vacuum housing and for the mass analyzer, and by separating the two sections with a relatively small (low conductance) aperture. In this way the gas load to the mass analyzer section is greatly reduced, resulting in an acceptably low pressure. Higher pressures may be tolerated in the ion source vacuum housing since the ions travel a much shorter distance there than in the mass analyzer.

A differentially pumped mass spectrometer vacuum system is shown in Fig. 4. Ion source pressure P_1 is governed by the gas load Q_1 and the conductance C_1. Ion source vacuum housing pressure P_2 is also set by Q_1, but here the conductances C_2, C_3, and C_4 must all be considered. The gas load is divided between C_2 and the series conductances of C_3 and C_4; so P_2 is given by

$$P_2 = Q_1/[C_2 + (C_3^{-1} + C_4^{-1})^{-1}] \tag{6}$$

Substituting a gas load of 2.5×10^{-2} Torr liters/sec for Q_1 and conductance values of 50 liters/sec for C_2 and C_4 and 10^{-1} liters/sec for C_3, the pressure in the ion source vacuum housing becomes about 5×10^{-4} Torr. Note that the conductance C_2 has the most influence on P_2. This means that the gas load on the source pump, Q_2, is much greater than on the analyzer pump, Q_3. Since the gas load on the mass analyzer, Q_3, is the same as the gas load on the analyzer pump, the analyzer pressure is given by

$$P_3 = P_2(C_3/C_4) = 10^{-6} \quad \text{Torr} \tag{7}$$

Because of this, differentially pumped systems with high total gas loads can produce results similar to a singly pumped systems with low gas loads.

II. Inlet Systems

Since mass spectrometers must operate under high vacuum conditions, it is necessary to have some form of interface to introduce the sample from atmospheric pressure into the vacuum system. There are a great variety of interface or inlet systems in existence; the most commonly encountered ones are described in the following sections.

A. Batch Inlet

The batch inlet consists of an evacuated sample reservoir from which a gas phase sample leaks through a small conductance into the ion source. A gaseous or high vapor pressure (≥ 1 mm Hg) liquid sample is injected through a silicone rubber septum into a heated expansion volume. A molecular leak meters the flow of sample from the sample reservoir into the ion source (Fig. 5). The expansion volume or reservoir is typically a 1-liter chamber

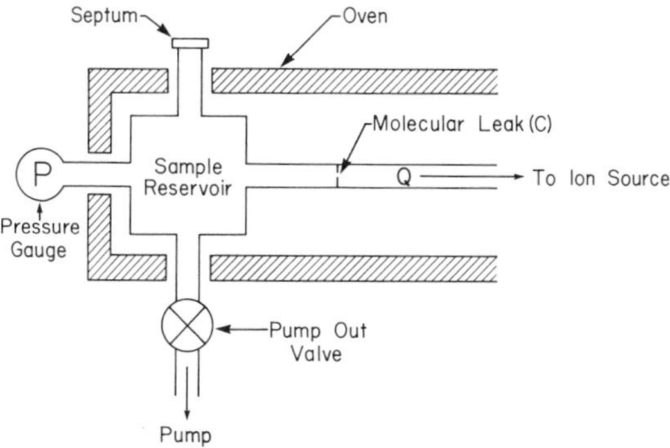

Fig. 5 Batch inlet for gaseous samples.

constructed of stainless steel, although glass can be used to reduce the decomposition of more labile samples. Reservoir pressures on the order of 10^{-2} Torr and molecular leak conductances of about 0.1 cc/min are typical.

A properly designed batch inlet produces a very stable, reproducible sample flow. This is particularly desirable for producing high quality ion abundance data for obtaining reference mass spectra or for evaluating mass spectrometer sensitivity. If the leak is operating under molecular flow conditions (mean free path ≫ orifice diameter), the quantity of gas passing through the orifice will be given by

$$\text{moles/sec} = PC/\text{RT} \tag{8}$$

Equation (8) assumes that the pressure is much lower on the ion source side of the leak than on the reservoir side and that the probability of a molecule in the reservoir passing through the leak is a function only of its pressure in the reservoir. If the conductance C is not known, it may be calculated by measuring the decrease in intensity of a mass spectral peak over an appropriate time interval and inserting the appropriate values into the equation

$$C = V \ln(P_0/P)/t \tag{9}$$

where V is volume, P_0 the initial pressure, and P the pressure at the end of the time interval t.

B. Direct Insertion Probe

The direct insertion probe is for the introduction of low volatility samples directly into the ion source. This is done by placing the sample in a small glass capillary at the tip of the probe, inserting the probe through a vacuum lock into the ion source, then heating the probe tip to volatilize the sample (Fig. 6). Sample flow from the probe tip is less stable than that from the batch inlet and does not provide as effective a means for evaluating basic spectrometer sensitivity. The flow rate, however, is sufficiently stable to provide good reference spectra for many compounds. The direct probe is a relatively quick and easy means for introducing low volatility samples. This type of sample, however, tends to be thermally labile and consequently tends to contaminate the ion source and reduce performance.

C. Gas Chromatograph

Recently the most versatile and widely applied inlet in mass spectrometry has become the gas chromatograph (GC). When these techniques are combined, the result is a significantly more powerful analytical tool than the two

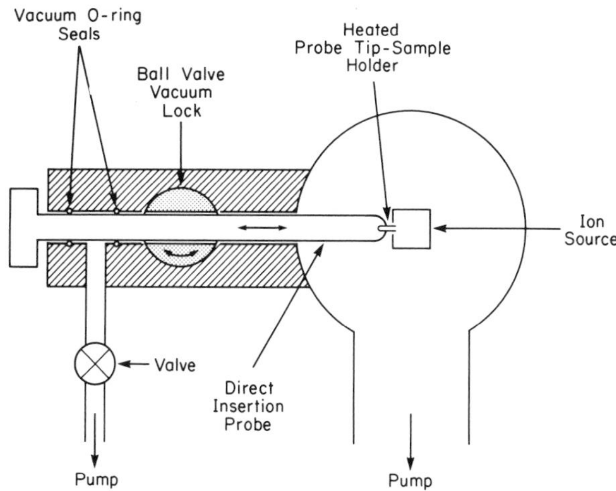

Fig. 6 Direct insertion probe for solid samples.

taken separately. Simultaneously the components of complex mixtures may be separated and identified.

1. *The GCMS Interface*

Because GC analysis requires a carrier gas flow rate of 30–70 cc/min, direct introduction of the total effluent into the mass spectrometer is usually impractical. The simplest way to reduce the high gas load is to place an effluent splitter between the two instruments to interface them. With only 1% of the total effluent diverted to the mass spectrometer, the gas load is reduced to about 10^{-2} Torr liters/sec and an acceptable pressure can be obtained in the ion source vacuum housing. However, effluent splitting reduces sensitivity because sample as well as carrier is removed. To overcome this problem with effluent splitting, several types of enriching devices have been developed such as the jet separator, the effusion separator, and the semipermeable membrane separator. Two considerations are of importance in these separator designs: enrichment ratio and sample transmission. Enrichment ratio is the increase in sample-to-carrier gas ratio produced by the separator, while sample transmission is the fractional quantity of sample passed to the mass spectrometer. Ideally the separator should produce maximum enrichment while maintaining maximum sample transmission.

(*a*) *The Jet Separator* The jet separator (Ryhage, 1964; Ryhage *et al.*, 1965) makes use of the fact that diffusion rate is a function of molecular weight. As the GC effluent is forced through an orifice (jet) into an evacuated region, the lighter carrier gas molecules diffuse away from the center axis more rapidly than do the heavier sample molecules (Fig. 7). Coaxial with the jet is a skimmer which has an orifice to accept the sample-rich center of the molecular beam. Since diffusion is inversely proportional to the square root of molecular weight, it is an advantage to use the smallest practical molecule, helium, as the carrier gas.

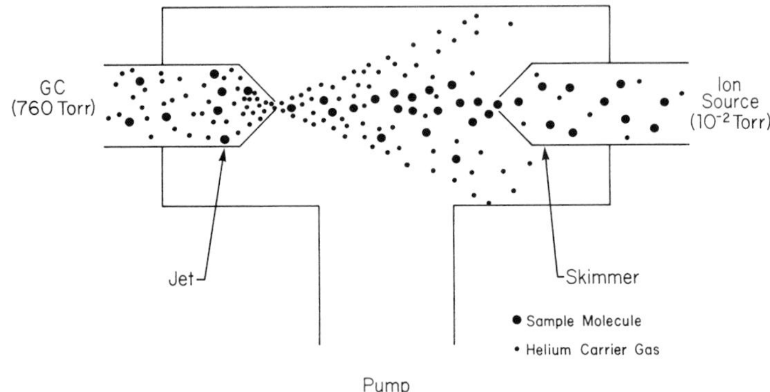

Fig. 7 Jet separator.

A major advantage of the jet separator is that it may be constructed entirely from glass, thereby greatly reducing the number of "active sites" for sample adsorption or decomposition. Also, it is applicable to all types of samples with enrichment ratio affected primarily by molecular weight. With a single stage jet, the transmission ratio may be as high as 0.9 and the enrichment ratio as much as 10. For a mass spectrometer with low pumping speed, two jets can be used in series to improve the enrichment ratio at the expense of some sample transmission.

Orifice diameter, spacing, and alignment are critical (+0.001 in.) in obtaining optimal operation. Consequently, care must be taken to prevent particulate material such as column packing from becoming lodged in the jet orifice. Also, operation is maximized only over a limited range of carrier gas flow rates. The disadvantages of the jet separator include the molecular weight bias inherent with its principle of operation although this is generally more of a theoretical than practical problem.

(*b*) *Effusion Separator* The effusion separator (Watson and Biemann, 1964), like the jet separator, is based on a diffusion related principle (Fig. 8). The rate at which gas passes through porous media is a function of its partial pressure P_i and molecular weight M. The quantity of gas, Q, to flow through the pores is given by

$$Q = P_i(M)^{-1/2}k \tag{10}$$

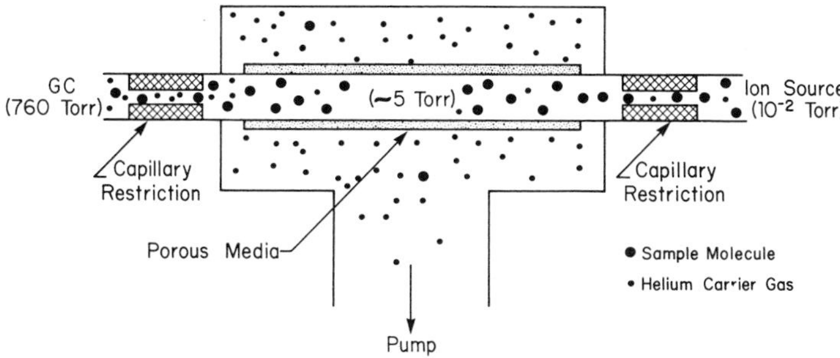

Fig. 8 Effusion separator.

where k is a constant. Consequently, the carrier gas which has a high partial pressure and low molecular weight will tend to pass through the pores and not enter the ion source.

The performance of this separator is maximized for some specific set of flow conditions and changes in GC carrier flow will decrease performance. The flow of gas is generally adjusted by altering the lengths and/or diameter of capillary restrictions at the entrance and exits of the separator. It may be constructed entirely of glass; however, scintered glass of uniform pore size is difficult to obtain, making commercial production a problem. Although high quality ceramic and porous metal materials are available, they tend to have a substantially greater number of reactive sites for adsorption and decomposition. On the other hand, the effusion separator does not require the highly precise mechanical alignment necessary for the jet separator.

(*c*) *Semipermeable Membrane Separator* The semipermeable membrane separator technique is based on conductance differences between sample and carrier through media such as silicone membrances (Fig. 9). The conductance

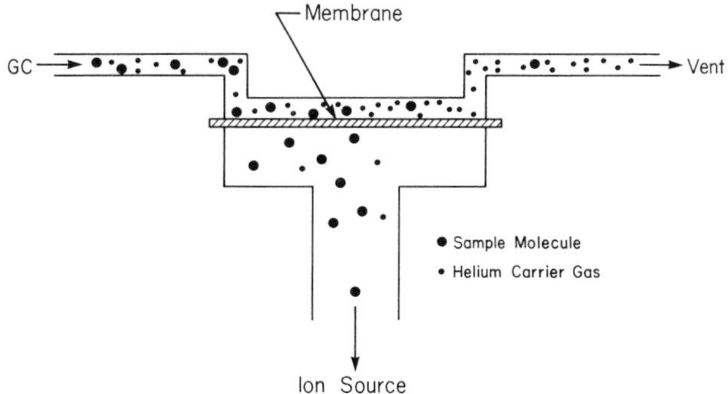

Fig. 9 Semipermeable membrane separator.

or permeability p of a compound through a membrane is given by

$$p = SDA(\Delta P/d)t \quad (11)$$

where S is the solubility and D is the diffusion coefficient of the compound in the membrane material. The terms A, d, ΔP, and t represent the area, thickness, pressure drop across, and exposure time to the membrane, respectively. For membranes constructed of silicone rubbers, p for most organic compounds is higher than for inert carrier gases by about a factor of 1000. This is due primarily to differences in solubility rather than diffusion, so mass dependence is not a major factor either in the enrichment ratio or in transmission. Membrane separators are extremely useful for mass spectrometers which are particularly sensitive to gas load since carrier gas partial pressures may be kept as low as 10^{-7} Torr while maintaining sample transmission at 40%.

However, membrane separators may produce peak broadening due to high dead volume, high background due to bleeding of the membrane material, and active sites from the metal used in its construction. Further, no membrane is likely to be uniformly highly permeable to all compounds except the carrier gas; therefore, it is wise to direct the separator vent flow into a flame ionization detector or other conventional GC detector in an attempt to monitor "lost" peaks.

2. *Mass Spectrometer Considerations*

(*a*) *Mass Spectra* Due to the dynamic nature of sample load generated by the GC, special considerations must be made to ensure that useful mass spectra are produced. A GC peak emerging from a packed column may be only 5–10 sec wide and, during this brief period, a characteristic mass spectrum must be generated. If the spectrum is taken while sample flow is rapidly increasing or decreasing, the ion intensities will be distorted relative to each other. The best way to reduce distortion is to scan rapidly so that sample load changes only a few percent during the scan. Fortunately, most compounds volatile enough to pass through a GC have molecular weights less than 600. Consequently, a mass scanning rate of about 500 atomic mass units (amu) per second is acceptable. Faster scanning rates would reduce distortion somewhat but would also reduce sensitivity. In fact, occasionally it is necessary to scan slowly to increase sensitivity and accept a distorted spectrum. Sensitivity may also be increased by reducing spectral resolution to coincide with the maximum expected mass.

(*b*) *Selected Ion Monitoring* While the sensitivity of GCMS is considered in terms of obtaining a complete mass spectrum (see Section I.C), it is possible to achieve a much lower detection limit by monitoring less than the full mass spectrum. By using long integration times at a limited number of masses, the signal-to-noise ratio (S/N) can be greatly enhanced. It is necessary that the sought-for compound have a known characteristic mass or masses. Identification is based on retention time as well as on the ion abundance data generated. The mass spectrometer is programmed to focus a selected mass at the detector for a specified period of time and to integrate the ion current over that period. At the end of the integration period, the focus is switched to the next selected mass in the sequence. When the ion current of the last mass has been integrated, the focus is switched back to the first mass and the cycle repeated. The maximum S/N would be obtained for data collected at only a single mass; each additional mass monitored decreases the S/N ratio and reduces sensitivity. If all masses were selected, a full mass spectrum would result and any sensitivity advantage would be entirely lost. This technique is known by a variety of terms including multiple ion monitoring, selected ion recording, multiple ion detection, mass fragmentography, and selected ion monitoring (SIM), the term which will be used here.

Data produced using the SIM technique is typically plotted as ion current versus time, resulting in records extremely reminiscent of gas chromatograms. In fact, when the mass spectrometer is used in the SIM mode, it is reasonable to consider the GC as the primary instrument and the mass spectrometer merely as the detector. The quality of the data is dependent to a very large degree on the quality of the chromatography and, even though the mass spectrometer is an extremely versatile detector due to its high degree of specificity, it will not compensate for poor chromatography or inadequate sample preparation. This is particularly true when extremely small quantities of sample are analyzed such as in some environmental and biomedical situations. The SIM technique has been used to great advantage in those cases where complex mixtures containing trace amounts of sample, sometimes as little as a few picograms (10^{-12} grams), must be analyzed.

(*c*) *The Role of the Computer* With GCMS, massive amounts of data may be generated. A typical analysis might consist of 450 spectra each with 750 masses. For an unaided operator to try to keep up with this sort of data generation is a formidable job in real time data handling. To ease the task of assigning masses, measuring ion current intensities, keeping track of which mass spectrum corresponds to which GC peak, small dedicated computers have been interfaced to GCMS systems with great success. Not only will the computer provide all the required bookkeeping services, but it will also perform area integrations for quantitative analysis, subtract background spectra from sample spectra, plot normalized data, and help with spectral interpretation. The advantages of the computer are all based on its ability to rapidly store and process large quantities of data. A computer is generally a worthwhile addition to a GCMS; however, it will approximately double the system price. Further discussion of computer systems is included in Sections V.D.2 and V.D.3.

III. Ion Sources

A. Electron Ionization

1. *Ionization Process*

Ions are formed in electron ionization (EI) by directing a beam of energetic electrons through a region containing a gaseous sample. When an electron and a molecule undergo collision, energy from the electron is imparted to the ground state molecule via a Franck–Condon mechanism, raising the molecule to some higher electronic and vibrational energy level. Since ion source pressure is low (less than 10^{-2} Torr), collisional deexcitation is unlikely. The most probable ways for the excited molecular species to lose energy are through electron losses, photon losses, and fragmentation. Fragmentation occurs as vibrational energy builds up along weak bonds, causing them to rupture. This breaking of bonds is very dependent on molecular structure, and from the appearance of characteristic ion fragments, the structure of the parent molecule can be deduced. The production of structurally related ion fragmentation patterns by electron ionization has resulted in its being the most commonly used and most highly developed ion source in mass spectrometry.

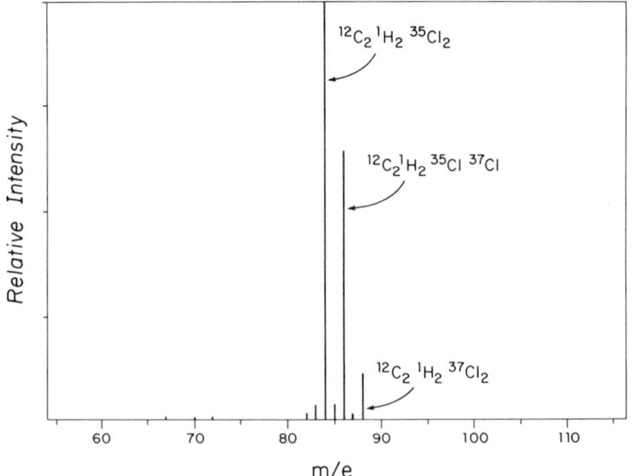

Fig. 10 Isotopic abundance pattern of the dichloromethane molecular ion, $CH_2Cl_2^{+\cdot}$

Schematically expressed, the ionization process for a hypothetical molecule ABCD is

$$ABCD + e^- \longrightarrow (ABCD^*)^- \longrightarrow ABCD^{+\cdot} + 2e^-$$

where $ABCD^{+\cdot}$ is the molecular ion. The positive ions are generally used for analysis because they are generally the most abundant species. The stability of the molecular ion depends on the strengths of the individual bonds involved and on the amount of internal energy present. If no or little excess internal energy is present, the mass spectrum will consist of the isotopic abundance pattern for the specific combination of elements present in the sample molecule (Fig. 10). Isotopic abundances for some commonly occurring elements are given in Table II.

TABLE II

Isotopic Abundance of Some Commonly Occurring Elements

^{1}H	99.985	^{2}H	0.015				
^{12}C	98.892	^{13}C	1.108				
^{14}N	99.635	^{15}N	0.365				
^{16}O	99.76	^{17}O	0.04	^{18}O	0.20		
^{28}Si	92.18	^{29}Si	4.71	^{30}Si	3.12		
^{32}S	95.06	^{33}S	0.66	^{34}S	4.20	^{36}S	0.14
^{35}Cl	75.4	^{37}Cl	24.6				
^{79}Br	50.57	^{81}Br	49.43				

More commonly the molecular ion possesses sufficient excess internal energy to undergo decomposition to form a neutral species plus a fragment ion, which may in turn undergo still further decomposition:

$$ABCD^{+\cdot} \longrightarrow ABC^+ + D^\cdot \qquad (ABC^+ \longrightarrow A^+ + BC)$$

$$ABCD^{+\cdot} \longrightarrow AB^+ + CD^\cdot$$

$$ABCD^{+\cdot} \longrightarrow A^+ + BCD^\cdot$$

The neutral species generated may be either molecules or radicals but, because they carry no charge, they are ignored by the mass spectrometer. The electron beam energy, or ionization voltage, which is required to produce a given fragment is called the appearance potential for that fragment (Fig. 11). At low ionizing voltages (6–20 V), the relative abundance of the molecular ion is greater due to the lesser degree of fragmentation. In the 60- to 80-V range there is generally little change in abundance with ionizing voltage and useful fragmentation patterns are produced.

Interpretation of electron ionization spectra is based primarily on the way different compounds fragment. For an unbranched hydrocarbon, such

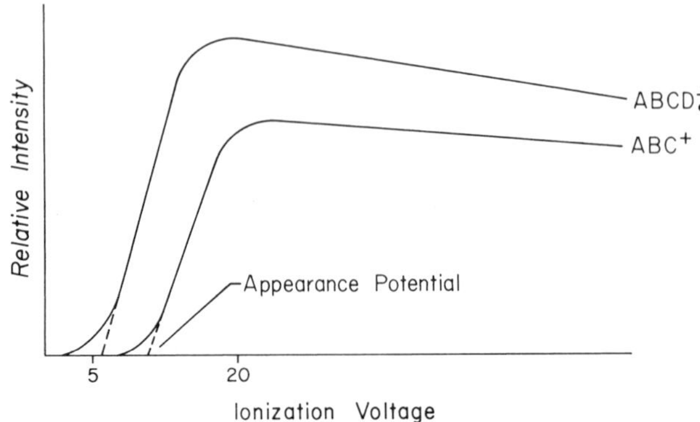

Fig. 11 Appearance potential.

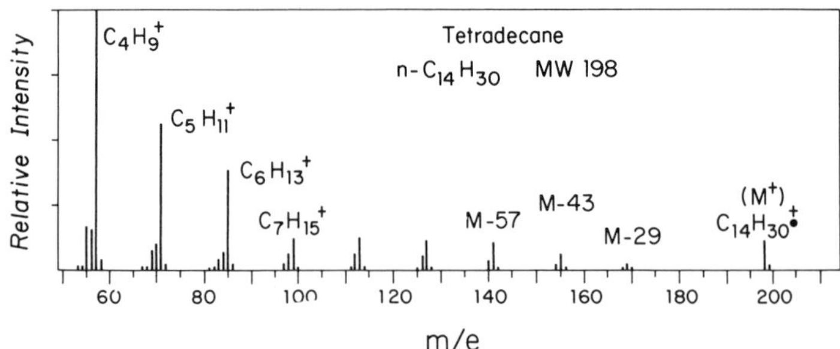

Fig. 12 Electron ionization spectrum of *n*-tetradecane.

as tetradecane (Fig. 12), the most probable fragmentation routes are through successive losses of C_nH_{2n+1}. Table III lists some of the common mass losses from other types of compounds, along with the compound or functional group inferred by the loss. Common losses are associated primarily with the high mass portion of the spectrum, while species called common fragment ions are associated with the low mass portion. Common fragment ions are stable species usually resulting from some functional group on the parent molecule. Several common fragment ions are given in Table IV. A third type of ion produced by certain molecular structures results from the rearrangement of the original ion. This type of ion is much less common than the preceeding two types. A good example is the γ-hydrogen rearrangement in carbonyl compounds known as the McLafferty rearrangement:

$$\left[\text{R, H, O, R}'\right]^{+\bullet} \xrightarrow{-\mathrm{RCH{=}CH_2}} \left[\text{H, O, R}'\right]^{+\bullet}$$

TABLE III

Common Neutral Losses from Molecular Ions

Mass	Entity Lost	Possible Source Compound and Functional Groups[a]
M − 14	CH_2	Homologous series
M − 15	CH_3	$—CH_3$
M − 16	O	$ArNO_2$, —SO—
M − 16	NH_2	$—CONH_2$
M − 18	H_2O	ROH, RCOH, RCOR
M − 27	HCN	ArCN
M − 28	C_2H_4	Ar—O—C_2H_5, $RCO_2C_2H_2$, RCO-n-C_3H_7
M − 29	CHO	—COH
M − 29	C_2H_5	$RCOC_2H_5$, Ar-n-C_3H_7, $—C_2H_5$
M − 30	NO	$ArNO_2$
M − 31	CH_3	$—CO_2CH_3$
M − 41	C_3H_5	$—CO_2C_3H_5$
M − 42	CH_2CO	$—COCH_3$, $ArCO_2CH_3$, $ArNHCOCH_3$
M − 43	CH_3CO	$—COCH_3$
M − 43	C_3H_7	$—C_3H_7$
M − 45	CO_2H	$—CO_2H$
M − 45	OC_2H_5	$—CO_2C_2H_5$
M − 46	NO_2	$ArNO_2$
M − 57	C_4H_9	$—C_4H_9$

[a] Ar, aromatic group; R, Nonaromatic group.

TABLE IV

Some Common Fragment Ions

Mass	Structure	Possible Source Functional Group
30	$CH_2{=}\overset{+}{N}H_2$	$H_2NCH_2—$
31	$CH_2{=}\overset{+}{O}H$	$HOCH_2—$
57	$C_4H_9^+$	$C_4H_9—$
71	$C_5H_{11}^+$	$C_5H_{11}—$
73	$(CH_3)_3Si^+$	$(CH_3)_3Si—$
77	$C_6H_5^+$	$C_6H_5—$
91	$C_7H_7^+$	$C_6H_5CH_2—$
105	$C_8H_9^+$	$CH_3C_6H_4CH_2—$
105	$C_6H_5CO^+$	$C_6H_5CO—$

Such an example is evident at mass 74 in the spectrum of methyl myristate (Fig. 13). In this spectrum, also note the M − 31 peak from the loss of $—OCH_3$ and the M − C_nH_{2n+1} series from the alkane portion of the molecule.

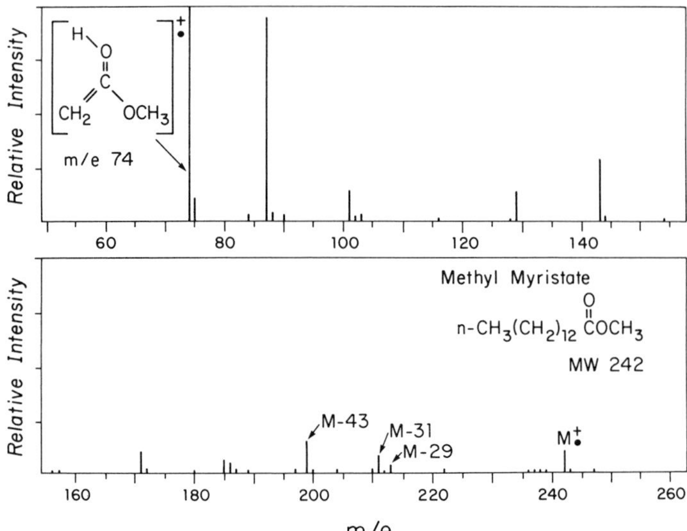

Fig. 13 Electron ionization spectrum of methyl myristate.

One of the most informative ions in the mass spectrum is the molecular ion (M^+) resulting from the unfragmented parent molecule. Three criteria can be used to help identify it if it is present:

(1) it is the highest mass peak in the spectrum;
(2) it is associated with certain common losses; and
(3) it has an odd number of electrons.

The odd electron aspect is important in that relatively few fragments in the spectrum will contain an odd number of electrons; rearrangement ions being the other major source. A useful clue in identifying odd electron ions is provided by the mass of the ions. If the ion has an even mass, it is an odd electron ion unless it contains an odd number of nitrogen atoms. This means that the molecular ion will occur at an even mass unless there are an odd number of nitrogen atoms present in the molecule. Once the molecular ion is identified, an elemental composition in agreement with the molecular weight can be hypothesized. This is done, keeping in mind the isotopic pattern of the molecular ion. From the molecular formula the number of rings plus double bonds in the molecule may be calculated. For the molecule $C_xH_yN_zO_n$,

$$\text{number of rings + double bonds} = x - y/2 + z/2 + 1 \qquad (12)$$

For pyridine, C_5H_5N, the number of rings plus double bonds is four, three for the π-electrons and one for the ring. At this point in the identification it is convenient to write out possible structures for the compound and see which, if any, might be expected to generate the observed spectrum. The molecular ion abundance is usually high in aromatic compounds but decreases with increased aliphatic character and increased branching. Tertiary alcohols, for example, have no significant molecular ion.

2. *Ion Source Design*

There are three items essential to the EI source: a source of electrons, a sample inlet, and an ion optical system to transport the ions into the mass analyzer. The exact configuration of these depends on the type of mass analyzer and sample inlet involved. A typical source for a magnetic system with a gas chromatograph or batch inlet is shown in Fig. 14. An electron current of several hundred microamperes is produced by heating a rhenium filament. The electrons are accelerated toward the block to provide the desired ionization energy and are apertured as they enter the ion source chamber to provide a well-defined beam. A weak magnetic field parallel to the direction of travel of the electrons is provided for two reasons. It forces the electrons to maintain a well-defined path, so that the ions are produced in a small, localized cylindrical volume in the ion source. This facilitates focusing of the ions into the mass analyzer. The magnetic field also causes the electrons to spiral as they travel, increasing the probability

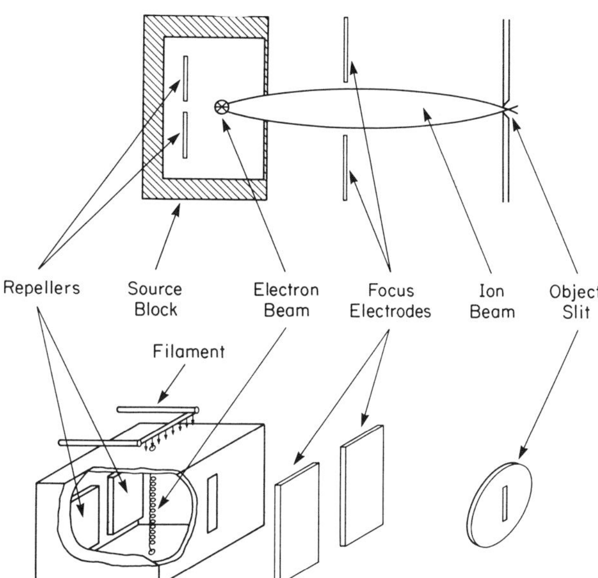

Fig. 14 Electron ionization source.

that a collision will take place. The electrons which fail to undergo collision are collected or trapped at the far side of the ion source and the current produced is used to regulate electron emission from the filament. Stable electron emission must be maintained in order to produce a stable ion beam. Ions are removed from the ion source by the repeller field and the resulting ion beam is focused into the mass analyzer by the focus electrodes.

B. Chemical Ionization

1. *Ionization Process*

Chemical ionization (CI) (Munson and Field, 1966; Munson, 1971) is the term used to describe the ionization resulting from ion–molecule reactions, a process which predominates when ion source pressure is approximately 1 Torr. At this pressure collisional deactivation of excited species occurs, so energy may be transferred from ion to molecule. Physically the CI source is much like an EI source but with lower conductance. The source is "pressurized" with a reactant gas which becomes ionized by the electron beam. The ionized reactant gas molecules undergo collisions with neutral reactant gas molecules, each time losing energy until a relatively stable ion species, the reactant ion, is produced. With methane, a commonly used reactant gas, two significant reactant ions are produced via the reactions

$$CH_4{}^+ + CH_4 \longrightarrow CH_5{}^+ + CH_3$$

$$CH_3{}^+ + CH_4 \longrightarrow C_2H_5{}^+ + H_2$$

The density of reactant ions ($CH_5{}^+$ and $C_2H_5{}^+$) becomes so high that statistically these ions are most likely to cause ionization of a sample molecule. Clearly, this is true only if the partial pressure of sample in the source is low compared to the reactant gas. Ions produced by CI tend to reflect traditional energy arguments used for predicting chemical reactions. Thus, for a hydrocarbon sample molecule, CI results in the loss of a proton to yield the molecular ion minus one, $(M - 1)^+$, an exothermic reaction. The $(M - 1)$ ion is often called the pseudo- or quasi-molar ion.

$$C_{18}H_{38} + CH_5{}^+ \longrightarrow C_{18}H_{37}^+ + CH_4 + H_2, \qquad \Delta H = -27 \text{ kcal/mole}$$

$$C_{18}H_{38} + C_2H_5{}^+ \longrightarrow C_{18}H_{37}^+ + C_2H_6, \qquad \Delta H = -25 \text{ kcal/mole}$$

This form of ionization is much less energetic than that produced with a 70-eV (1600-kcal/mole) electron beam, so generally fewer fragment ions are produced. For hydrocarbons some CI fragment ions with the formula $C_nH_{2n} + 1$ would be observed.

The abundance of CI fragment ions depends primarily on reactant gas selection and pressure. For example, methane produces somewhat more fragmentation than isobutane which has lower energy reactant ions. Also, pseudomolecular ions are produced by processes in addition to proton loss. For example, cycloalkanes and aromatic compounds generally produce an $(M + 1)^+$ pseudomolecular ion. This can be confusing if it is not interpreted in conjunction with other data such as an EI spectrum. The wide acceptance of chemical ionization is based on the intensity of the molecular ion produced by EI relative to that of the pseudomolecular ion produced by CI. Many compounds do not produce an EI molecular ion, but most produce an intense CI pseudomolecular ion. Secobarbital is a good example (Fig. 15).

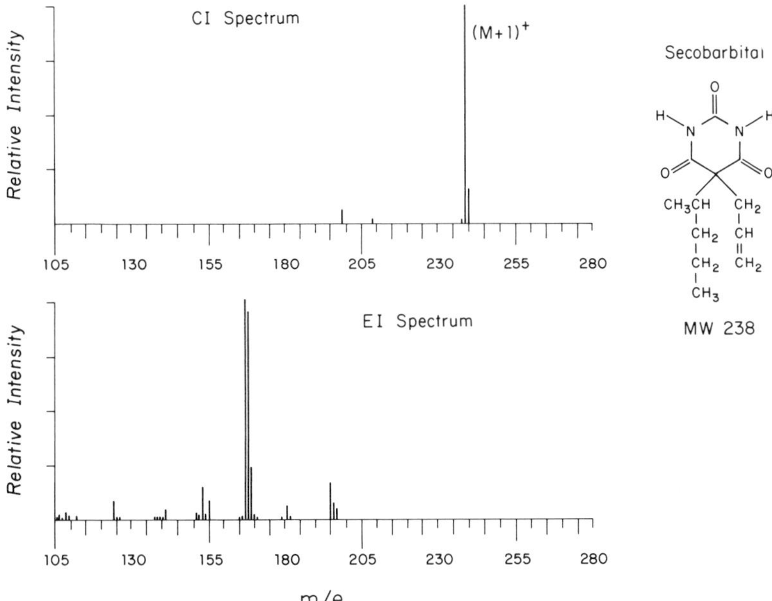

Fig. 15 Comparison of the CI and EI spectra of secobarbital.

2. *Ion Source Design*

Basically, CI and EI sources are similar. However, because high pressure is required in order that ion–molecule reactions predominate ion production, the CI source must have a lower conductance. Otherwise, the high gas load required for the ionization process would generate high pressure in the mass analyzer.

Although single-purpose CI sources can be constructed, it is desirable to have the combined capability of EI and CI in the same mass spectrometer. The simplest way to alter the source pressure is to "switch" the reactant gas load in a moderately low conductance source "on" for CI and "off" for EI using a valve. It is possible also to "switch" ion source pressure mechanically by increasing conductance for EI and decreasing it for CI (Fig. 16). An ion source incorporating both features would be an advantage with GCMS systems in which fairly high gas loads exist even for EI.

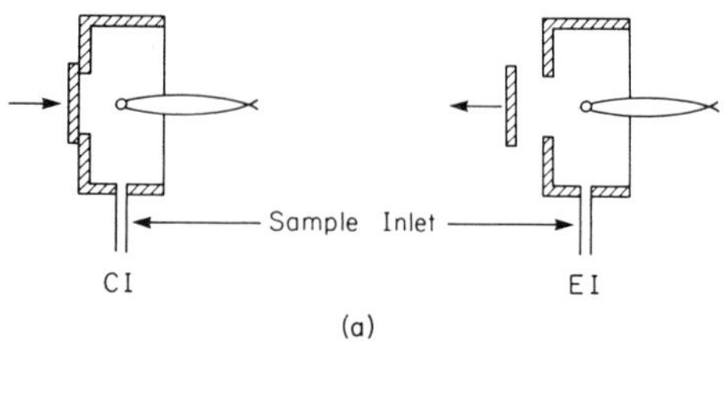

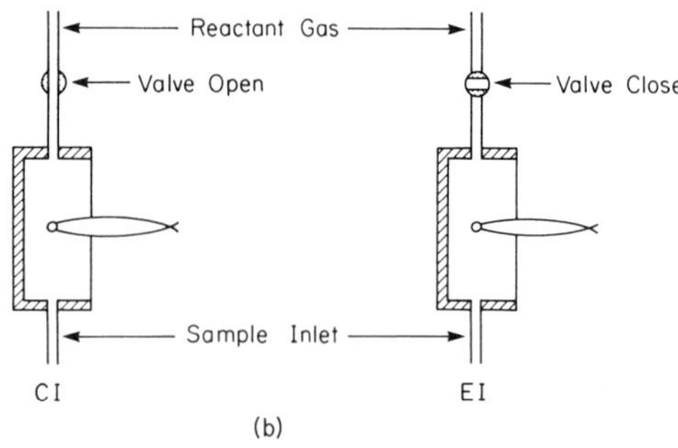

Fig. 16 Two types of switchable CI/EI ion sources: (a) variable conductance; (b) variable gas load.

C. *High Voltage Spark Source*

Spark source mass spectrometry (SSMS) is an extremely worthwhile and useful technique. It is unmatched in its ability to generate simultaneous multielemental analysis for concentrations ranging from a few parts per billion to several percent on a few milligrams of sample.

1. *Ion Process*

When an electrical spark is generated between two electrodes, a portion of the electrode material is eroded away from the surface and is ionized (Fig. 17). First, an electrical field in excess of 10^8 V/cm is generated at the tips of microprotrusions on the electrode surface causing a few electrons to be drawn away from the electrode. This process is called field electron emission. The field-emitted electrons ionize residual gas molecules which

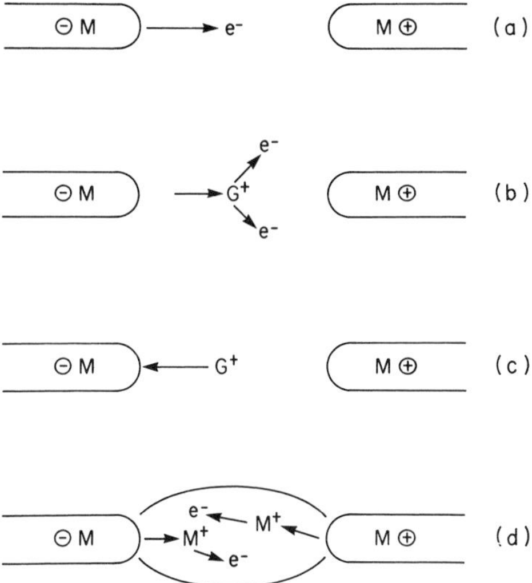

Fig. 17 Production of ions in a spark source: (a) field electron emission; (b) ionization; (c) sputtering; (d) plasma.

are then accelerated toward the electrode by the high electrical field gradient. As the ions strike the electrode, kinetic energy is transferred to the localized area of impact. This localized energy pulse results in the ejection or "sputtering" of more electrode material into the spark gap. The spark gap rapidly becomes electrically conductive due to the presence of a large number of gas phase electrons and ions being generated. From this plasma, ions are drawn into the mass analyzer. Due to the high thermal and electron temperatures generated by this process, the chemical integrity of any compounds present in the sample is destroyed and the mass spectrum is composed primarily of the singly charged, monoatomic ions of the elements in the electrode. The process is so energetic that all elements are ionized with essentially equal probability. This means that, merely by correcting for isotopic abundance, elemental concentrations can be calculated directly from ion current measurements.

Interpretation of spark source spectra is relatively simple, especially when the original sample was one of high purity such as a silicon crystal intended for use in the electronics industry. The spectrum is inspected for characteristic elemental isotopic patterns. When a pattern is recognized, the ion current ratio of impurity-to-matrix (silicon) is determined and an atomic concentration calculated. Detection limits as low as a few parts per billion are possible Although the interpretation process is similar for more complex samples such as minerals or biological tissues, the amount of effort involved in recognizing isotopic patterns is greatly magnified due to the presence of interfering ion species and the overlapping of isotopic patterns. While singly-charged, monoatomic ions are the most prevalent, some doubly, triply, and higher order charged species are also generated. Consequently, an analysis for traces of silicon in a stainless steel (Fe, Cr, Ni) matrix would be difficult due to the interference at $^{28}Si^{+}$ with $^{56}Fe^{2+}$. Also, there are some polyatomic (molecular) species generated which may result in interferences. An example of this would be detecting traces of sulfur in a silicate matrix where $^{16}O_2^{+}$ would interfere with $^{32}S^{+}$.

2. *Ion Source Design*

Both ac and dc discharges have been used to generate ions but the pulsed radio-frequency (rf) spark is the most common (Fig. 18). A 100-MHz, 100-kV signal is applied between the sample electrodes 100 times a second for periods of 100 μsec to produce a 1% "duty cycle." Higher duty cycles generate excessive electrode temperature which results in the loss of the more volatile elements. Because of the type of energy involved, ions generated by the spark source have a large spread in kinetic energy (hundreds of volts) and exit angle (about 45°) when they leave the ion source. The consequences of these two facts will be discussed in Section IV.

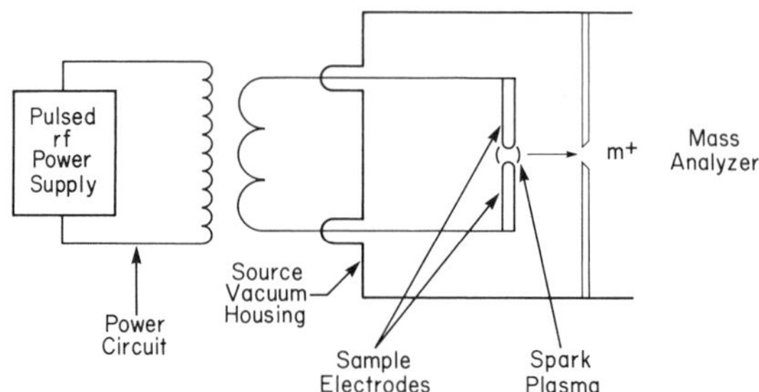

Fig. 18 The radio-frequency spark ion source.

Several restrictions exist on the nature of samples suitable for SSMS: first, they must be solids, and second, they must be electrically conductive or semiconductive. If they are not, they must be made conductive. This may be done with powdered samples by mixing with either graphite or silver powder and then compacting the powder into a suitable shape using a press. Electrode holders and source ion optics must be changed after each run to prevent contamination from causing "memory" effects. This means that the ion source vacuum housing must be vented and pumped down for each analysis.

D. *Ion Bombardment*

Ion bombardment is a technique used to generate ions from the surface of a solid by directing a beam of energetic ions (5–25 keV) at the surface which is eroded or "sputtered" away (Evans, 1972). Some of the eroded material leaves the surface in the form of ions and may be mass analyzed to help characterize the surface composition. Two mechanisms are useful in describing the ionization process. The first involves transfer of kinetic energy from ions in the primary beam to the first few atomic layers of a conductive sample surface (Fig. 19). This results in ejection of electronically excited, neutral material. Direct ejection of ionized species is unlikely as they would be neutralized by conduction band electrons at the sample surface. When the electronically excited, or metastable, species is free from interactions with the surface, it may eject an Auger electron and become an ion for mass analysis. The second ionization process predominates with electrically insulating sample surfaces. Here conduction band electrons are not available and ions may be ejected directly from the surface without being neutralized.

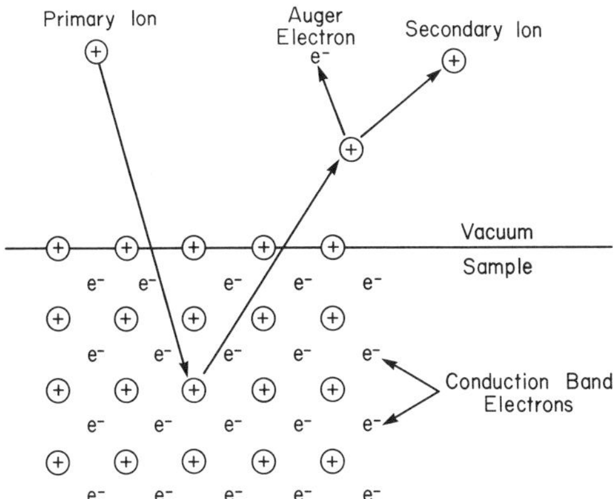

Fig. 19 Secondary ion production from a solid surface.

The high energies involved in the ion bombardment process results in the production of elemental rather than compound related information. The energy is lower than that encountered with the pulsed rf spark source and thus the expected spectral differences are observed: polyatomic ion species are more abundant while multiply charged ions are much less abundant.

With this technique, the sampling process rather than the mass spectrum delineates the unique analytical capability since only the uppermost few atomic layers of a sample are analyzed. The elemental context of the surface may be determined with sensitivities better than one part in 10^6 (1 ppm) for many elements. Also, as the primary ion beam erodes the sample surface and deeper layers are exposed, an in-depth elemental or isotopic analysis can be produced with depth resolution on the order of 100 Å. This type of information is of value in the study of thin films used in the electronics industry. Alternatively, the primary ion beam may be focused to a small spot (about 1 μm in diameter) and moved about on the sample surface. This "ion microprobe" is used to produce localized elemental information from the surface. If the beam is rastered across the sample surface, an image of elemental distribution on the surface can be generated in much the same way that a picture is generated on a television screen. The lateral resolution of the image is dependent on primary ion beam diameter at the present stage of development. If both depth and lateral analytical capabilities are utilized, a three-dimensional elemental analysis is possible.

Vaccum conditions in secondary ion mass spectrometry (SIMS) systems, particularly in the sample region, are critical with respect to possible contamination or modification of the sample surface. For example, on a clean metal sample surface, a reactive molecule such as oxygen at a pressure of 10^{-6} Torr, can deposit a full monolayer in about 1 sec. Consequently, a vacuum of about 10^{-8} to 10^{-10} Torr is desirable.

E. Other Sources

1. *Field Ionization*

Gas phase chemical compounds may be ionized if they are exposed to an electrical field gradient of about 10^8 V/cm (Beckey, 1971). Such a field can be created by placing a mechanically sharp edge, such as a steel razor blade, opposite a slit and applying a high voltage between the two (Fig. 20). Electron loss from the molecule is via quantum mechanical tunneling, so the resulting

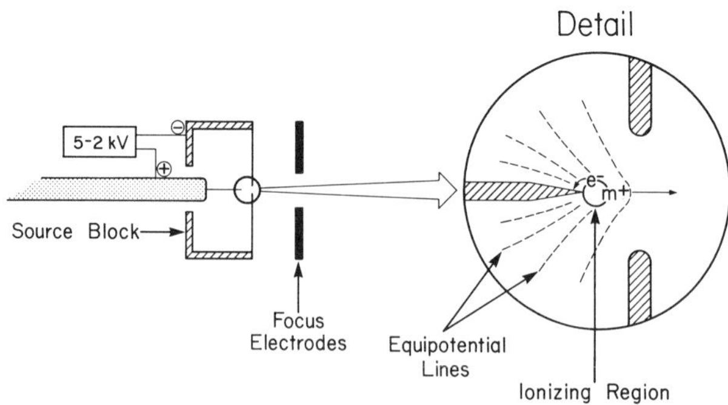

Fig. 20 Field ionization source.

ion has a minimum amount of excess electronic and vibrational energy. This leads to the production primarily of molecular ions with a minimum of fragmentation. The field ionization (FI) source has not gained general acceptance as a routine ionization method for several reasons: it is generally at least a factor of 10 less sensitive than EI or CI sources, the emitter elements tend to be fragile and easily destroyed by arcs in the source, and the same kind of information can be obtained more easily with chemical ionization. The principle behind FI is significant and one should be aware of its existence.

2. *Field Desorption*

The field desorption (FD) source (Winkler and Beckey, 1972) is very similar to the FI source, except that the FD emitter is a small wire on which many highly pointed needles are grown. A solution containing the sample is applied to the emitter exterior to the ion source vacuum housing and the solvent evaporated. The emitter is inserted through a vacuum lock into the ion source and accurately positioned. A few milliamperes of current are passed through the emitter to warm it, causing the sample molecules to migrate to the high field regions (10^8 V/cm) at the needle points where they are ionized or "field desorbed." FD spectra are at present not easily interpreted, but since the technique provides the ability to generate mass spectra from extremely low volatility, highly labile compounds, it is an area of considerable research interest.

3. *Photoionization*

Since the ionization potential of most molecules falls between 7 and 19 eV, it is possible to produce ionization with a beam of uv light in the wavelength range of 177 to 78 nm. Photoionization (PI), like EI, proceeds via a Franck–Condon mechanism, and the resulting mass spectra are very similar. Photoionization is used primarily for studies of the ionization process since the ionization energy can be well defined by the light source. However, this wavelength region falls in the vacuum ultraviolet and, below 105 nm, no window materials are available. Consequently, the technical complexity involved in source design has kept PI from becoming a major analytical tool.

4. *Atmospheric Pressure Ionization*

As the name implies, atmospheric pressure ionization (API) does not require a vacuum environment (Horning *et al.*, 1974a,b). Rather, ions are created in a high pressure environment and then passed through a very small (25-μm diameter) conductance into the vacuum system of the mass analyzer (Fig. 21). Because of the high pressure, ionization is achieved primarily via ion–molecule reactions, and the resulting spectra are similar to those produced by chemical ionization. The electron source used to ionize the "reactant" species must survive a much higher pressure than that encountered in the CI source. Two types of electron source have been used: ^{63}Ni, a β emitter, and a corona discharge. The technique is in its early stages of development and is not yet a standard analytical tool.

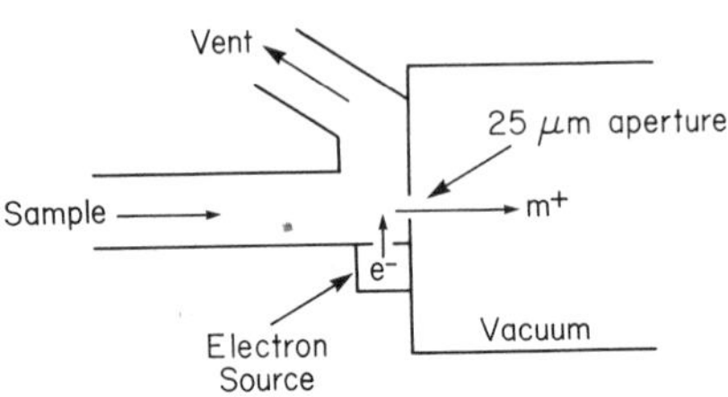

Fig. 21 Atmospheric pressure ion source.

5. *Californium Ion Source*

When a high energy particle from the decay of ^{252}Cf passes through an organic layer deposited on a thin nickel substrate, a very sharp, localized thermal spike results (Macfarlane *et al.*, 1974; Torgerson *et al.*, 1974) (Fig. 22). In a process possibly similar to ion bombardment, ions are emitted from the organic layer. The spectra produced are composed primarily of a quasi-molecular ion similar to those of chemical ionization, although at present the ionization process is not well understood. The ability to produce mass spectra from condensed phase organic compounds without the necessity of first volatilizing them is particularly attractive to those in the biomedical fields where many labile compounds are encountered. However, much work will be required to establish what application the californium source will have in the field of mass spectrometry.

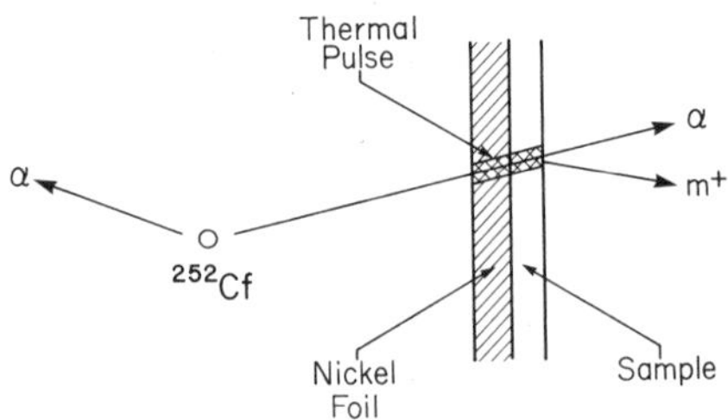

Fig. 22 Californium ion source.

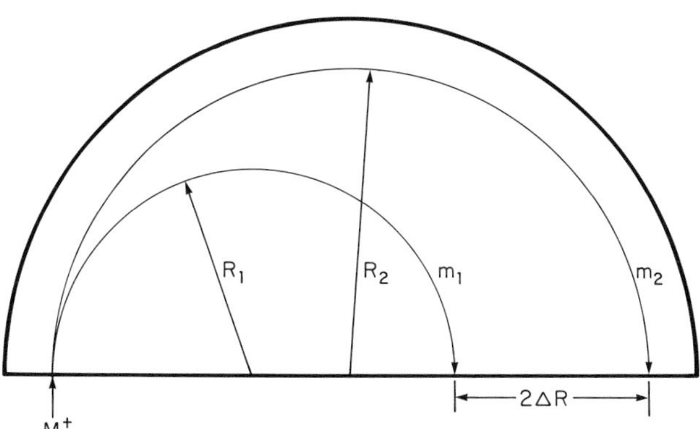

Fig. 23 The 180° magnetic sector mass analyzer.

IV. Mass Analyzers

In the design of a mass spectrometer, a suitable configuration of electric and/or magnetic fields must be selected. Before the advent of large scale computing systems, the only method available to perform this task was through the mathematical solution of the functions that describe ion trajectories. These solutions resulted in a variety of mass analyzer designs, some of which are still very capable performers. For a given type of design, improved performance was usually achieved by increasing the size of the instrument, a procedure that also increased cost. In recent years, computer programs have been developed with which the ion optician can quickly trace the trajectory of an ion through a series of electric and/or magnetic fields. The computer model of a mass analyzer is easily modified through programming changes and produces reasonable estimates for both sensitivity and resolution. This is done by taking into account not only field strengths and shapes, but also expected angular and energy distributions of the ion produced by the ion source. Using this method, an optical system which will produce the desired performance for a minimum of cost can be rapidly selected.

A. Single Focusing Mass Analyzer

1. *The 180° Sector*

The 180° sector was first introduced by A. J. Dempster (1918) (Fig. 23). If an ion of charge e is accelerated through a potential V_a, its resulting kinetic energy is given as

$$eV_a = \tfrac{1}{2}mv^2 \tag{13}$$

If this ion enters a magnetic field H perpendicular to its direction of travel, it will be deflected into a circular orbit of radius R, where centripetal and centrifugal forces are equal:

$$Hev = mv^2/R \tag{14}$$

When Eq. (13) is combined with Eq. (14) and rearranged to eliminate v, it becomes the basic magnetic analyzer equation:

$$m/e = H^2R^2/2V_a \tag{15}$$

When Eq. (15) is differentiated, it yields

$$2\,\partial R/R = (\partial m/m) + (\partial V_a/V_a) - (2\,\partial H/H) \tag{16}$$

indicating the effects of small changes in the various parameters on the radius of focus. Also, if ∂V_a and ∂H are small, Eq. (16) becomes

$$2\,\partial R = R(\partial m/m) = D \tag{17}$$

which describes the dispersion D of the analyzer. Although the mass dispersion provided by the 180° sector is good, the linewidths which result are poor even when the magnetic field and the accelerating voltage are perfect ($\partial H/H = 0$ and $\partial V_a/V_a = 0$). This results from the inability to focus ions with an angular divergence α from the ion beam center line (Fig. 24). Thus, a limit is imposed on the image width and, consequently, on the resolution.

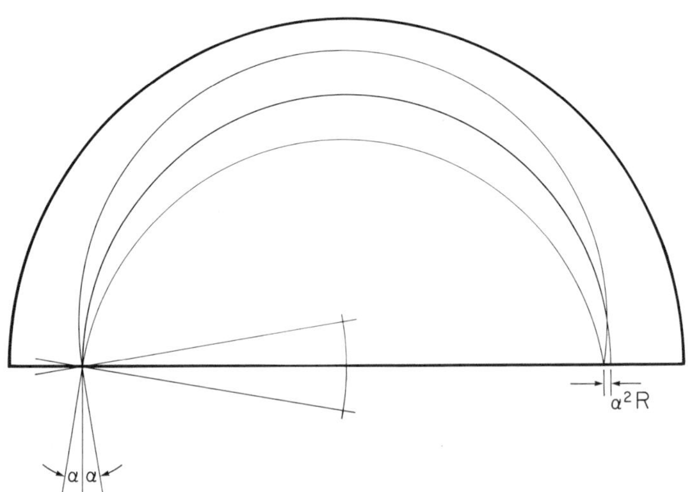

Fig. 24 Illustration of the angular aberration, α^2R, in a 180° magnetic sector.

2. *The 60° Sector*

The 60° magnetic sector was developed and introduced in the 1930's by A.O.C. Nier (1940) and has proven to be a very popular basic design (Fig. 25). This design represents a major improvement over the 180° sector in several ways. Directional or α focusing is provided so that resolution can be improved. Both the ion source and the detector are located away from the magnetic field, which eases their design requirements. With the ion source removed from the magnetic field, it becomes possible to pump the system differentially to obtain more favorable analyzer pressure. Finally, a smaller, less expensive magnet can be used. Manufacturers are still incorporating modified 60° sectors into new instrument designs.

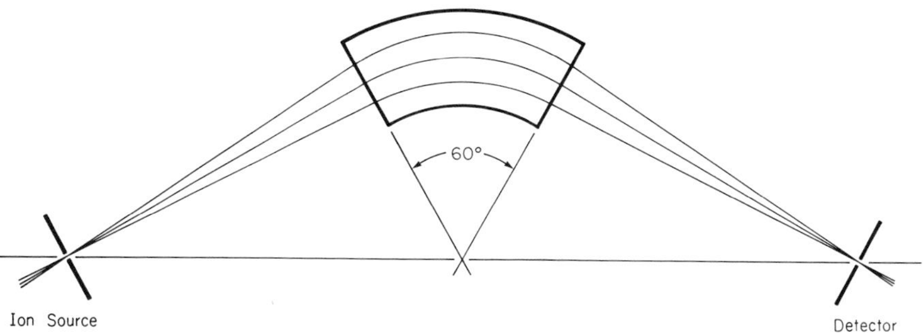

Fig. 25 The 60° magnetic sector mass analyzer.

B. Double Focusing Mass Analyzer

In mass analyzers that employ only a magnetic sector, no provision can be made for focusing an ion beam with a significant spread in initial kinetic energy, $\partial V_a/V_a \neq 0$ in Eq. (16). The only method to improve resolution under these conditions is to increase the ion accelerating voltage V_a such that $\partial V_a/V_a$ is reduced in magnitude. Alternatively, an energy filter can be added to the optical system to improve resolution. Such a filter can be formed using a radial electric field or electrostatic analyzer as shown in Fig. 26. Here the radius (in cm) traced by an ion is given by

$$R_e = 2V_a d/V_e \tag{18}$$

where V_a is the ion accelerating voltage, V_e the potential difference, and d the distance between the two plates. The width of the exit slit will determine the energy spread of the filtered ion beam. To obtain true double focusing, however, the radial electrostatic field must be configured with a magnetic

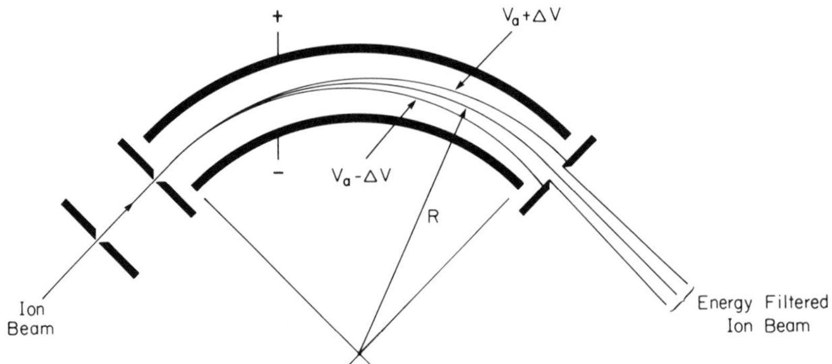

Fig. 26 The electrostatic energy filter.

field such that the angular (α) and energy (β) aberrations of the ion beam essentially cancel each other. Simply placing an electrostatic analyzer in tandem with a magnetic sector does not constitute double focusing.

1. *Mattauch–Herzog Geometry*

When Mattauch and Herzog worked out the general equations for double focusing in the 1930s (Mattauch and Herzog, 1934; Mattauch, 1936), they discovered a combination of field shapes with which double focusing could be obtained along a planar surface. With this optical configuration a glass photographic plate could be used to simultaneously record the full mass spectrum. A $\pi/4\sqrt{2}$ degree electrostatic sector and a 90° magnetic sector

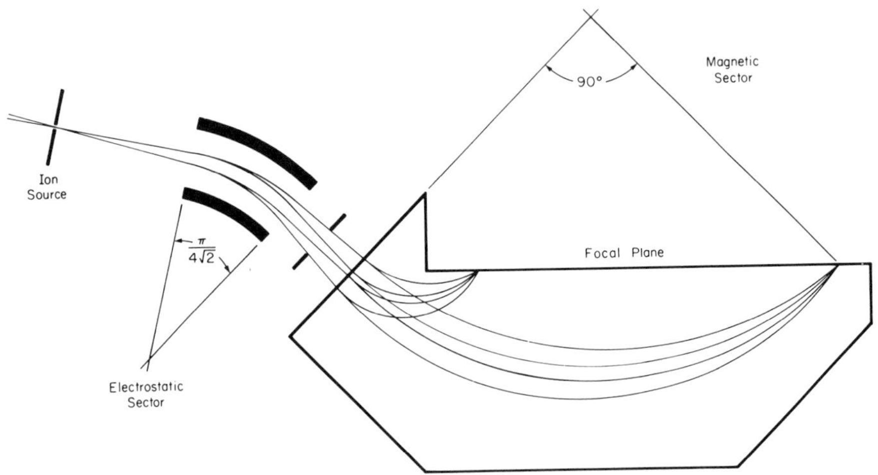

Fig. 27 Mattauch–Herzog double-focusing mass analyzer geometry.

was used as shown in Fig. 27. This geometry, modified to allow the photoplate to be placed outside the magnetic field, has been successfully applied to commercial instruments, and resolving powers in excess of 50,000 have been achieved.

This geometry has two notable applications, both based on the simultaneous recording provided by the photographic plate. A nondistorted mass spectrum can be recorded even when the ion beam is fluctuating wildly and the initial kinetic energy spread ($\partial V_a/V_a$) is large. This is the case with the spark source and essentially all SSMS work is done using this type of analyzer. The second application involves high resolution GCMS where sample load is changing rapidly, making analysis time very short. With this analyzer, a nondistorted spectrum can be obtained at high resolution by using the photographic plate.

2. *Nier–Johnson Geometry*

The Nier–Johnson geometry (Johnson and Nier, 1953), which was developed using an adaptation of the Mattauch–Herzog equations, provides double focusing at a single point (Fig. 28). Successive mass-to-charge ratios are focused at the collector slit by changing the magnetic field strength. If the ion accelerating voltage is adjusted to scan the mass scale, the electrostatic field must be adjusted proportionally in order to keep the beam in focus. In terms of performance, the Nier–Johnson mass analyzer can produce resolutions in excess of 100,000 while maintaining good sensitivity.

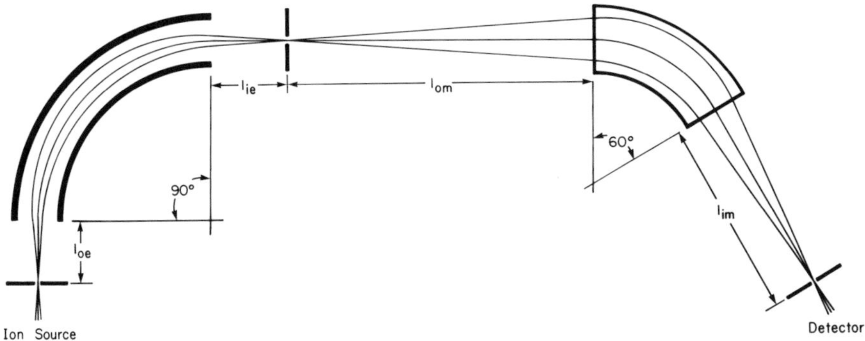

Fig. 28 Nier–Johnson double-focusing mass analyzer geometry.

It should be mentioned that the general solutions to the double focusing problem are, in fact, approximations. Mattauch and Herzog's calculations were first order approximations and, although they have been improved upon by others to yield second order solutions, no absolute general solution has been produced.

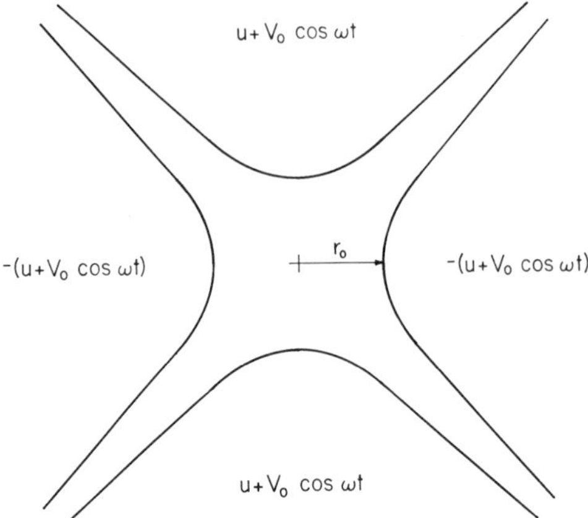

Fig. 29 Quadrupole field potentials.

C. *Quadrupole Mass Analyzer*

A variety of nonmagnetic mass analyzers exist, of which the quadrupole is by far the most widely used. This analyzer is based on the principle, introduced by Paul and Steinwedel (1953), that involves resonant ion oscillation in an oscillating electric field. The hyperbolic field is shaped by four rods in cross section in Fig. 29. Until recently, cylindrical rods were used to approximate the ideal hyperbolic field; however, several instruments have now been introduced with hyperbolic rods. The rods are all operated at the same absolute potential (Fig. 30), but with adjacent rod potentials opposite in sign. The applied potential has a dc component U and a radio

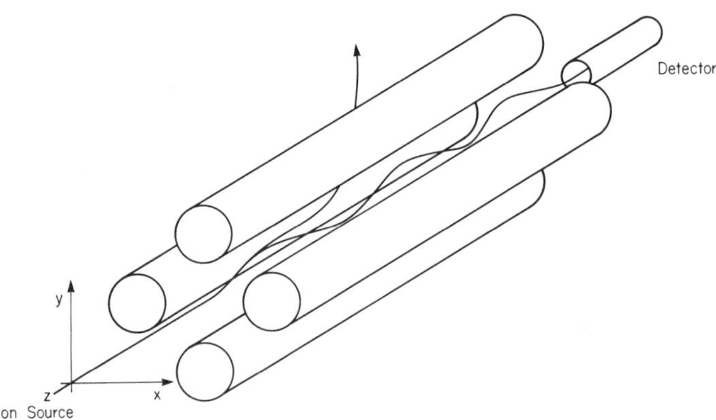

Fig. 30 Quadrupole mass analyzer.

frequency component, $V_0 \cos \omega t$, which forms a potential inside the rods described by

$$I = (U + V_0 \cos \omega t)[(x^2 - y^2)/2r_0^2] \tag{19}$$

where x and y are the distances from the z axis and $2r_0$ is the distance between opposite rods. Consequently, the motion of a charged particle, m/e, moving along the z axis will be given by

$$\ddot{x} + (e/mr_0^2)(U + V_0 \cos \omega t)x = 0 \tag{20}$$

$$\ddot{y} - (e/mr_0^2)(U + V_0 \cos \omega t)y = 0 \tag{21}$$

By substituting $\xi = \frac{1}{2}\omega t$, $a = 4eU/mr_0^2$, and $q = 2eV_0/mr_0^2\omega^2$ into Eq. (20) and (21), the orthogonal set of equations in x and y becomes

$$(d^2x/d\xi^2) + (a + 2q \cos 2\xi)x = 0 \tag{22}$$

$$(d^2y/d\xi^2) - (a + 2q \cos 2\xi)y = 0 \tag{23}$$

which are known as the Mathieu equations. The general solution of Eqs. (22) and (23) shows that the oscillatory amplitude of an ion may cover a range of a and q values and yet remain bounded or stable. In less mathematical terms, values of U and V_0 may be selected to maintain a selectable range of m/e values in stable oscillation as the ions travel down the length of the rods. For values outside the acceptable range, the ions are unstable and fly out of the analyzer. The stable m/e values depend on the dc and rf voltage values plus the frequency. It is most convenient to vary U and V_0 at a constant frequency setting to produce a scan linear in mass.

With quadrupoles, as with the magnetic sectors, there are limits to the angular and energy dispersion which may be focused. Also, a tradeoff in sensitivity and resolution is possible. Commercial quadrupoles are available which operate at resolving powers ranging from several hundred up to about 4000 and RPs up to 16,000 have been achieved. Basic sensitivity is comparable with magnetic sector instruments, and mass scanning rates of up to 500 amu/sec are readily achieved. This is faster than many magnetic sector instruments are scanned and tends to favor the quadrupole for GCMS applications.

D. *Time-of-Flight Mass Analyzer*

The velocity v of an ion, m/e, which has been accelerated through a potential V_a is given by

$$v = (2eV_a/m)^{1/2} \tag{24}$$

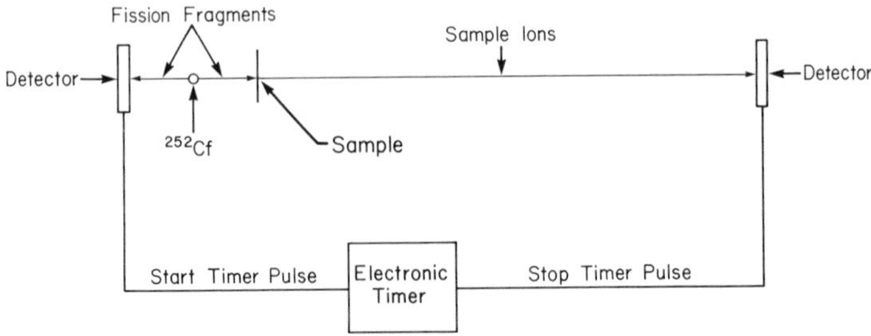

Fig. 31 Time-of-flight mass analyzer with a californium ion source.

Consequently, for two ions m_1 and m_2 which leave the ion source at time t_0 and travel the distance L, there is a mass-dependent time differential Δt in their transit through the analyzer.

$$\Delta t = \frac{L(\sqrt{m_1} - \sqrt{m_2})}{\sqrt{2eV_a}} \tag{25}$$

Although this mass analyzer system is conceptually straightforward, the practical difficulties encountered when using an EI source are significant. There are problems associated with producing sufficiently narrow ionization pulses, reducing initial kinetic energy spread, and lowering basic sensitivity due to a low duty cycle. These analyzers, however, may be scanned at 10,000 to 100,000 spectra per second depending on mass range, path length, and accelerating voltage. This fast scanning rate has made possible the observation of reactions which take place on a 100-μsec time scale. Also, when time-of-flight (TOF) is used with the ^{252}Cf ion source, which is inherently a pulsed ion source, the difficulty in defining t_0 is greatly reduced. The nuclear decay of ^{252}Cf is such that two high energy particles (fission fragments) are emitted simultaneously; one is used to start a timing circuit while the other ionizes the sample (Fig. 31). The arrival of an ion at the detector is then stored as a count in a multichannel analyzer as a function of time and a TOF mass spectrum is produced.

V. Data Collection

A. Photographic Plate

When a high energy particle strikes a photographic emulsion, silver grains in the emulsion are exposed. The location of the exposure constitutes the mass scale and the degree of darkening, the intensity scale. This detection scheme is used almost exclusively with the instruments of the Mattauch–Herzog geometry, in which ions are focused along a planar surface. Although the photographic plate is simple in concept, it is rather cumbersome to work with. The emulsion tends to outgas considerably and must be preconditioned in a separate vacuum chamber before it is introduced into the mass spectrometer. The exposure-to-abundance function is linear only over a limited range of exposure values and is variable from plate to plate and occasionally along the length of an individual plate. Finally, a relatively tedious developing procedure is required, and the data cannot be inspected until well after the mass spectrum has been taken. It is interesting to note that all commercial Mattauch–Herzog mass spectrometers are provided with a means of electrical detection which may be used when the photographic procedure is not required by the application or when immediate readout is desirable.

B. *Faraday Collector*

There are two basic forms of electrical detection used with ion beams: the Faraday collector and the electron multiplier. The Faraday collector consists of a cup-shaped trap from which the flow of current is measured directly (Fig. 32). The collector is constructed with a depth greater than its width to reduce the loss of secondary electrons. Any escaping secondary electrons resulting from ion bombardment would produce an error in the measured ion current. Electron suppression is used to reduce secondary losses further as well as to exclude any secondary electrons which arise as the ion beam strikes the mass resolving slit. Since it is electrically difficult to amplify and measure electrical currents less than 10^{-13} A, Faraday collector systems are limited in sensitivity. In spite of the sensitivity limitation imposed by this detector, it finds considerable application in dual collector systems where small physical size and uniform response are important.

C. *Electron Multiplier*

The electron multiplier is used when high gain and high frequency response are required. This system takes advantage of secondary electrons produced

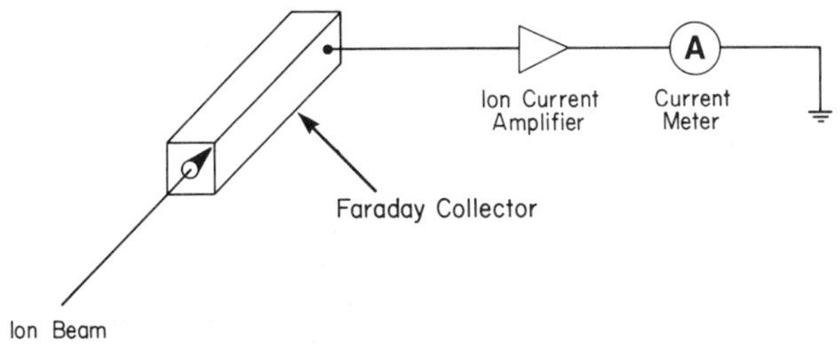

Fig. 32 Faraday ion collector.

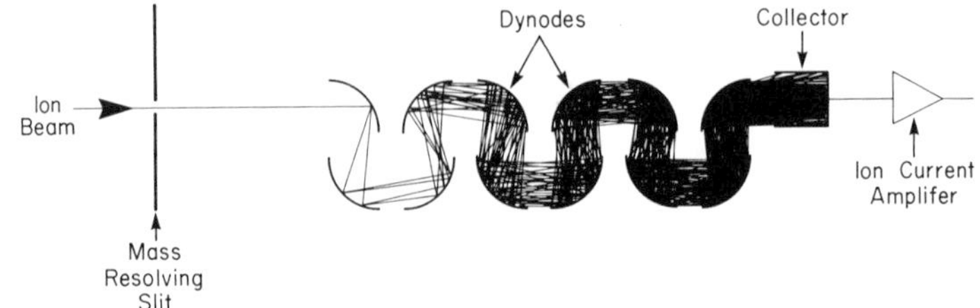

Fig. 33 Twelve-stage electron multiplier.

when a high energy particle strikes a metallic surface or dynode (Fig. 33). The quantum yield of electrons is greater than one, so that by properly shaping the dynodes and applying electrical fields, a cascading effect results which produces a significant net gain in current at the final dynode. Gain is adjusted by changing the voltage applied along the dynode string. Higher voltages produce higher energy bombardment, and thus greater secondary electron yield, up to a point of saturation. At the end of the dynode string there is a collector similar in function to the Faraday collector. With the electron multiplier, however, the effective ion current is increased by up to 10^8 so amplification limitations are not a problem. In fact, single ions can be detected at rates up to 10 to 100 kHz. The electron multiplier does produce a significant noise level at high gain, but fortunately the current pulses which result are generally of lower magnitude than ion current pulses and may be discriminated against by using an appropriate thresholding circuit. Thus, by using ion counting techniques, both high gain and low noise are possible. However, ion counting systems require more complex and expensive electronics to produce a spectrum than does an analog system.

D. Recording Methods

1. *Oscillographic Recorder*

Due to the high rates at which data are produced with electrical detection, special recording methods must be employed. The oscillographic recorder has proven to be the most suitable (Fig. 34). Records are made by exposing light sensitive paper to a light beam deflected by a small mirror attached to a galvanometer. Often several galvanometers are used to increase the dynamic range of the recording system. Frequency responses up to 5 kHz are possible, so even GCMS data may be readily recorded by this means. Sometimes one galvanometer is used to print a mass scale or mass marker automatically on the record. Although these markers are not absolutely accurate, they greatly reduce the time and effort that goes into establishing the mass scale manually.

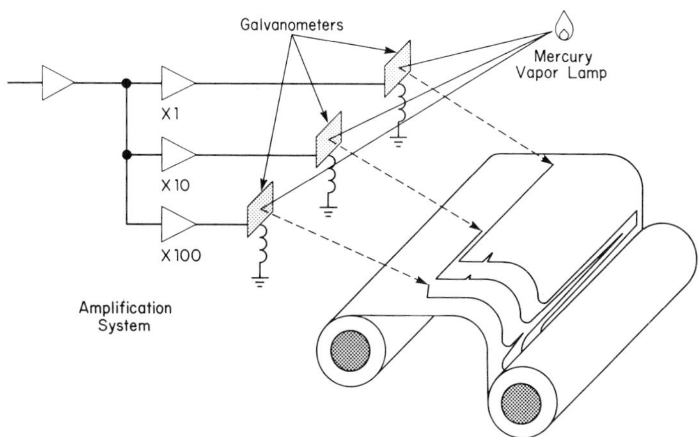

Fig. 34 Oscillographic recorder.

2. *Magnetic Tape*

When excessive quantities of data are anticipated, such as with GCMS analysis of complex mixtures, recording data on a magnetic tape is more practical than recording on an oscillographic recorder. The tape may be processed by a remote computer and, depending on software sophistication employed, mass scale marking, data normalization, and background subtraction may be performed. This can be a practical approach to computerized data reduction if a dedicated on-line computer system is not available.

3. *On-Line Computer*

With high resolution experiments or GCMS, extremely large quantities of data may be generated. If efficiency is of sufficient importance, an on-line computer is the recording means of choice. The major limitation is, of course, cost, since the price of an appropriate data system may exceed the price of the GCMS with which it is intended to operate. This is not the case with high resolution mass spectrometers, however, where basic instrument cost is considerably higher. However, in both situations the computer data system is becoming a standard and integral part of the total system.

The first task of the computer system is to establish a mass scale by analyzing a standard reference compound. Materials routinely used for this purpose are high molecular weight perfluorinated compounds, such as perfluorokerosene (PFK), perfluorotributyl amine (PFTBA or FC43), and trisperfluoro-heptyl-*S*-triazine. These compounds are volatile and yield rather simple spectra with abundant peaks over a wide mass range. Perfluorotributyl amine has peaks up through m/e 614 and the base peak of the triazine is at *m/e* 866. The computer software is designed to "recognize" the reference spectrum and generate a mass scale which is a function of the scanned parameter.

There are two basic techniques for data collection (Fig. 35). In the first system, the mass spectrum is scanned while the computer monitors the resolved ion beam. Points of maximum ion intensity are assigned mass values using the mass scale established during calibration, and the intensity is recorded along with the mass. The second system uses the established mass calibration scale and actively drives the mass spectrometer to measure ion intensity at the expected location of each mass. The differences between the two techniques are subtle but real. With the first system, the ion intensity value will closely represent true ion abundance, a desirable feature for spectral interpretation. It does, however, spend time taking data between and on sides of peaks where there either is no data or where the data has a larger noise component than the peak maximum. As a result, sensitivity suffers compared to the second system which is driven so as to make all intensity measurements at the peak maximum. If, however, the sample ions have a substantial mass defect, the intensity measurements using the second system will be derived from the side of the peak or, conceivably, the peak could be missed entirely. Also, the second method is not at all practical for obtaining high resolution data since a peak center is not determined from which to assign the exact mass. In practice, either system can be made to work well at low resolution.

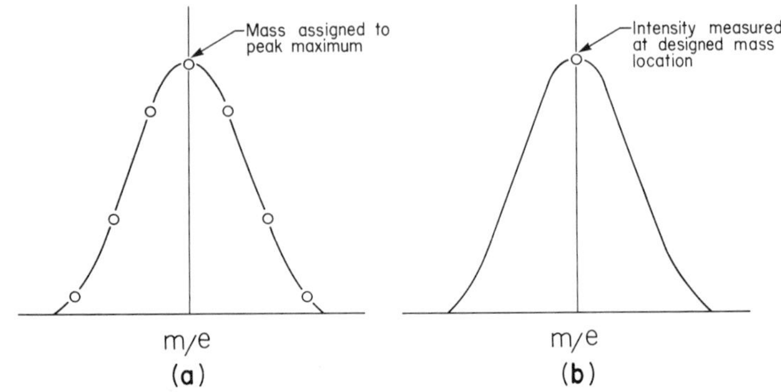

Fig. 35 Computer data collection techniques: (a) first system; (b) second system.

Once data has been acquired, a variety of software routines which can reduce the data to a convenient form are available. With high resolution data, the most probable elemental compositions of the ions are automatically selected. For low resolution data, sample spectra may be displayed either in

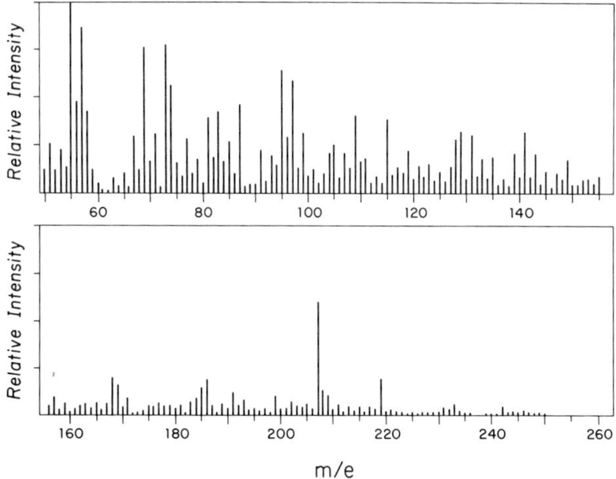

Fig. 36 EI spectrum of 400 pg of methyl myristate including background.

tabular or in graphical form with or without the subtraction of background spectra. The effect of background is shown in Fig. 36, a GCMS spectrum of 400 pg of methyl myristate including background. This same spectrum, but with the background subtracted, is shown in Fig. 13. With GCMS data a gas chromatogram may be reconstructed by plotting the total number of ions in each scan versus scan number or time (Fig. 37). If a specific compound is being sought, the intensity of characteristic ion in the mass spectrum of the compound may be plotted as a function of time. This reconstruction of a single mass chromatogram (Fig. 38) is particularly useful when the compound

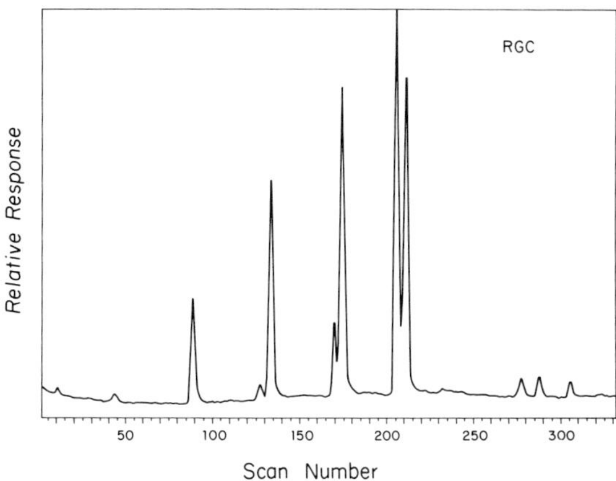

Fig. 37 Computer reconstructed gas chromatogram.

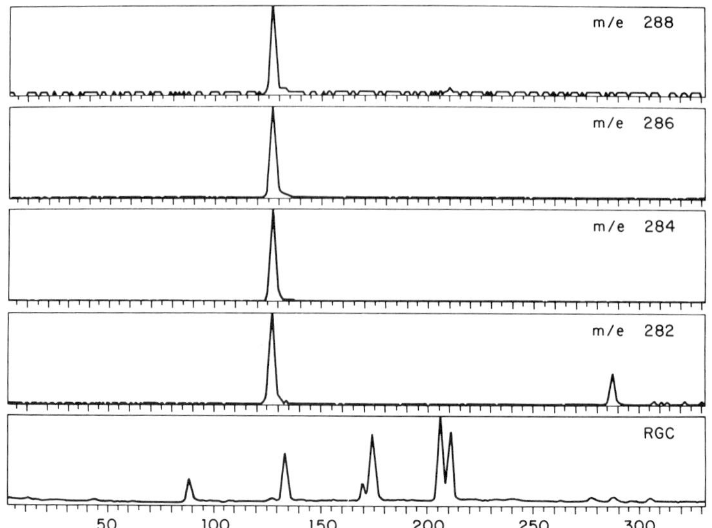

Fig. 38 Mass chromatograms for isotopic peaks in the molecular ion of hexachlorobenzene.

being sought is present at very low concentration or is one of a multitude of GC peaks. Areas under chromatographic peaks may be integrated for quantitative purposes or to determine isotope ratios. Finally, spectra may be searched against a library of known spectra to determine what compounds produce similar spectra. These libraries can easily exceed 10,000 stored spectra and reasonable identification can often be carried out using them. This is particularly true with certain well-documented groups of compounds, such as pesticides or drugs. Caution must always be exercised with this simplified approach to identification. The library search is no better than the stored spectra, and it should be considered as an aid to, rather than a crutch for, identification purposes. The limits to what can be done using the computer based data system have not yet been approached and the cases included here are only examples of what can be done.

VI. Conclusion

Mass spectrometers consist of four basic components, which, when combined, have the potential for producing answers to a great many analytical questions. Perhaps a greater variety of analytical problems can be successfully solved using mass spectrometry than by any other single technique. A sampling of these problems will be presented in the chapter by Fenselau. Due to the diversity of the hardware encountered in mass spectrometry, however, it is not possible to have the best aspects of all components combined into a single general purpose instrument since the result would be a completely ineffective compromise. Any set of mass spectrometer components must be selected to produce a given type of information. Each additional type of information capability added to an instrument will take its toll in some form of compromise, be it performance, cost, or complexity. For example, an instrument originally intended to measure carbon isotope ratios in CO_2 samples as accurately and precisely as possible would have to be greatly compromised if it were also expected to function as a high sensitivity GC detector where air leaks would be expected to alter the CO_2 background level. The good analyst will learn to accept a certain degree of compromise, fully utilize an instrument's strong points, and work around its weak ones.

Bibliography

Ahearn, A. J., ed. (1966). "Mass Spectrometric Analysis of Solids." Elsevier, Amsterdam.

Barrington, A. E. (1963). "High Vacuum Engineering." Prentice-Hall, Englewood Cliffs, New Jersey.

Beynon, J. H. (1967). "Mass Spectrometry and Its Applications to Organic Chemistry." Elsevier, Amsterdam.

Blanth, E. W. (1966). "Dynamic Mass Spectrometers." Elsevier, Amsterdam.

McFadden, W. H. (1973). "Techniques of Combined Gas Chromatography/Mass Spectrometry: Applications to Organic Analysis." Wiley, New York.

White, F. A. (1968). "Mass Spectrometry in Science and Technology." Wiley, New York.

Williams, D. H., and Howe, I. (1972). "Principles of Organic Mass Spectrometry." McGraw-Hill, New York.

References

Aston, F. W. (1919). *Phil. Mag.* **38**, 707.

Aston, F. W. (1927). *Proc. Roy. Soc.* (*London*) **A115**, 487.

Beckey, H. D. (1971). "Field Ionization Mass Spectrometry." Pergamon, Oxford.

Dempster, A. J. (1918). *Phys. Rev.* **11**, 316.

Evans, C. A., Jr. (1972). *Anal. Chem.* **44(13)**, 67A.

Horning, E. C., Carroll, D. I., Dzidic, I., Haegele, K. D., Horning, M. G., and Stillwell, R. N. (1974a). *J. Chromatogr. Sci.* **12**, 725.

Horning, E. C., Horning, M. G., Carroll, D. I., Dzidic, I., and Stillwell, R. N. (1974b). *Anal. Chem.* **45**, 936.

Johnson, E. G., and Nier, A. O. (1953). *Phys. Rev.* **91**, 10.

Macfarlane, R. D., Torgerson, D. F., Fares, Y., and Hassel, C. A. (1974). *Nucl. Instrum. Methods* **116**, 381.

Mattauch, J. (1936). *Phys. Rev.* **50**, 617.

Mattauch, J., and Herzog, R. (1934). *Z. Phys.* **89**, 786.

Munson, M. S. B. (1971). *Anal. Chem.* **43(13)**, 28A.

Munson, M. S. B., and Field, F. H. (1966). *J. Am. Chem. Soc.* **88**, 2621.

Nier, A. O. (1940). *Rev. Sci. Instrum.* **11**, 212.

Paul, W., and Steinwedel, H. (1953). *Z. Naturforsch.* **8A**, 448.
Ryhage, R. (1964). *Anal. Chem.* **36**, 759.
Ryhage, R., Wikström, S., and Waller, G. R. (1965). *Anal. Chem.* **37**, 435.
Thomson, J. J. (1911). *Phil. Mag.* **10**, 225.
Torgerson, D. F., Skowronski, R. P., and Macfarlane, R. D. (1974). *Biochem. Biophys. Res. Commun.* **60**, 616.
Watson, J. T., and Biemann, K. (1964). *Anal. Chem.* **36**, 1135.
Winkler, H. U., and Beckey, H. D. (1972). *Org. Mass Spectrom.* **6**, 655.

Applications of Mass Spectrometry

Catherine Fenselau

Department of Pharmacology and Experimental Therapeutics
The Johns Hopkins University
School of Medicine
Baltimore, Maryland

I. Introduction

This chapter discusses the kinds of information that mass spectrometry can provide about compounds which may be broadly considered as organic and will apply to samples from the pharmaceutical, chemical, and foods and

ISBN 0-12-430801-5

flavors industries, to forensic, toxicologic, and environmental residue analysis, to clinical, biochemical, and pharmacologic research, as well as to research in synthetic, physical organic, and natural products chemistry.

Samples at the current analytical frontier are often characterized by submicrogram availability, occurrence in multicomponent mixtures, polar functional groups, and thermal instability. Polar functional groups of course render the compounds lipophobic and nonvolatile.

Severe demands are imposed on the mass spectrometer by such samples. High sensitivity is clearly desirable, and it has been pointed out many times that mass spectrometry provides more information from less sample than any other instrumental technique. Certain kinds of mass spectral analyses have been reported in which subpicogram (10^{-13} g) levels of sample were used.

The requirement for high resolution separation of mixtures has been met by combining a gas chromatograph with a mass spectrometer. This very effective instrumental combination has the potential to produce vast quantities of data. Computerized data acquisition and processing of spectra have thus facilitated the analysis of multicomponent mixtures. Derivatization techniques and new ionization techniques are now making possible mass spectral analysis of more and more classes of nonvolatile or thermally fragile compounds.

This chapter commences with a discussion of approaches for obtaining structural information from mass spectra. Subsequent sections present the four areas of application for mass spectrometry: structure elucidation, pattern matching, isotope analysis, and quantitation. Examples are cited in each category to illustrate strengths, limitations, and the different kinds of information provided by the various mass spectral techniques.

II. Reading the Spectrum

The dimensions and the quality of the spectrum must be evaluated before a fragmentation pattern can be interpreted.

A. Dimensions of the Spectrum

In a mass spectrometer, ions are separated and analyzed according to the ratio of their mass to charge, *m/e* (Colby, this volume). In greater than 95% of the ions formed, one electron is removed, so that the charge has unit value. Thus it is reasonable to call the data record a mass spectrum and not an *m/e* spectrum. Ion current is recorded as a function of the *m/e* ratio.

The spectrum is best presented as a bargraph or histogram, in which ion abundance or peak intensity is plotted along the abcissa and *m/e* values along the ordinate (e.g., Fig. 1). If no ions are detected at a given *m/e* value, no peak is found at that value. In most analytical mass spectrometers the actual ion current is not recorded, but rather ions are detected via an electron multiplier. In addition to a variable amplification, the ion current is also a function of sample size and instrumental variables. Thus, the intensities of the signals recorded usually have no absolute units, but rather their significance lies in their intensities relative to one another. Most often a spectrum is normalized by plotting the largest or base peak with an intensity of 100%, and the intensities of all other peaks are measured relative to that of the base peak. Alternatively, the intensities of all peaks (in any units) may be summed and each characterized as carrying a certain percent of the total ion current, $\% \sum_x^y i$. Occasionally, intensity is plotted on a logarithmic or square root scale. In any case, ions of some *m/e* values will be formed with relatively great abundance and others not at all. Each spectrum of *m/e* values and intensities is highly characteristic of a given compound.

A good mass spectrometer has a dynamic range of 1 to 10,000, through which relative ion intensities may be reproducibly recorded. This dynamic range is often abbreviated in reporting analytical data on a relative intensity scale of 1 to 100 (see Fig. 1), although it lends itself well to tabular presentation (Table I) or to a lograithmic abcissa. Often a critical portion of the

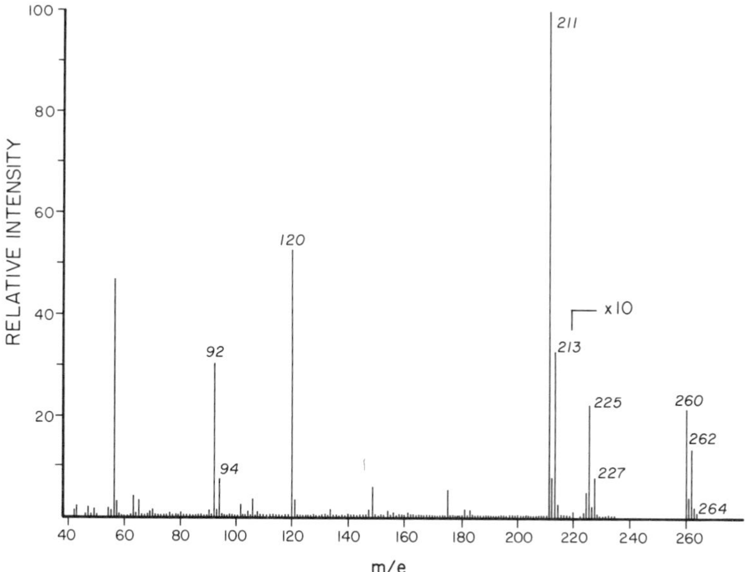

Fig. 1 Electron impact mass spectrum of cyclophosphamide.

TABLE I

Electron Impact Mass Spectrum of Cyclophosphamide

Mass	% RI	$\% \sum_{50}^{270}$	Mass	% RI	$\% \sum_{50}^{270}$	Mass	% RI	$\% \sum_{50}^{270}$
50	0.03	0.01	101	0.10	0.03	158	0.13	0.04
51	0.03	0.01	102	2.39	0.66	159	0.05	0.01
52	0.02	0.01	103	0.30	0.08	161	0.81	0.22
53	0.03	0.01	104	0.87	0.24	162	0.34	0.09
54	1.72	0.47	105	0.15	0.04	163	0.04	0.01
55	1.36	0.37	106	3.41	0.94	168	0.23	0.06
56	46.90	12.91	107	0.08	0.02	170	0.08	0.02
57	3.16	0.87	108	0.96	0.27	173	0.01	0.00
58	0.60	0.17	109	0.07	0.02	175	5.59	1.54
59	0.03	0.01	110	0.06	0.02	176	0.22	0.06
60	0.10	0.03	111	0.10	0.03	177	0.01	0.00
61	0.17	0.05	112	0.10	0.03	178	0.02	0.01
62	0.34	0.09	113	0.02	0.01	181	1.39	0.38
63	4.15	1.14	114	0.13	0.04	183	1.42	0.39
64	0.73	0.20	115	0.03	0.01	185	0.17	0.05
65	3.34	0.92	116	0.01	0.00	189	0.25	0.07
66	0.02	0.01	117	0.06	0.02	190	0.01	0.00
67	0.24	0.07	118	0.20	0.05	194	0.05	0.01
68	0.56	0.15	119	0.43	0.12	195	0.05	0.01
69	0.86	0.24	120	52.82	14.54	197	0.41	0.11
70	1.49	0.41	121	3.47	0.96	199	0.12	0.03
71	0.22	0.06	122	0.46	0.13	201	0.01	0.00
72	0.12	0.03	123	0.02	0.01	204	0.02	0.01
73	0.15	0.04	124	0.22	0.06	207	0.13	0.04
74	0.17	0.05	125	0.05	0.01	210	0.77	0.21
75	0.09	0.02	126	0.23	0.06	211	100.00	27.52
76	0.75	0.21	127	0.03	0.01	212	8.27	2.28
77	0.07	0.02	128	0.39	0.11	213	32.89	9.05
78	0.43	0.12	129	0.06	0.02	214	2.56	0.70
79	0.22	0.06	132	0.19	0.05	216	0.19	0.05
80	0.68	0.19	133	0.24	0.07	218	0.15	0.04
81	0.10	0.03	134	1.21	0.33	220	0.05	0.01
82	0.12	0.03	135	0.10	0.03	224	0.48	0.13
83	0.26	0.07	136	0.01	0.00	225	2.21	0.61
84	0.12	0.03	137	0.15	0.04	226	0.20	0.06
85	0.06	0.02	138	0.01	0.00	227	0.75	0.20
86	0.13	0.04	140	0.41	0.11	228	0.06	0.02
88	0.11	0.03	141	0.03	0.01	229	0.01	0.00
89	0.09	0.02	142	0.43	0.12	231	0.06	0.02
90	1.23	0.34	144	0.09	0.02	234	0.01	0.00
91	0.24	0.07	145	0.19	0.05	239	0.10	0.03
92	30.48	8.39	146	0.02	0.00	242	0.04	0.01
93	1.29	0.36	147	1.31	0.36	260	2.10	0.58
94	7.42	2.04	148	0.21	0.06	261	0.34	0.09
95	0.27	0.07	149	5.64	1.61	262	1.30	0.36
96	0.11	0.03	150	0.18	0.05	263	0.19	0.05
97	0.06	0.02	151	0.03	0.01	264	0.11	0.03
98	0.06	0.02	154	1.06	0.29	265	0.01	0.00
99	0.09	0.02	156	0.98	0.27			

spectrum, usually the high mass range, may be amplified by a factor of 10 or 100 in a conventional histogram (e.g., Fig. 1) exploiting the full dynamic range. The dynamic range of the mass spectrometer is often truncated by computerized data processing systems.

The mass range that can be scanned varies with the particular instrument. A typical mass range is *m/e* 12 to *m/e* 900. However, it is only necessary to scan the range of the expected *m/e* for a sample. Similarly, the entire mass range is not often plotted in a spectrum, but rather the range in which structurally significant ions are detected. For many samples the spectrum may be initiated at the low mass end at *m/e* 40 with no loss in critical structural information. Thus, air peaks at *m/e* 32 and 28, water, and other instrumental background peaks are excluded from the spectrum.

Mass spectroscopists use a mass scale in which an atom of carbon weighs 12.000 Daltons (atomic mass units). The masses of atoms most commonly encountered in organic, biomedical, and environmental compounds are presented in Table II. The mass spectrometer also is capable of separating

TABLE II

Atomic Weights and Approximate Natural Abundance of Selected Isotopes

Isotope	Atomic Weight (^{12}C = 12.000000)	Natural Abundance (%)
^{1}H	1.007825	99.985
^{2}H	2.014102	0.015
^{12}C	12.000000	98.9
^{13}C	13.003354	1.1
^{14}N	14.003074	99.64
^{15}N	15.000108	0.36
^{16}O	15.994915	99.8
^{17}O	16.999133	0.04
^{18}O	17.999160	0.2
^{19}F	18.998405	100
^{28}Si	27.976927	92.2
^{29}Si	28.976491	4.7
^{30}Si	29.973761	3.1
^{31}P	30.973763	100
^{32}S	31.972074	95.0
^{33}S	32.971461	0.76
^{34}S	33.967865	4.2
^{35}Cl	34.968855	75.8
^{37}Cl	36.965896	24.2
^{79}Br	78.918348	50.5
^{81}Br	80.916344	49.5
^{127}I	126.904352	100

the isotope species of the same atom. (Indeed, the greatest contribution of mass spectrometry to science is probably the first demonstration of the existence of isotopes.) Thus, molecular weights measured by this technique are not averaged for isotopic populations, but are monoisotopic. For example, the molecular weight of HCl by mass spectrometry is 35.977 and 37.974, reflecting the presence of isotopic $H^{35}Cl$ and $H^{37}Cl$ and not 37.538, the value found in most molecular weight tables.

Although mass spectrometers are available which can measure the m/e values for HCl to three decimal places, in most instruments these ions are observed at the nominal m/e values of 36 and 38. In fact, the masses of commonly encountered isotopes fall sufficiently close to whole numbers that ions are observed in most mass spectrometers at integral mass numbers (see Fig. 1). Peaks are encountered occasionally at half masses, and these represent doubly charged ions; for example, a peak at 91.5, represents a doubly charged ion of mass 183 ($m/e = 183/2$). Occasionally weak diffuse signals are observed which can extend through several mass values. These metastable peaks represent ions that have decomposed in the flight tube after leaving the ionization chamber.

B. Quality of the Spectrum

One feature to look for in the initial evaluation of a spectrum is the presence of ^{13}C isotope peaks accompanying the ^{12}C ions. The natural abundance of ^{13}C is variable, but is about 1.1% relative to that of ^{12}C. Thus the molecular ions of a compound such as benz(*a*)pyrene, which contains 20 atoms of ^{12}C, will be accompanied by ions which are 1 mass unit heavier and contain 1 atom of ^{13}C and 19 atoms of ^{12}C. The abundance of the $^{13}C^{12}C_{19}H_{12}^{+\cdot}$ species will be about 22% relative to that of the $^{12}C_{20}H_{12}^{+\cdot}$ species. This is illustrated in Fig. 2. Although the theoretical abundances may be calculated of species containing more than one ^{13}C atom, their abundances are very low.

The heavier isotopes of N and O occur with such low natural abundance that they do not contribute any substantial ion currents to most spectra. However, isotopic species of Si, Cl, Br, S, Fe, and many other elements are quite clearly distinguishable in a mass spectrum. The unique capability of the mass spectrometer to analyze isotopes will be discussed in Section V.A. The occurrence of satellite isotope peaks at higher masses and of peaks at lower masses representing ions which have lost one or two hydrogen atoms gives rise to several peaks in a group instead of a single peak for a major ion.

Spectra are occasionally encountered in the literature in which no isotope peaks are shown. This usually reflects a decision on the part of the investiga-

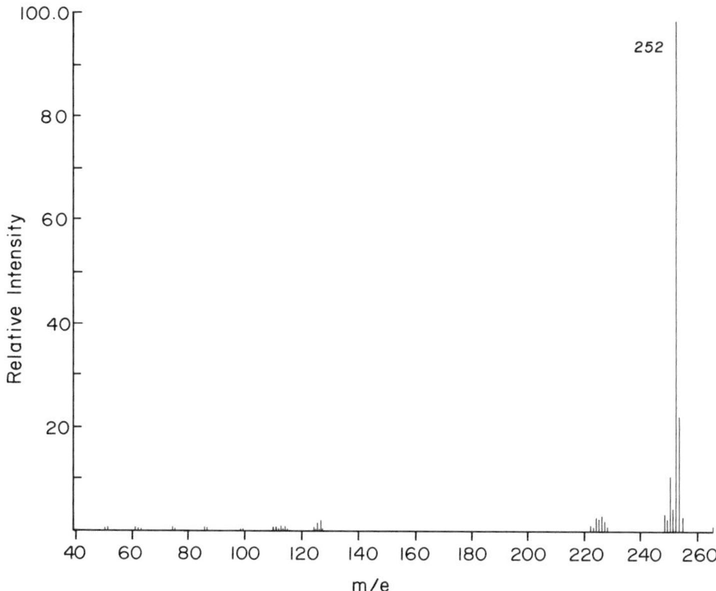

Fig. 2 Electron impact mass spectrum of benz(*a*)pyrene.

tor not to report the entire spectrum, but to report only peaks he thinks are important. If he has omitted the isotope peaks, one wonders what else he has left out that might be of interest. Instead of heavy handed subjectivity on the part of the investigator, the absence or incorrect intensities of some of the isotope peaks can also reflect a malfunctioning computerized data processing system or a careless draftsman. Klaus Biemann has suggested another explanation for the absence of ^{13}C isotope peaks in a spectrum—the use of isotopically pure ^{12}C samples. The absence of isotopic peaks is usually a reporting problem.

In one's own laboratory, spectra that have strong peak groups, weak peaks, and blank spaces with no signals (see Figs. 1 and 2) are sought. If a peak occurs at every mass, and all have about equal intensities, then the sample is either too small to be characterized or it is too impure. Instrumental background can contribute peaks to a spectrum. Background spectra are often run and subtracted from sample spectra.

Another characteristic of a good spectrum is a definite termination point at the high mass end. One hopes to find a peak group of reasonable relative intensity, beyond which no more peaks are recorded. If the peaks in a spectrum become gradually weaker at increasing m/e values, trailing off into the sunset as it were, then it will be hard to assign a molecular ion, if indeed one has been detected.

If a peak from the high mass end of the spectrum is to be assigned as the molecular ion peak, it must be separated from lower mass fragment ions by mass differences which correspond to the loss of a reasonable neutral fragment Thus, lower peaks at mass intervals of 15 (CH_3) and 18 (H_2O) Daltons provide corroboration for the assignment of a molecular ion. The occurrence of peaks at mass intervals of 8, 13, or 25 mass units (for example) at the high mass end of a spectrum suggests that the sample is a mixture. Of course the molecular ion may not always be observed, and the highest mass ions in the spectrum may be due to fragment ions.

The question of sample purity may also be addressed by scanning a series of spectra as the sample elutes or evaporates into the mass spectrometer. If more than one compound is present, some fractionation will usually occur, which will be reflected in changes in the relative intensities of the peaks contributed by the different compounds. Pyrolysis of the sample in the instrument can often be detected as it proceeds by running sequential scans. The requirements that peaks contributed by a single compound retain their relative intensities throughout the analysis period has been developed into algorithms for computerized deconvolution of overlapping peaks on a GCMS.

The mass spectrum of a sample should be reproducible from day to day, from one sample preparation to another, and, with less precision, from instrument to instrument. It is difficult to define acceptable interinstrument reproducibility. Basically, the major and minor peaks should remain as such. Peaks should not be present in one spectrum with medium or strong intensity and absent in another. Obviously, operating conditions should be duplicated as closely as possible.

C. *The Fragmentation Pattern*

Once a spectrum of good quality has been obtained, it can be used for whatever purpose the investigator has in mind. Except for simple pattern matching, most applications will involve some degree of interpretation. The first step in interpreting a mass spectrum is usually the identification of the molecular ion peak, or the recognition of its absence. The second step usually is the definition of the fragmentation pattern. A large and active literature exists in which correlations are reported of fragmentation patterns with structures of known compounds. Most sophisticated laboratories employ stable isotope labels, accurate mass measurements, and metastable ions to define fragmentation patterns and also mechanisms of decomposition processes. In general, specific fragmentation patterns have been found to be characteristic of given structural elements, and especially of functional groups. Among the most simple of these is the loss of water usually observed in the spectrum of compounds that contain aliphatic hydroxyl groups. Fragmentation patterns are often altered when several structural elements or functional groups are combined in a molecule, and one usually has to evaluate each family of compounds anew.

Once the fragmentation patterns and characteristic ions have been discerned in a spectrum, the delicate process begins of moving on to the structure of the sample. Several approaches have been described in the literature (Biemann, 1962; Budzikiewicz *et al.*, 1967; McLafferty, 1973). While these approaches share many features in common, that developed by Djerassi and co-workers may be termed "mechanistic," and that of McLafferty "arithmetic" (Budzikiewicz *et al.*, 1967). In defining the "shift technique," Biemann (1962) early recognized that most samples are already known to be members of a given structural family and that it is of great value to compare spectra of related compounds from this family.

This section will address the delineation of fragmentation patterns, particularly in the context of comparing the unknown spectrum to spectra of related known compounds. Specific examples of structure elucidation will be presented in a subsequent section.

1. *The Molecular Ion Peak*

As has been previously stated, the prime consideration is usually the assignment of the molecular ion peak. Candidates in interpreting a spectrum are usually sought in the high mass end of the spectrum. Note in Figs. 1 and 2 that the peaks all occur in groups and not individually, because of the isotope species already mentioned, and also because some fragment ions lose 1, 2, or even 3 hydrogen atoms to form less abundant and lighter ions. Usually one or two peaks in the group are much more intense than the others and these represent the candidates for the molecular ions and the most significant fragmentation ions. Thus the peak at m/e 252 in the spectrum of benz(a)pyrene (Fig. 2) represents a molecular ion, and is accompanied by higher mass isotope peaks and a lower mass $M - \mathrm{H}$ peak.

McLafferty (1973) and others argue that some idea of the elemental composition of a compound may be gained by examining the relative abundance of its molecular ions. Thus a molecular ion with a 13% $M + 1$ peak can have no more than 12 carbon atoms. If some of the ion current in this peak arises from other sources, of course, the actual carbon count will be lower. Usually, when the sample being analyzed is polar, or when it is introduced on the direct probe, $M + \mathrm{H}$ ions will make small contributions to the $M + 1$ peak, along with ^{13}C isotope species, and the upper limit deduced for the carbon content of the molecular formula can be so high as to be useless. Nonetheless, this potential source of important information should always be explored. This approach to carbon content is usually not employed for fragment ions because of the high probability of other fragment ions overlapping with isotope peaks.

Some frequently encountered elements occur with isotope distributions so unique that the isotope pattern permits their rapid identification in an unknown. For example, the isotopes of bromine, ^{79}Br and ^{81}Br, occur with an abundance ratio of nearly 1:1. This means that a compound containing 1 bromine atom will be ionized to 2 major molecular ions, 2 Daltons apart, with nearly 1:1 abundance. Carbon-13 isotope peaks will also be observed, of course. The patterns for 2 and more bromine isotopes may be calculated by polynomial expansion (Biemann, 1962). Similarly, the ratio of abundances of naturally occurring ^{35}Cl and ^{37}Cl (approximately 3:1) produces a distinct pattern in ions containing chlorine atoms. The number of peaks observed for a polybrominated or polychlorinated ion is always one more than the number of bromine or chlorine atoms present. The isotope patterns of Si and S (Table II) often reveal the presence of these elements. And, of course, the presence of many other elements will be reflected in their isotope patterns, e.g., ^{54}Fe:^{56}Fe 1:15 and ^{10}B:^{11}B 1:4.

As has been mentioned, the molecular ion peak must occur at mass differences from the next two or three peaks (or peak groups really) that correspond to reasonable fragment losses. Reasonable masses can include 1 (H), 15 (CH_3), 16 (O), 18 (H_2O), 20 (HF), 27 (HCN), 31 (CH_3O), 34 (H_2S), etc. (4,5). They do not include masses 5–13 or 21–25 in most organic compounds. Methylene groups are rarely eliminated from a molecular ion, and a peak 14 mass units below the molecular ion most likely represents a homologous impurity.

The nitrogen rule holds for organic compounds that contain C, H, O, N, Si, S, halides, or P. This guideline states that the molecular weight of a compound containing an odd number of nitrogen atoms will be an odd number of Daltons, and that of compounds containing no nitrogen or an even number of nitrogen atoms will be an even mass number. This correlation can be very valuable for spotting compounds containing 1, 3, or 5 nitrogen atoms.

2. *High Mass Ions and Neutral Species Eliminated from the Molecular Ion*

Often the most informative portion of the spectrum is that in the upper third or half of the mass range. Here interpretation depends on calculating the mass differences between fragment ions and the molecular weight, and considering what neutral species have been lost in primary decomposition processes. Primary processes reflect the structure of the molecule more directly than later proceeses in multistep cleavage sequences. Tables are available (Williams and Howe, 1972) of neutral fragments that are frequently lost in primary processes. It should also be emphasized that high mass peaks should be considered to be important, even if their relative intensity is quite low. Often the high mass regions in a spectrum are plotted on an expanded scale (e.g., $\times 10$ in Fig. 1).

In the spectrum in Fig. 1, peaks at m/e 225 and 227 correspond to ions formed by the loss of 35 and 37 mass units from the molecular ions. Loss of an atom of chlorine is an obvious interpretation, which can be correlated with the structure of the sample. In other cases, neutral molecules are eliminated as well as neutral radicals.

Even when the mass difference between the molecular ion and a fragment ion is substantial, it is often constructive to consider the fragmentation in terms of the neutral fragment as well as the charged species. For example, peaks at 386, 412, and 429 in the spectra of the trimethylsilylated methyl esters of Δ^9-tetrahydrocannabinol glucuronide (molecular weight 720), diethylstilbestrol glucuronide (molecular weight 746), and morphine glucuronide (molecular weight 763) are all $M - 334$ peaks, despite their varying m/e values, and are all formed by loss of the sugar acid moiety, a fragmentation pattern characteristic of phenol-linked glucuronides. (The structures shown in scheme **1** have not been established, but allow one to keep track of the fragmentation process.)*

COOCH$_3$ OTMS OTMS OTMS → OTMS +

$M - 334$

$+ C_{13}H_{26}O_6Si_2$

1

It is also useful to view a fragmentation process not as the loss of a methyl group, or of an ethyl group, etc., but as cleavage of a particular bond in the molecule. Often the same bond is cleaved in a group of related molecules, although the masses of the charged and neutral species are different. This can be illustrated in the fragmentation patterns of two derivatized dipeptides, N-acetyl glycinylalanine methyl ester, and N-acetyl valinylleucine methyl

* Generally accepted symbols are used here for the purpose of "electron book keeping" (Budzikiewicz *et al.*, 1967), and do not imply a firm knowledge of the mechanisms of fragmentation or of the structures of ions. These symbols include $^{+\cdot}$ to denote an ion radical, ↑ to denote a two electron shift (usually a heterogeneous bond cleavage) and 1 to denote a one electron shift (usually a homogeneous bond cleavage). In addition, the charge is shown as localized in most fragmentation pathways described here. While this is an approximation to the real molecular charge distribution it is generally accepted as a good approximation and a useful device for keeping track of electrons and bond cleavages.

ester (scheme **2**). Although the masses of the ions and neutral species formed differ greatly, the same amide bond is broken in each case, in a cleavage characteristic of amides as a class.

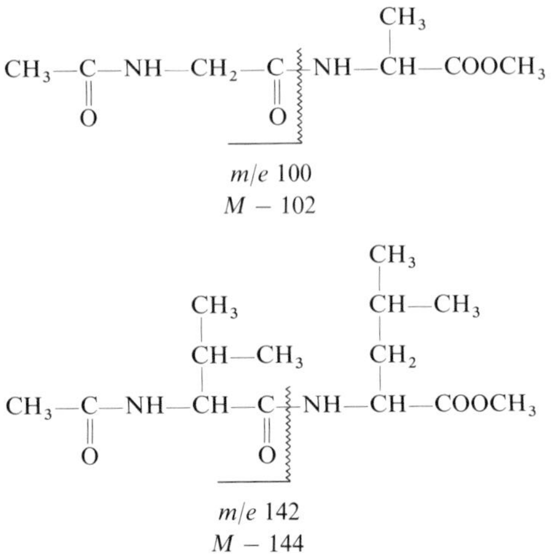

$$CH_3{-}\underset{\underset{O}{\|}}{C}{-}NH{-}CH_2{-}\underset{\underset{O}{\|}}{C}\,\wr\,NH{-}\overset{\overset{CH_3}{|}}{CH}{-}COOCH_3$$

m/e 100
M − 102

$$CH_3{-}\underset{\underset{O}{\|}}{C}{-}NH{-}\overset{\overset{\overset{CH_3}{|}}{CH-CH_3}}{\underset{}{CH}}{-}\underset{\underset{O}{\|}}{C}\,\wr\,NH{-}\overset{\overset{\overset{\overset{CH_3}{|}}{CH-CH_3}}{|}}{\overset{CH_2}{\underset{}{CH}}}{-}COOCH_3$$

m/e 142
M − 144

2

The value of this mechanistic approach to interpretation cannot be emphasized enough, because it is frequently not employed and because it is just as frequently the key to deducing the structure of a sample. It will be emphasized subsequently in discussing specific examples of structure elucidation.

3. *Information from the Low Mass Ions*

There are two general ways in which information may be obtained from the lower half of the spectrum. One is to recognize abundant class characteristic ions, and the other is to trace homologous series of ions to estimate the extent of aliphatic hydrocarbon portions of unknown molecules.

A small number of intense peaks have been found to be associated with specific functional groups or structural features (scheme **3**). Thus the presence

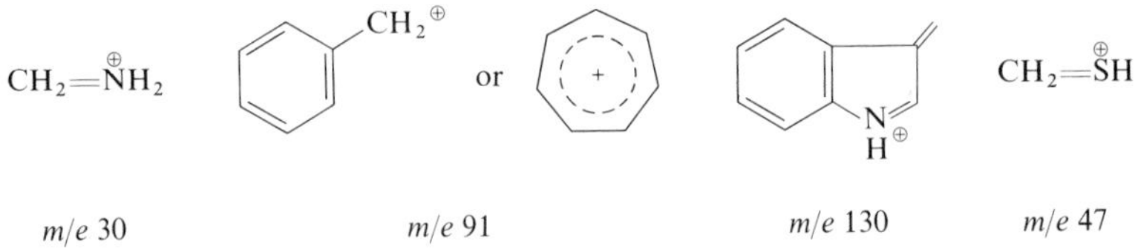

$CH_2{=}\overset{\oplus}{N}H_2$ — *m/e* 30

$C_6H_5CH_2^{\oplus}$ or (tropylium, +) — *m/e* 91

(indole, $\overset{\oplus}{N}H$) — *m/e* 130

$CH_2{=}\overset{\oplus}{S}H$ — *m/e* 47

3

of the intense peak at *m/e* 30 is usually associated with an aliphatic amine, *m/e* 91 with a benzyl moiety, *m/e* 130 with the indole nucleus, 47 with sulfides.

If a sample has been trimethylsilylated (scheme **4**), an intense peak is almost

$$\underset{\text{(TMS)}\ m/e\ 73}{CH_3-\overset{CH_3}{\underset{CH_3}{Si^{\oplus}}}} \qquad \underset{m/e\ 147}{CH_3-\overset{CH_3}{\underset{CH_3}{Si}}-O=\overset{\oplus}{Si}(CH_3)_2}$$

4

always observed at m/e 73. An intense peak at 147 reflects the presence of two or more trimethylsiloxy groups. Several peaks may be considered together as evidence for certain functional groups. Thus peaks at m/e 105 and 77 strongly suggest the presence of a phenylketone (scheme **5**).

$$C_6H_5-C(=\overset{+\cdot}{O})-X \longrightarrow \underset{m/e\ 105}{C_6H_5-C\equiv O^{\oplus}} \longrightarrow \underset{m/e\ 77}{(C_6H_5)^+}$$

5

To qualify as an indicator peak in the sense described here, ions must always be formed by compounds in a certain class, and must not be formed readily by other kinds of compounds. Mass-43 ions are formed by most methyl ketones, but they are also formed by many acetates and by most molecules with any aliphatic hydrocarbon component. Thus they are not indicative of any particular functional class. Lists are available of ions and structural elements commonly associated with peaks at lower m/e values (McLafferty, 1973; Williams and Howe, 1972; Spiteller, 1966). Many more indicator ions can be identified that are not in the tables, associated with less frequently encountered classes of compounds. For example, intense peaks at m/e 299 and 315 suggest that a sample is a trimethylsilylated phosphate ester.

Homologous series of ions 14 units apart can provide a good estimate of the chain length of aliphatic portions of a molecule. These homologs may occur in the C_nH_{2n+1} series, or an XC_nH_{2n} series, where X can be a variety of groups. The lowest members of the series are sometimes missing in the spectrum, and peak intensities may fluctuate as one progresses through the series.

In a number of reference compounds widely used for making accurate mass measurements, hydrogen atoms have been replaced by fluorine atoms. Homologous series occur in the spectra of these perfluorinated samples with intervals of 50 Daltons, corresponding to CF_2 units.

III. Structure Elucidation

The most intellectually challenging application of mass spectrometry is the elucidation of structures of unknown samples. It should be pointed out that a structure cannot usually be obtained from the mass spectrum of the sample alone. A mass spectrum is used most profitably in conjunction with all other chemical and spectroscopic information that is available for the sample. Leads from other spectroscopic methods about the identities of specific functional groups present in the molecule are particularly useful in interpreting the mass spectrum in order to deduce more general information. Stereochemistry and some aspects of bonding cannot be deduced reliably from a mass spectrum, and in the end, rigorous identification still requires synthesis of an authentic sample for comparison.

In practice, one rarely encounters a completely unknown sample. Most often the compound under study is the product of a chemical reaction, the metabolite of an administered drug, the by-product of a plant process, or a sample isolated by a protocol that excludes everything but one class of compounds, e.g., amino acids. The most widely used approach to the identification of an unknown sample is to obtain and interpret spectra of related compounds to obtain an understanding of the fragmentation pattern of that particular family or class of compounds, and to examine the changes in the masses of ions formed as the sample fragments along the same lines.

One of the most elegant mass spectral approaches to structure elucidation and one that is used surprisingly little, is the element map. Accurate mass measurements (Colby, this volume) may be made on the molecular ion in low resolution; however, the probability that isobars may be formed with the same nominal mass is higher for fragment ions, and the reference compound necessary for accurate mass measurements may contribute ions of the same masses. Thus, if accurate mass measurements and probable elemental compositions are to be measured for all ions, it is best done with a high resolution instrument. An element map of deoxy-dihydro-N_6-methylajmaline is given in Fig. 3 along with notation by Biemann *et al.* (1964). Once the elemental compositions of all the ions are assigned, these are arranged according to the heteroatoms they contain. As the notation on Fig. 3 indicates, this permits facile recognition of the elemental composition of various portions of the molecule. For example, the column on the extreme right lists only two entries for ions which contain 2 nitrogen and 1 oxygen atom. One is, of course, the molecular ion, whose carbon content is indicated as 21 atoms and hydrogen content as 30 atoms. The last number in the entry reports the deviation (in millimass units) of the mass found from that calculated for the elemental composition given. That deviation is zero in the lower right-hand entry. The asterisks denote intensity. The other entry in the right-hand column reveals that a saturated 4-carbon unit may be lost without losing any heteroatoms.

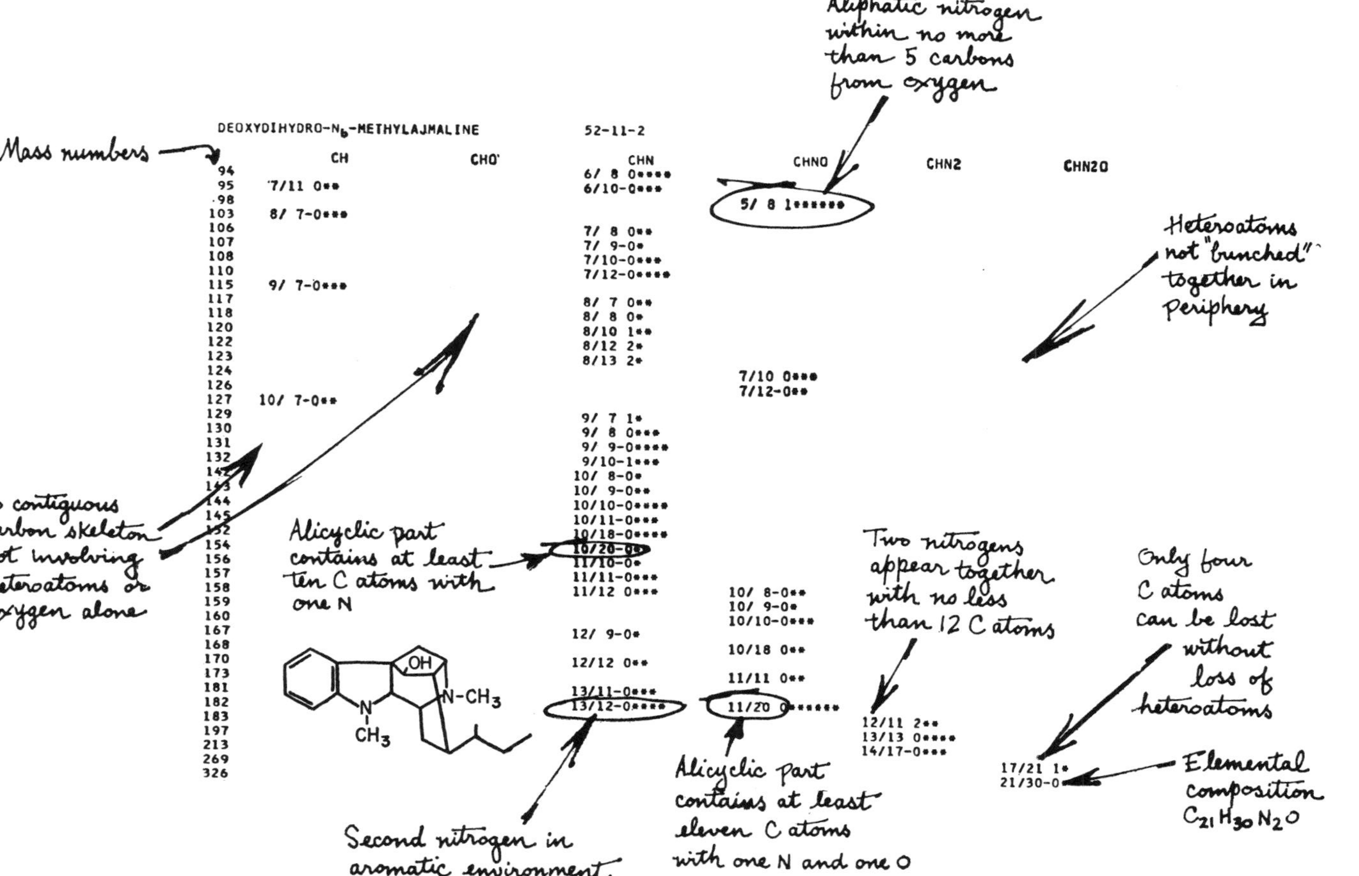

Fig. 3 Element map of deoxy-dihydro-N_b-methylajmaline. [Reproduced by permission from *Chem. Eng. News* July 6, pp. 42–44 (1964).]

A third approach to identifying an unknown is to try to match its spectrum or features of its spectrum to those in a reference library. This may be a library of a limited size, as is used in toxicologic and forensic work, or one may work with nonspecific collections of 30,000 to 50,000 spectra (see Section IV). In the latter case, certainly, the searching must be computerized. Since identical spectra are not obtained from instrument to instrument, or even from week to week in the same laboratory, flexible guidelines have to be established as to what constitutes a match. It is also important to realize that partial structural features can be deduced by library searching even if the exact compound is not in the library.

A. Molecular Weight

The most important single piece of information which may be obtained from a mass spectrum is the molecular weight of the compound, or its elemental composition if accurate mass measurements can be made. This kind of information has classically been deduced from combustion analyses, which require a relatively large amount of a pure compound. Mass spectrometry has the great advantage with samples which may be impure and available only in very small amounts. In contrast to x-ray crystallography, mass spectral samples need not be crystalline. An outstanding example of the value of mass spectrometry for elucidating molecular weights (and of the ambiguity of elemental combustion analysis for assignment of empirical formulas) is the study of the structure of the antibiotic picromycin. This compound was isolated and its structure was characterized by Brockman and co-workers in 1951. The structure was based on extensive chemical work and a very careful combustion analysis. The combustion analysis obtained by these workers is shown in Table III, along with the elemental composition they assigned to the compound. Their formula corresponded to a molecular weight of 469 Daltons. In 1968 Muxfeldt in the United States working in collaboration with Brockman, and also Richards in England, both obtained mass spectra of picromycin (Muxfeldt *et al.*, 1968; Rickards *et al.*, 1968) and the molecular weight was assigned as 525. Because Muxfeldt obtained a high resolution mass spectrum, his accurate mass 525.3302 led him to a revised empirical formula which is shown in the Table III. Subsequent chemical work led to the identification of the portion of the molecule which had not been included in the original structure and to correction of the formula and structure. The combustion analysis and mass spectral molecular weights observed are presented in Table III along with the theoretical values for both the old and the new structures of picromycin. The elemental composition observed by combustion analysis is closer to the theoretical composition

TABLE III

Analyses for Original and Revised Pikromycin Formulas[a]

	Calculated for Original Formula $C_{25}H_{43}NO_7$	Calculated for Revised Formula $C_{28}H_{47}NO_8$	Observed Analyses
Combustion analysis	C 63.94 H 9.23 N 2.98	C 63.96 H 9.02 N 2.66	C 63.86 H 9.02 N 2.92
Molecular weight by high resolution mass measurement	469.3028	525.3298	525.3302

[a] Taken from Fenselau (1972).

values for the incorrect structure, and all three sets of values are very similar. However, the mass spectrum clearly differentiates between the possibilities.

Not all compounds produce molecular ions on electron impact and Henry Fales has observed publicly that the more important a sample is, the less likely its spectrum is to contain a molecular ion. There are several reasons why molecular ions may not be detected. The sample may be involatile. It may be thermally unstable. It may fragment completely on ionization. One or more of these properties is usually associated with more polar compounds and with compounds of molecular weights in excess of 1000. Several approaches may be taken to remedy this situation. The sample may be altered by derivatization or chemical modification to produce a more volatile species, a more stable compound, or a sample less prone to fragmentation. In addition, a number of ionization techniques are available, alternate to electron impact, which produce ions from involatile samples, and/or which produce ions with the transfer of less excess energy than electron impact. If less energy is transferred to the samples, fewer bonds break and more molecular species can be detected. In another chapter in this volume, Colby describes chemical ionization, field ionization, field desorption, fission fragment ionization using ^{252}Cf, and also ionization by various means at atmospheric pressure (Colby, this volume). Chemical ionization (CI) encompasses proton transfer ionization, charge exchange ionization, and also negative and positive ion attachment (Munson, 1971; Dougherty *et al.*, 1972). Different reagent gases provide some chemical selectivity in their ionization reactions, and most reagent gases provide sensitivity comparable to that of electron impact. The most important feature of chemical ionization is that fragmentation is usually greatly reduced and molecular ion species (most commonly $M + H$ ions) are observed with high abundance. This makes chemical ionization especially useful in assays employing selected ion monitoring. It is important to remember that the extent of fragmentation is directly related to temperature in a chemical ionization source.

In an interesting variation on chemical ionization, sources have been developed in several laboratories in which compounds are analyzed at atmospheric pressure (Horning *et al.*, 1977; Colby, this volume). A variety of energy sources has been evaluated for ionization including corona discharges and radioactive emitters. Volatile samples may be introduced directly from the gas chromatograph and less volatile samples may be injected in solution.

Although field ionization (Beckey, 1971) has been found to provide less sensitivity than chemical ionization, it does achieve low energy ionization with reduced fragmentation and enhanced intensity of molecular ions.

The apparent ionization of some molecules in the solid phase detected in field ionization spectra led to the development of field desorption (FD) ionization, a technique in which the sample is ionized exclusively in the solid state and the requirement for volatility is obviated (Beckey and Schulten, 1975; Holland *et al.*, 1976). Although field desorption spectra of pure samples are readily obtained, difficulties are sometimes encountered in extending the technique to analysis of unknown compounds, especially those isolated from biological milieu. Impurities usually present in such samples cause them to desorb under different conditions than pure reference compounds, and these conditions are often unpredictable. Perhaps the largest obstacle encountered has been the problem of contamination by metallic cations, particularly sodium and potassium. If a large amount of sodium salt is present in a sample, the spectrum of the sample will be obscured and mainly sodium cations and cluster ions will be detected. These difficulties may be mediated by more careful sample preparation and by careful instrumental control. The extent of fragmentation in an FD spectrum can be controlled by controlling the electrical current in the emitter on which the sample is coated (Maine *et al.*, 1976).

Figures 4 and 5 contain the CI (isobutane) and FD spectra of cyclophosphamide, which may be compared with its electron impact spectrum in Fig. 1. In the CI spectrum, $M + 1$ ions are the most abundant species detected. Chemical ionization induced fragmentation usually involves the elimination of small neutral molecules, and the loss of HCl in Fig. 4 is exemplary. The FD spectrum reveals primary fragmentation similar to that observed on electron impact. Although the conditions used to obtain the CI and FD spectra shown here were chosen to enhance the abundances of molecular ion species, fragmentation could be increased by raising the temperature in the CI source or the current in the FD emitter wire.

Another technique for ionizing involatile samples in the solid state involves their bombardment with fission fragments from ^{252}Cf (Colby, this volume). The fission fragment is allowed to penetrate a thin metal sheet with the analyt-

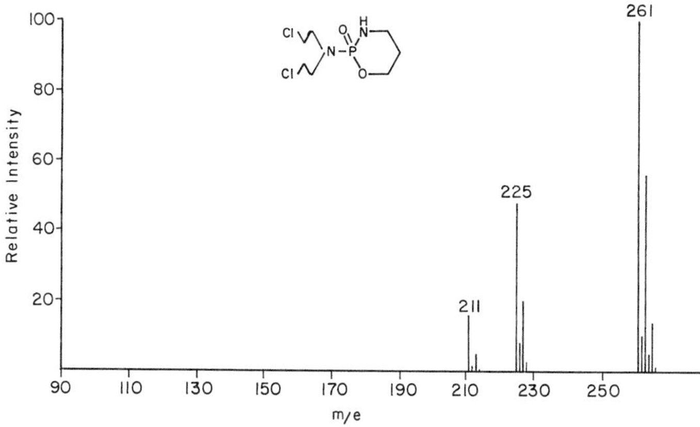

Fig. 4 Chemical ionization mass spectrum of cyclophosphamide.

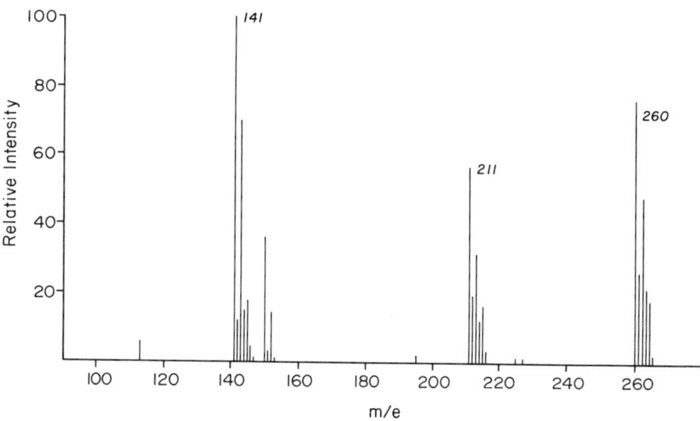

Fig. 5 Field desorption mass spectrum of cyclophosphamide.

ical sample coated on the other side. Ionizing fission events occur at discrete time intervals and many hours, may be required to generate enough ions from the analytical sample to compose a mass spectrum. One marked advantage of this ionization technique is that most of the sample can be recovered from the nickle foil, in contrast to other mass spectral ionization techniques. A spectrum of tetrotatoxin and chiriquitoxin is shown in Fig. 6, obtained by MacFarlane and Torgerson (1976) who developed the technique. The molecular ions are indeed detected in abundance. It was reported that a satisfactory mass spectrum of tetrotatoxin could not be obtained using any other ionization technique.

Once the molecular weight or empirical formula of the sample is established, the classical calculations for degrees of unsaturation (rings and

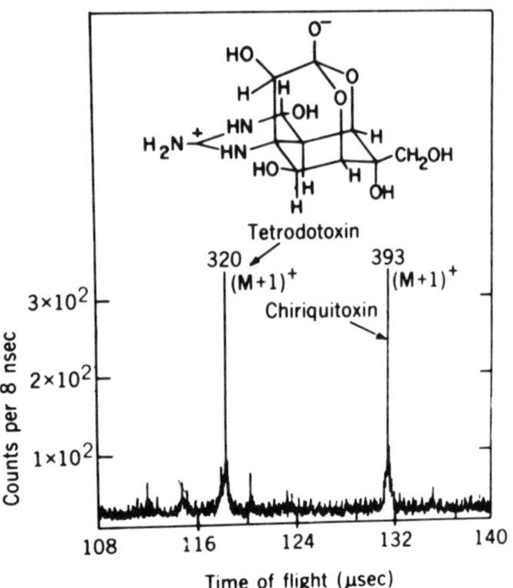

Fig. 6 Californium-252 plasma desorption spectrum of tetrodatoxin and chiriquitoxin. [Reproduced by permission from R. D. Macfarlane and D. F. Torgerson, *Science* **191**,920–925 (1976).]

double bonds) may be applied. Often an estimate of the number of active hydrogens (OH, NH, SH) may be obtained by exchanging H for D from D_2O in a chemical ionization source, or by counting the number of derivative groups that can be added. If the sample must be trimethylsilylated to be analyzed mass spectrometrically, the number of trimethyl groups that have been added may be determined by counting the increments of 9 Daltons introduced when the sample is derivatized with d_9-trimethylsilylating reagent. For example, the molecular weight of tris-(trimethylsilyl)-1-ethylthioethyl deoxyadenosine increases from 555 to 582 when three d_9-trimethylsilyl groups are added in place of the normal trimethylsilyl groups (scheme **6**).

TMS, TMSO—CH$_2$, OTMS — M^+ 555

TMS-d_9, d_9-TMSO-CH$_2$, OTMS-d_9 — M^+ 582

6

B. *Related Classes*

The most straightforward application of mass spectrometry to structure elucidation comes when one knows that the unknown compound belongs to a particular class of compounds. This would be the case in analyzing the products of many chemical reactions, metabolites obtained from drugs, some by-products of manufacturing processes, components of fractions which have been purified so as to include only certain classes of compounds, for example, amino acids or glucuronides. In this circumstance, the spectra of other members of the class may be obtained and examined to provide guidelines for interpreting the spectrum of the unknown sample. Biemann has called this approach the "shift technique" and first demonstrated its utility in several classes of alkaloids (Biemann, 1962).

1. *Drug Metabolism*

Drug metabolism is an area in which the sample (metabolite) can always be related to a known parent drug. An example may be drawn from the structure elucidation of animal metabolites of the antitumor drug cyclophosphamide. The electron impact mass spectrum of cyclophosphamide has already been presented (Fig. 1). There are three molecular ions corresponding to the presence of 2 chlorine atoms in the molecule, at *m/e* 260, 262, and 264 in relative intensities of approximately 9:6:1. The major peaks in this spectrum are the pair *m/e* 211 and 213. These peaks have a relative intensity of about 3:1, and thus correspond to ions that contain 1 chlorine atom. Their formation is indicated in scheme **7** as requiring cleavage of the carbon–carbon bond in the chlorothyl side chain. Another important fragmentation

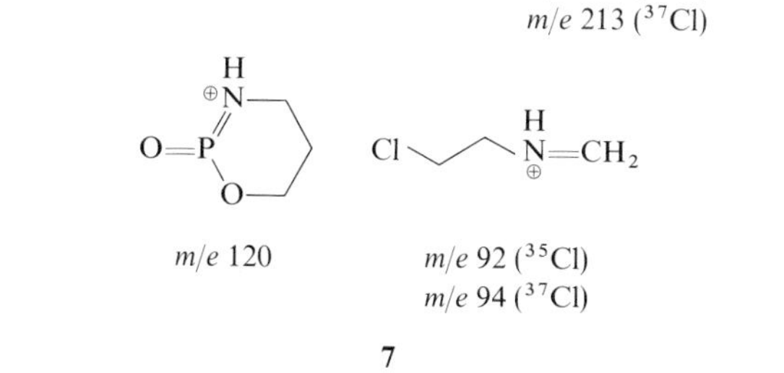

m/e 211 (^{35}Cl)
m/e 213 (^{37}Cl)

m/e 120

m/e 92 (^{35}Cl)
m/e 94 (^{37}Cl)

7

occurs with bond cleavage of the nitrogen–phosphorous ion and generates ions of mass 120. These ions comprise the heterocyclic portion of the molecule as is indicated in the scheme. A third set of peaks in the spectrum is that at *m/e* 92 and 94. This pair of peaks has a 3:1 relative intensity and, as is indicated in the scheme, correspond to a portion of the mustard moiety of the original drug.

In Table IV the fragmentation pattern of cyclophosphamide is summarized, along with the analogous information drawn from the spectra of three metabolites, A, B, and C. In deducing the structure of metabolite A, three molecular ions are confirmed, with the relative intensity pattern characteristic of 2 chlorines. Thus this metabolite still retains both chlorine atoms from the original drug. However, the molecular weight has been increased by 14 mass units. The addition of 14 mass units to a drug metabolite usually reflects replacement of a hydrogen atom by a methyl group or a replacement of 2 hydrogen atoms by oxygen in a carbonyl group. $M - 49$ ions are formed by loss of $\cdot CH_2Cl$ in the fragmentation of metabolite A, and have masses of 225 and 227 Daltons (Table IV). These ions are 14 mass units heavier than the analogous ions in the spectrum of the parent drug. Cleavage of the N–P bond generates ions comprising the heterocyclic moiety which in this case weighs 134 mass units. Thus the 14 mass unit increment is located in the heterocyclic portion rather than the nitrogen mustard portion of the molecule. The pair of peaks at *m/e* 92 and 94 are present, again suggesting that no change has taken place in the mustard portion of the molecule. Metabolite A was eventually identified as the 4-keto analog of cyclophosphamide (see scheme **8**), based on its infrared spectrum and also by comparison of its physical properties with a synthetic standard (Struck *et al.*, 1971).

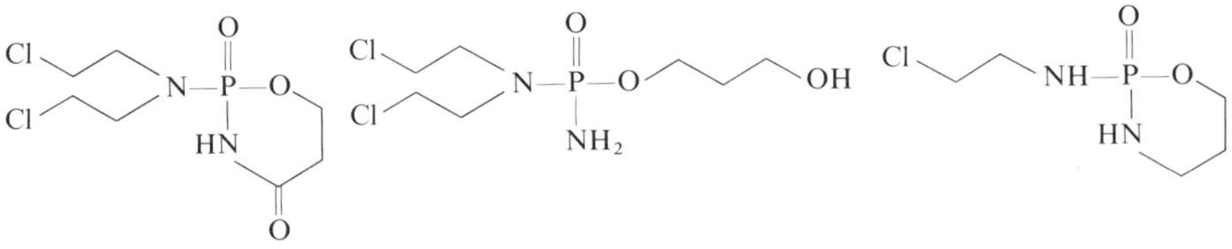

Cl Cl N–P(=O)–O HN O (A)   Cl Cl N–P(=O)–O OH NH₂ (B)   Cl NH–P(=O)–O HN (C)

8

The molecular ions of metabolite B, shown in Table IV, reflect the presence of 2 chlorine atoms. This time the molecule weighs 18 mass units more than cyclophosphamide. The addition of 18 mass units is usually caused by the addition of water to a molecule. This change does not appear to have taken place in the nitrogen mustard portion of the molecule, because elimination of $\cdot CH_2Cl$ leads to a pair of peaks at *m/e* 229 and 231, 18 mass units higher than the corresponding peaks in the cyclophosphamide spectrum, and be-

TABLE IV

Important Peaks in the Spectra of Cyclophosphamide and Some Metabolites

	M^{+} (2 Cl Pattern)	$M - 49$ (1 Cl) Base Peak	$M - 140$	92, 94
Cyclophosphamide	260, 262, 264	211, 213	120	Yes
Metabolite A	274, 276, 278	225, 227	134	Yes
Metabolite B	278, 280, 282	229, 231	138	Yes
Metabolite C	198, 200	149	120	No

cause the pair of peaks at 92 and 94 are still observed in the spectrum. More to the point, the heterocyclic ions, which weigh 120 mass units in the spectrum of cyclophosphamide, are now observed to have a mass of 138. Thus, water has been added to the heterocyclic moiety. In fact, this metabolite was eventually demonstrated to be an open-chain hydroxy compound formed by addition of the elements of water (Bakke *et al.*, 1972).

The entries in Table IV for metabolite C look different from those for the other three compounds. To begin with, only two molecular ions are present, suggesting that 1 chlorine atom has been eliminated in whatever biotransformations have taken place. Similarly, the loss of CH_2Cl from the molecular ion leads to a single ion of mass 149 with no chlorine present. The heterocyclic ions are observed to have a mass of 120, suggesting that the changes have occurred in the mustard moiety. The pair of peaks at m/e 92 and 94 are not observed in the spectrum. The molecular weight is considerably lighter than that of the parent drug cyclophsophamide, and arithmetical analysis of the possibilities should suggest that metabolite C is the N-dealkylated analog of cyclophosphamide (scheme **8**), in which one of the beta-chloroethyl chains has been removed from the mustard nitrogen. This example should illustrate that if one works within a defined family or related group of compounds, one can reasonably expect the same basic fragmentation to occur throughout the set.

2. *Peptide Sequencing*

Another example of the application of mass spectrometry to structure elucidation of a sample known to belong to a certain class of compounds is that of peptide sequencing. In this case the sample is almost always known to be a peptide beforehand, and preliminary information has usually been obtained, including identification of the amino acids liberated by total hydrolysis. Another boundary limit in this kind of work is provided by the knowledge that only 20 or so amino acids will commonly be encountered in peptide sequencing. As was previously indicated (scheme **2**), the fragmentation pattern of peptides as a class features cleavage in the amide groups that link the monomeric acids.

The identification of two related pentapeptides from rat brain, which have potent opiate agonist activity was based primarily on the mass spectrum of the mixture (Hughes *et al.*, 1975.) of the two. The amino acid composition after hydrolysis with hydrochloric acid at elevated temperatures was found to be glycine 36.5, methionine 12.3, tyrosine 16.9, phenylalanine 22.5, and leucine 4.2 nmoles. The peptide mixture was derivatized by acetylation of amine groups and by permethylation to increase the volatility of the pentapeptides and also to enchance fragmentation of the amide bonds so that sequence information could be more easily deduced from the spectrum. The mass spectrum (Fig. 7) was interpreted as characterizing two pentapeptides, each containing the sequence tyrosine, glycine, glycine, phenylalanine as shown in scheme **9**. Methionine was assigned to the carboxy terminal of one peptide and leucine to the other.

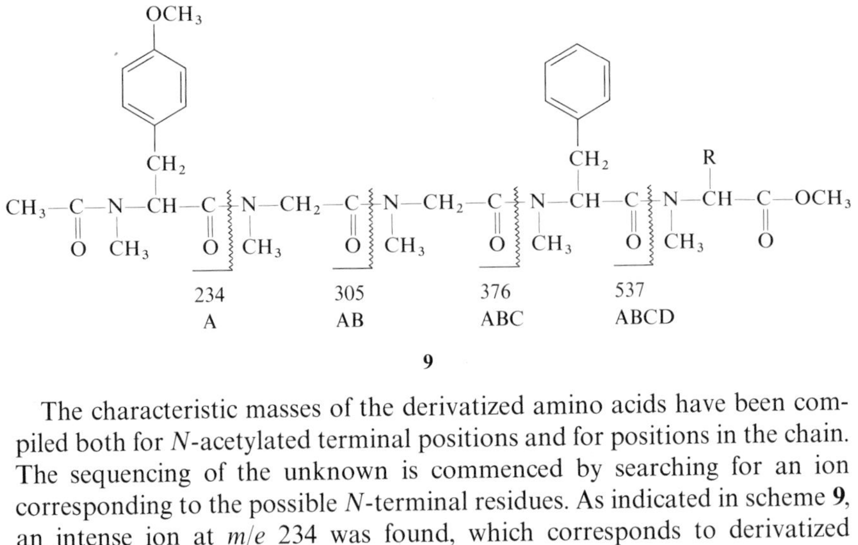

$$CH_3-\underset{\underset{O}{\|}}{C}-\underset{\underset{CH_3}{|}}{N}-\underset{|}{\overset{\overset{\overset{OCH_3}{|}}{C_6H_4}}{\overset{|}{CH_2}}}{CH}-\underset{\underset{O}{\|}}{C}\,\wr\,\underset{\underset{CH_3}{|}}{N}-CH_2-\underset{\underset{O}{\|}}{C}\,\wr\,\underset{\underset{CH_3}{|}}{N}-CH_2-\underset{\underset{O}{\|}}{C}\,\wr\,\underset{\underset{CH_3}{|}}{N}-\overset{\overset{C_6H_5}{|}}{\overset{CH_2}{|}}{CH}-\underset{\underset{O}{\|}}{C}\,\wr\,\underset{\underset{CH_3}{|}}{N}-\overset{\overset{R}{|}}{CH}-\underset{\underset{O}{\|}}{C}-OCH_3$$

234 A | 305 AB | 376 ABC | 537 ABCD

9

The characteristic masses of the derivatized amino acids have been compiled both for *N*-acetylated terminal positions and for positions in the chain. The sequencing of the unknown is commenced by searching for an ion corresponding to the possible *N*-terminal residues. As indicated in scheme **9**, an intense ion at m/e 234 was found, which corresponds to derivatized tyrosine. Trial addition of mass values for various peptides led to the identification of the peak at m/e 305, corresponding to derivatized tyrosyl glycyl units, which may be designated as AB (scheme **9**) in the ABCDE sequence. In this manner, the first four units of the peptide were identified. However, the C-terminal groups could not be characterized by extension of this sequence series.

In order to finish the sequence, and also to confirm the first four sequence assignments, a second series of ions was traced. This series is formed by the

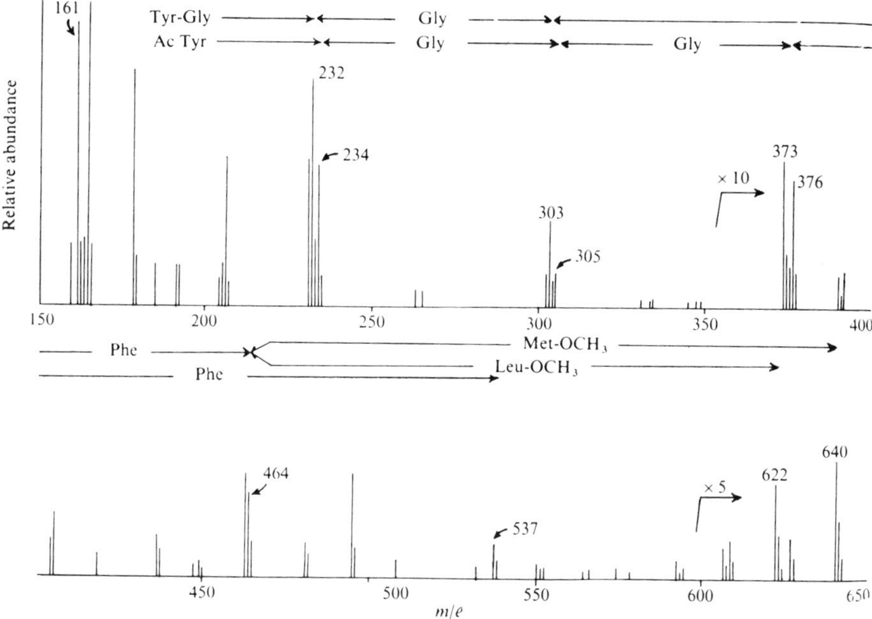

Fig. 7 Partial mass spectrum of the *N*-acetyl permethyl derivative of natural enkophalin. [Reproduced by permission from J. Hughes, T. W. Smith, H. W. Kosterlitz, L. A. Fothergill, B. A. Morgan, and H. R. Morris, *Nature* (*London*) **258**, 577–579 (1975).]

same cleavages shown in scheme **9** as has already been discussed, with the additional loss of *N*-methylacetamide from the *N*-terminal group. Thus the corresponding sequence ions are 73 mass units lighter in this second series. The ABCD sequence is confirmed by peaks at *m/e* 161, 232, 303, and 464, and the peaks at 640 and 622 indicate the presence of methionine and leucine at the carboxy termini in the two pentapeptides. Preliminary identification of the hydrolyzed amino acids made the interpretation of the mass spectrum easier and more reliable. These structures were ultimately confirmed by comparison with synthetic standards.

C. *Unknown Samples*

The most challenging application of mass spectrometry is probably its use in structure elucidation of completely unknown samples. Of course, it is rare to encounter a sample about which no other information is available. Information from other physical methods, such as ultraviolet spectroscopy, infrared spectroscopy, and nmr, should be used to augment mass spectral information in interpreting the structure. The most sophisticated computer programs envisioned to assist in structure elucidation by interpretation of mass spectra have provision for the inclusion of structural information from other spectroscopic or chemical techniques (Cheer *et al.*, 1976).

1. *Plasticizers*

One simple example of a complete unknown to whose structure elucidation mass spectrometry contributed is that of an artifact from a liver perfusion experiment. Toxicologists studying drug metabolism using a perfused liver system found that circulation of blood through their perfused liver led to the accumulation of small organic compounds, even before the drug to be studied had been added (Jaeger and Rubin, 1970). They purified a quantity of the organic material, and determined that it was acidic from its behavior in various extraction systems. The sample was treated with diazomethane (to esterify carboxylate groups) and its low resolution mass spectrum was measured. The spectrum is shown in Fig. 8. It is a simple spectrum, and the assignment of the peak at 252 to the molecular ion is consistent with the mass differences between 252 and lower peaks in the spectrum; that is, the loss of 31 mass units to form ions of mass 221 would be expected from a sample which is a methyl ester. High resolution mass measurements were made on the major peaks in the spectrum. These are shown in Table V, along with the elemental compositions assigned. The molecular ion is seen to be highly unsaturated. Mass-221 ions are indeed formed by loss of a methoxy radical, probably from a methyl ester, and the abundant ions of mass 163 are formed by loss of $C_3H_5O_3$ from the molecular ion. Infrared

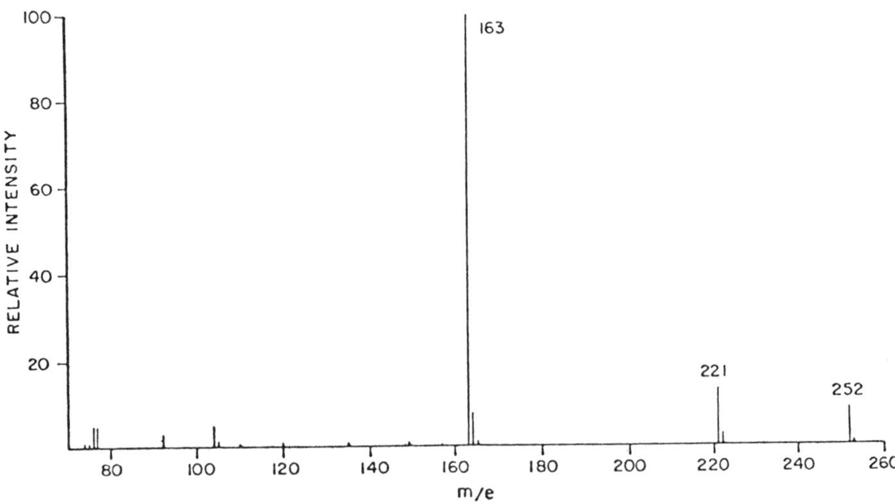

Fig. 8 Mass spectrum of the phthalate artifact from perfused liver.

TABLE V

High Resolution Accurate Mass Measurements on Major Peaks in Fig. 9

Observed	Calculated	Formula Assigned	Interpretation
252.06359	252.06338	$C_{12}H_{12}O_6$	M^+
221.04523	221.04499	$C_{11}H_9O_5$	M—OCH_3
163.03996	163.03951	$C_9H_7O_3$	M–$C_3H_5O_3$

spectroscopy suggested that two different kinds of carbonyl groups were present in this molecule. Ultraviolet spectroscopy confirmed the presence of an aromatic ring. A nuclear magnetic resonance spectrum revealed three different kinds of protons in the molecule, aromatic, methyl ester, and another, with relative populations of 4:6:2. The structure which was finally assigned to this unknown is that of the dimethyl ester of glycolyl phthalate (scheme **10**). The investigators subsequently determined that phthalate esters of this sort were present as plasticizers in the tubing used in the perfusion apparatus and were leached out by the circulating blood and metabolized in the liver. A rationale for formation of the major ions of mass 163 is shown in scheme **10**.

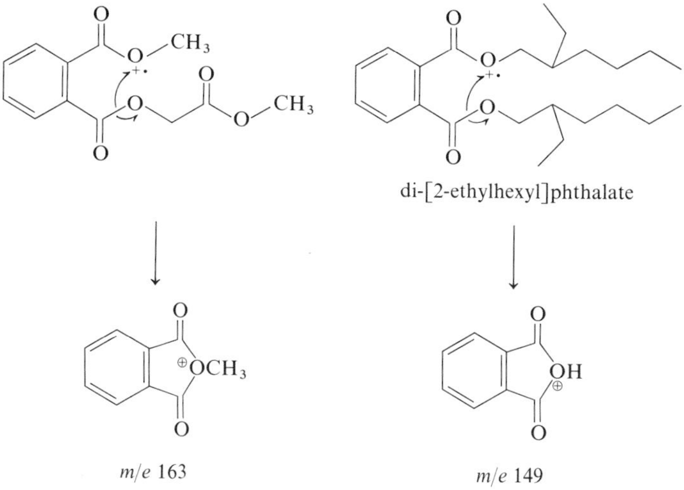

di-[2-ethylhexyl]phthalate

m/e 163

m/e 149

10

Phthalates are used in most plastic devices and containers, in some vacuum pump oil, and as stiffening agents in cardboard containers. Although no health risk has been unambiguously defined, they are almost ubiquitously encountered as environmental contaminants. They have been found in

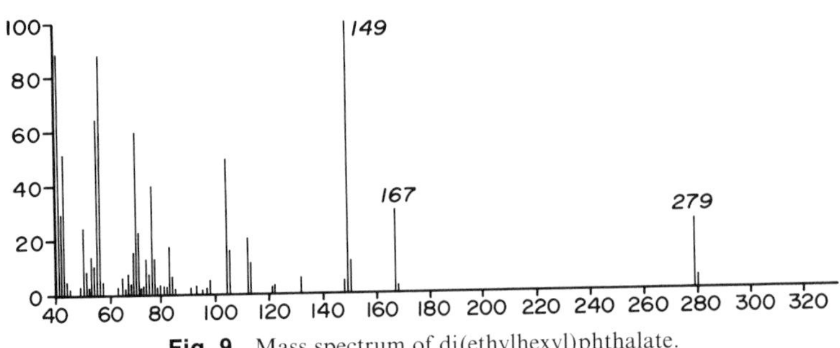

Fig. 9 Mass spectrum of di(ethylhexyl)phthalate.

municipal drinking water, in human and animal tissue, in blood stored in the blood bank, and in the atmosphere above new vinyl seat covers in automobiles. They are also frequently observed as contaminants in mass spectral samples.

The most commonly encountered phthalate is di(2-ethylhexyl)phthalate, whose mass spectrum is shown in Fig. 9. Early studies on the mechanisms of fragmentation of dialkylphthalates (Djerassi and Fenselau, 1965) support the formation of mass-149 ions of the structure shown in scheme **10**.

2. *Components of Smog in Rainwater*

Another example of the use of mass spectrometry to identify the structure of a completely unknown compound came from a study of organic smog components found in rainwater collected in Washington, D.C. (Saunders *et al.*, 1974). Organic compounds were purged from the water with helium gas, and the constituents of the mixture were identified by GCMS. The major component was found to be 3-methyl furan. This structure was assigned by computer-based comparison of the spectrum obtained with a library of more than 30,000 spectra maintained by the US Environmental Protection Agency. The structure was then confirmed by comparison with the gas chromatographic retention time and mass spectrum of an authentic standard. The investigators suggest that this compound originated not from auto exhaust but from photooxidation of unsaturated terpene hydrocarbons exuded in large quantities by forests in the nearby Appalachian Mountains.

3. *Insect Pheromones*

Although mass spectrometry provided critical information in all areas of natural product chemistry, it has been especially useful in the identification of insect pheromones. This is because there is often too little sample available to obtain other physical measurements, and because these compounds are often small open-chain aliphatic compounds whose fragmentation patterns relate well to those of textbook models.

Elucidation of the Douglas-fir tussock moth sex pheromone (see scheme **11**) provides an instructive example (Smith *et al.*, 1975). This work was carried out on about 240 μg of sample isolated from 6000 female moth abdominal tips. The functional groups present in the molecule were characterized by assaying whether each of a wide selection of chemical reactions destroyed biological activity or not. On this basis a carbonyl group and a double bond were assigned. A high resolution mass spectrum established the molecular weight as 308.3074 Daltons, and permitted elemental compositions to be assigned to major fragment ions in the spectrum (see Table VI). The incorporation of 4 deuterium atoms under basic conditions for aqueous exchange was confirmed mass spectrometrically, characterizing the carbonyl group as an unconjugated ketone. The product of catalytic reduction was found to have a GC retention time and mass spectrum identical to that of authentic 11-heneicosanone, and the rest of the spectrum was interpreted in the context of ketone fragmentation, as shown in scheme **11**.

$C_{10}H_{21}$ (ketone, O with H transfer) ... C_5H_{11} ⟶ $[CH_2{=}CH{-}CH{=}CHC_5H_{11}]^{+\cdot}$

M^{+} 308.3074
(Z)-6-heneicosen-11-one

m/e 124.1246

11

Although an argument for the position of the double bond was advanced, based on retention of charge on the hydrocarbon fragment (mass 124) of the frequently encountered ketone rearrangement process, this position was rigorously defined by comparison of the product of ozonolysis with an authentic ketoaldehyde. However, no clue could be obtained from the spectra regarding the stereochemistry about the double bond. Both isomers had to be synthesized for comparison of their physical properties and also their bioactivity.

This brings us to an important caveat about structure elucidation by mass spectrometry. Although there are usually reproducible differences in

TABLE VI

Major Fragments in the Spectrum of the Sex Pheromone of Douglas-Fir Tussock Moth

Observed	Calculated	Formula Assigned
308.3074	308.3079	$C_{21}H_{40}O$
169.1587	169.1592	$C_{11}H_{21}O$
167.1412	167.1435	$C_{11}H_{19}O$
124.1246 (base peak)	124.1252	C_9H_{16}

peak intensities between the spectra of geometric isomers or of diastereomers, it is usually not possible to assign stereochemistry to an unknown without having spectra of all the candidate stereoisomers for direct comparison. This has been found to be the case whether one is using simple scanned EI or CI spectra, or more complicated techniques such as ion kinetic energy spectra of metastable ions, or collisionally induced dissociation.

IV. Pattern Matching

Pattern matching is perhaps the easiest way in which to use mass spectrometry. Here the spectrum of the sample is compared with the spectrum of a known compound without special regard for the fragmentation processes which lead to the peaks in the spectrum. Pattern matching can be done manually, and it is amenable to computerization in conjunction with library searching techniques. This kind of fingerprinting is used in quality control, in residue analysis, in air monitoring, in toxicology, in forensic chemistry, as well as in research. In addition to conventional scanned spectra, there are many other kinds of mass spectral measurements which can be fingerprinted. These include high resolution mass spectra, selected ion records, and mass chromatograms.

A. Scanned Spectra

Examples are taken from the areas of pharmacology, toxicology, forensic chemistry, and environmental chemistry.

1. *Antitumor Drugs*

The first example is taken from a study of the distribution of cyclophosphamide in body fluids (Duncan *et al.*, 1973). The mass spectrum of this widely used antineoplastic drug is shown in Fig. 1. Although the fragmentation pattern is well understood (and is discussed in Section III.B., it need not be understood to confirm the presence of the drug in an extract of human milk whose spectrum is shown in Fig. 10. In this case, milk from a patient receiving the drug was first extracted with chloroform. The drug in the chloroform solution was subsequently backextracted into aqueous solution. This provides considerable purification from the lipophillic endogenous components of milk. With minor differences in relative intensities, the two spectra (Figs. 1 and 10) are identical, and on the basis of this comparison the drug was reported to be secreted in milk. The manufacturer subsequently contraindicated its use in nursing mothers.

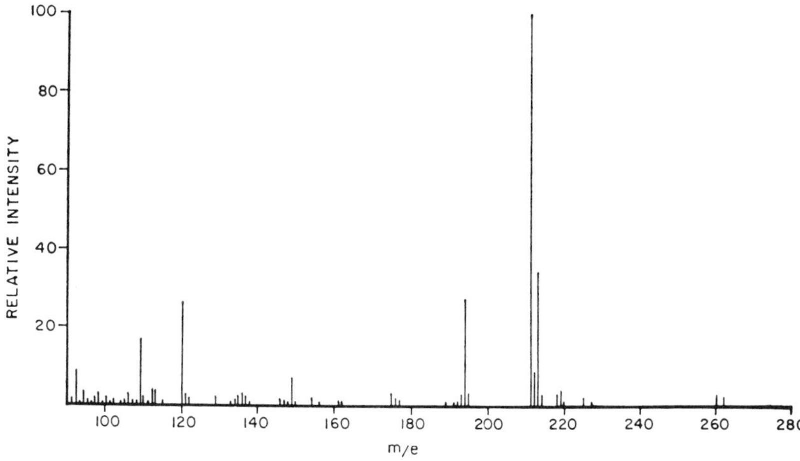

Fig. 10 Mass spectrum of an extract of milk from a patient receiving cyclophosphamide. [Reproduced by permission from J. H. Duncan, O. M. Colvin, and C. Fenselau, *Toxicol. Appl. Pharmacol.* **24**, 317–323 (1973).]

The spectrum of an extract from synovial fluid from a patient receiving the drug is shown in Fig. 11. In addition to the familiar cyclophosphamide peaks, this spectrum contains a number of other intense peaks and sources of some of these have been identified on the spectrum. These include cholesterol, palmitic acid, and diethylhexylphthalate, the last presumably from the plastic container in which the fluid was collected. Identification of cyclophosphamide on the basis of this spectrum may be less convincing, particularly to the nonspecialist, and this introduces the problem of fingerprinting components of mixtures.

Two of the more successful approaches for identifying impure compounds are the introduction of the sample into the mass spectrometer through a gas chromatograph, and the measurement of ions in high resolution. The latter method often permits the mixture to be introduced on the direct probe. High resolution spectra were obtained (using the direct probe) on extracts from synovial fluid and a number of other physiologic fluids which were being analyzed for cyclophosphamide. In Table VII the calculated exact masses of three different ions, all having nominal mass 213, are listed and also the value observed for the synovial fluid extract. These exact masses differ in the first or second decimal place, according to the empirical formula of the ion. The observed value matches the theoretical value for the fragment ion to within 0.003 mass units, the limit of precision usually considered acceptable for high resolution work of this kind. Few endogenous contaminants or artifacts due to collection procedures would contain the variety of heteroatoms, particularly the chlorine atom, necessary to produce an ion

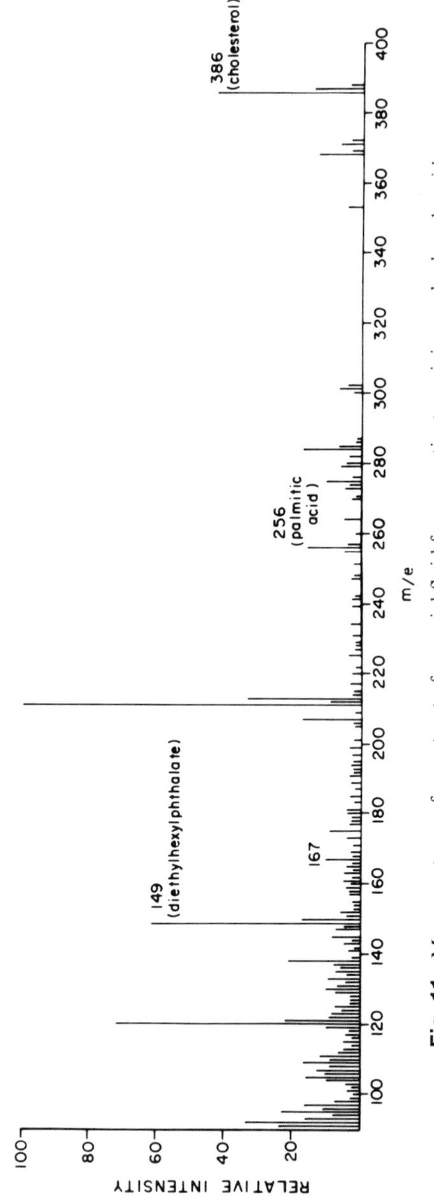

Fig. 11 Mass spectrum of an extract of synovial fluid from a patient receiving cyclophosphamide.

TABLE VII

High Resolution Measurements at *m/e* 213 on Extracts of Body Fluids Containing Cyclophosphamide

Theoretical	213.1854	$C_{13}H_{25}O_2$
	213.1643	$C_{16}H_{21}$
	213.0374	$C_6H_{13}N_2O_2P^{37}Cl$
Saliva	213.0385	
Sweat	213.0380	
Milk	213.0385	
9-hr serum	213.0381	
Spinal fluid	213.0387	
Synovial fluid	213.0327	
Urine	213.0381	

with the exact mass measured. Unique masses were also measured for other ions characteristic of cyclophosphamide, notably those with nominal masses 211, 120, 94, and 92. All of these ions were detected under high resolution in the extract, lending support to the identification of the compound. Because these exact measurements were made using a photoplate detector, less emphasis is placed on the relative intensities. Ideally, fingerprinting by high resolution should also involve an evaluation of the relative intensities of the ions.

2. *Drug Overdoses*

A second example of fingerprinting comes from the area of toxicology. Figure 12 shows a partial low resolution mass spectrum of levorphan obtained from the reference files of the US customs house in Baltimore, Maryland, and Fig. 13 the mass spectrum obtained from the extract of contents of a

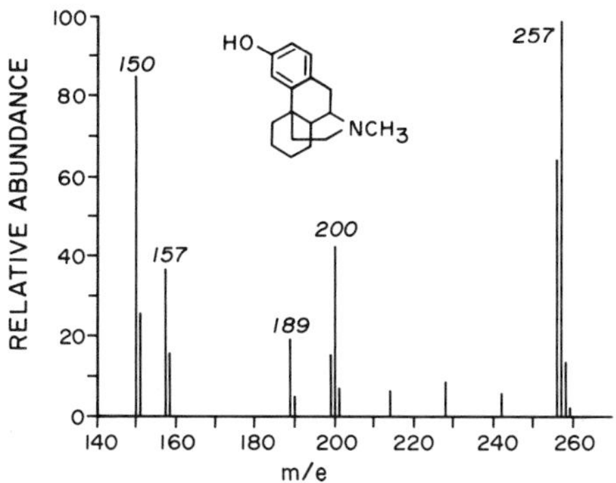

Fig. 12 Partial mass spectra of levorphan.

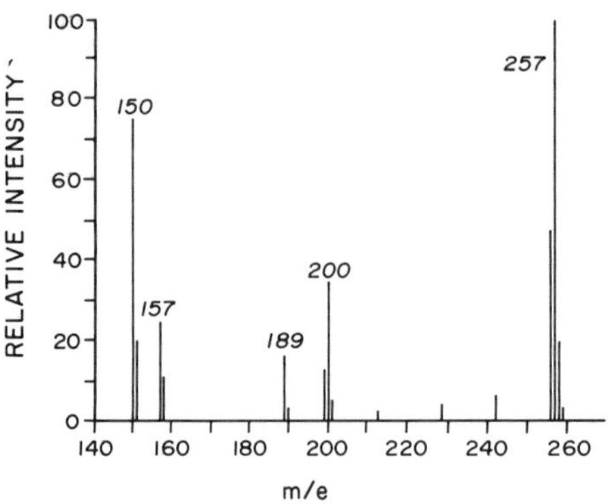

Fig. 13 Partial mass spectra of contents of a confiscated syringe.

syringe found with a man who died from a drug overdose. Levorphan is not a commonly abused drug, and probably was obtained in a drugstore robbery. The use of mass spectrometry to identify drugs in the urine or blood of victims of accidental and self-inflicted drug overdoses has been popularized by Fales and co-workers (Law *et al.*, 1971) and has been refined to a highly computerized technique by Costello Biemann, and co-workers (1974) at MIT. In the protocol of the latter group, the chloroform extract of gastric aspirate, blood, or urine is injected into a GCMS. The mass spectrometer is repetitively scanned every few seconds under computer control, and the computer searches a library of reference spectra for matches for each of the spectra scanned. Identification can be based both on chromatographic retention time and on mass spectral fingerprinting.

Mass spectrometry with computerized data processing and pattern matching is also being applied to other clinical problems, including profiling the changes in metabolic patterns associated with various disease states, and the characterization of different strains of bacteria.

3. *Narcotics*

Scientists, including those in the New Jersey state police laboratory, who are primarily concerned with identification of narcotic samples, have recommended the use of chemical ionization to simplify spectra. They point out that the ideal situation, when pattern matching with a limited number of possible unknowns, is to produce a single peak spectrum. Preferably, that single peak should correspond to molecular ions or, in the case of chemical ionization, to $M + \mathrm{H}$ ions, because the higher the mass of the ion, the lower the probability that it may be coincidentally contributed by a contaminant. Although it is not yet possible to obtain spectra which contain only the molecular ion peak group, chemical ionization does simplify spectra considerably, allowing potential narcotic samples (mixtures) to be analyzed easily by insertion of the sample on the direct probe, eliminating the need for time-consuming gas chromatographic introduction (Saferstein *et al.*, 1974).

4. *Tobacco Smoke Carcinogens*

Many interesting examples of fingerprinting mass spectra may be found in the area of environmental contaminants. An early example is a study of the components of tobacco smoke reported by Masuda and Hoffman (1969). In Fig. 14 is shown the reference spectrum of the bladder carcinogen 2-naphthylamine, converted to its pentafluoropropionamide derivative so that it may be introduced into the mass spectrometer via the gas chromatograph. This is to be compared with the spectrum obtained by GCMS analysis of the pentafluoropropionamide derivative of a sample obtained by blowing the smoke of 200 cigarettes through acidic methanol. Despite two chemical purification steps, and separation on the gas chromatograph, the sample still is visibly contaminated. However, the characteristic features of the

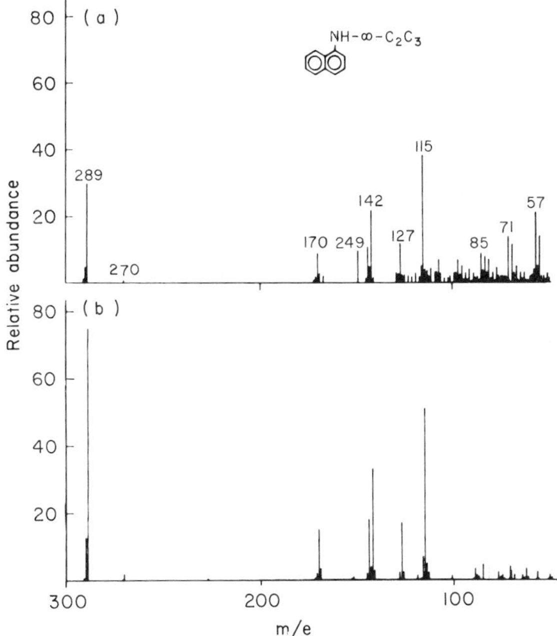

Fig. 14 Mass spectra of 2-naphthylamine (a) isolated from cigarette smoke and derivatized as the pentafluoropropionamide and (b) authentic 2-naphthylamine pentafluoropropionamide. [Reproduced by permission (Y. Masuda and D. Hoffman, *Anal. Chem.* **41**, 650–652 (1969)). Copyright © by the American Chemical Society.]

standard spectrum are visible in the spectrum of the sample, and this evidence along with the identity of retention times has been accepted as an identification of the bladder carcinogen in tobacco smoke. Although contamination of this sort is not unusual in challenging analytical problems, this example is particularly instructive, because the authors were willing to put it into the literature. More often, when a spectrum is not a good match with that of reference material, it is discussed but not actually presented.

B. Library Searching

Library searching or computerized fingerprinting of scanned spectra, was mentioned in an earlier section. Algorithms differ, for example, as to how many peaks must be compared for reliable matching, how to choose these peaks, how to handle variability in peak intensity, and whether to match the library spectra to the sample spectrum or the sample spectrum to the library spectra. Many environmental and biological samples are not completely resolved from overlapping peaks in the gas chromatograph, and the resulting spectra may be difficult to match accurately. Classical deconvolution techniques for gas chromatographic peaks depend heavily on peak shapes and symmetry. With computer-controlled repetitive scanning, one

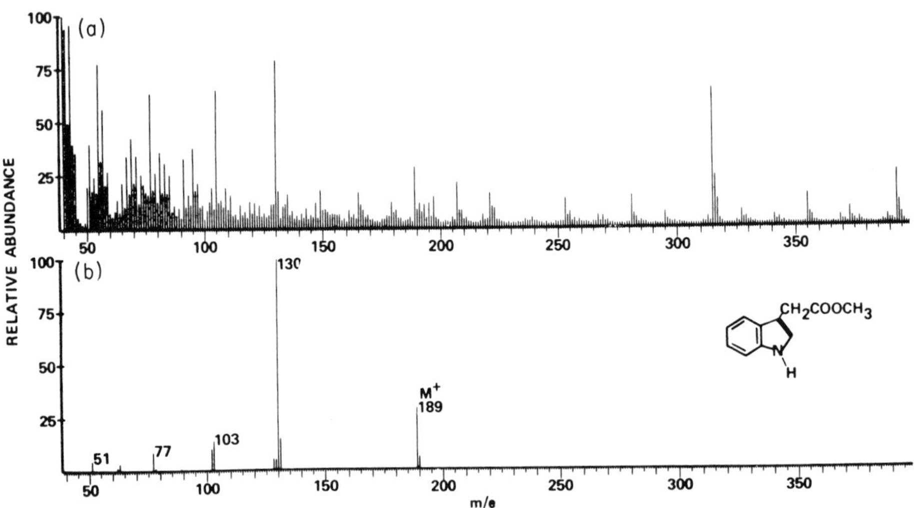

Fig. 15 (a) Spectrum of indole acetic acid-3-methyl ester from a GCMS analysis of human urine and (b) the same spectrum resolved by automated data processing. [Reproduced by permission (R. G. Dromey, M. J. Stefik, T. C. Rindfleisch and A. M. Duffield, *Anal. Chem.* **48**, 1368–1375 (1976)). Copyright © by the American Chemical Society.]

can focus on the masses of ions formed from the various components to effect spectral deconvolution. In the simplest form, one requires that all ions that are formed from the same precursor in the mixture must exhibit maximum intensities at the same time, i.e. in the same scan. Ions whose intensities maximize in other scans are considered to be formed from other components of the mixture and are subtracted from the spectrum. In Fig. 15 one can see the marked simplification of the spectrum that can result from this kind of data processing (Dromey *et al.*, 1976). Figure 16 shows how the gas chromatographic peaks can be substantially narrowed and deconvoluted when the resolved spectra are resumed for total ion chromatograms (Biller and Bieman, 1974). These resolved spectra are reported to provide greatly improved potential over spectra of impure samples for matching by library searching.

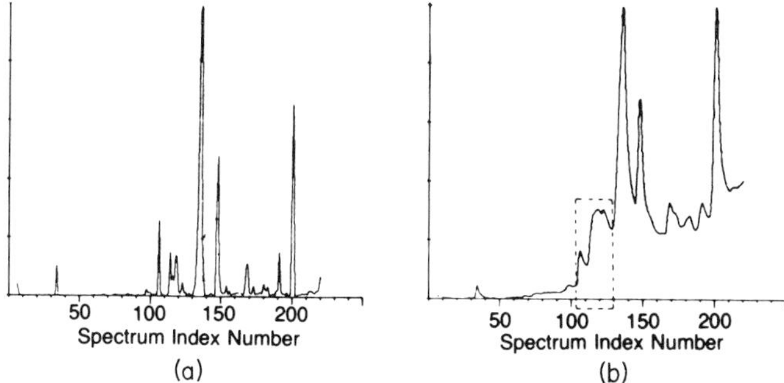

Fig. 16 Total ion current chromatogram of an extract from gastric lavage, reconstructed (a) from repetitive scans and (b) from resolved or deconvoluted repetitive scans. [Reproduced by permission from J. E. Biller and K. Biemann, *Anal. Lett.* **7**, 515–528 (1974)].

Library matching of scanned spectra, manually or with computer assistance, has a very old tradition in mass spectrometry. To this end, the American Society for Testing and Materials began to assemble spectra in the 1950s, the Mass Spectrometry Data Center was established at Aldermaston (UK) in the 1960s, and many individual collections have been assembled in subspecialty areas such as urinary acids, derivatized carbohydrates, and polychlorinated hydrocarbon residues. Spectral collections are available in book form (Cornu and Massot, 1975; Stenhagen *et al.*, 1974; "Eight Peak Index of Mass Spectra," 1975) and also on computer tape. Perhaps the largest data collection is that assembled by many groups but presently maintained by the Mass Spectrometry Data Center, the US National Institutes of Health, and the US Environmental Protection Agency, which contains more than 30,000 spectra. Among the important applications of this library has been its use in the identification of organic compounds in drinking water from various metropolitan areas of the United States.

Abuse of this powerful technique has also developed, with some laboratories reporting identification of unusual compounds supported by no experimental data, other than a statement to the effect that the identification was made by computer-based mass spectral library searching.

C. *Selected Ion Monitoring*

As has already been pointed out, pattern matching or fingerprinting can be done with selected ion records, as well as with scanned spectra. In this technique the instrument is not scanned through a range of m/e values, but rather is focused to continually record ions of a single m/e value or is switched rapidly between a small number of preselected masses. This kind of detection (called "peak stepping" at the time) was used before 1948 to monitor uranium isotope enrichment (Falkner, 1977). Its use with the combined GCMS was demonstrated in 1966 (Sweeley *et al.*, 1966).

The experimental record from this kind of detection presents the abundance of ions formed with the selected mass, as a function of time, and looks very much like a chromatogram. If the sample is introduced into the mass spectrometer via a gas chromatograph, then selected ion monitoring detects only those compounds eluted from the gas chromatograph that fragment in the mass spectrometer to ions of the preselected masses. Chromatographic eluents are not detected if they do not form ions of the masses being monitored.

In order to select the appropriate masses to monitor, one must have knowledge of the spectrum of the compound being sought. Thus one chooses ions which are characteristic of the compound, preferably the molecular ion or other high mass ions. It has been shown that the higher the mass of an ion, the smaller the probability of its being formed in the fragmentation of other compounds. Another consideration in the selection of ions to be monitored is that of intensity. Since the spectrum may contain ions whose abundances differ by as much as 4 orders of magnitude, one can obtain additional sensitivity by monitoring the more intense peaks. Reports in the literature of nanogram to picogram (10^{-9} to 10^{-12} g) sensitivity (reflecting longer dwell times than conventional scanning) and of exceptional specificity have brought selected ion monitoring—and mass spectrometry—to many new laboratories (Fenselau, 1977; Falkner *et al.*, 1975; Holmstedt and Palmer, 1973).

1. *Low Resolution*

An example can be taken from studies of the pharmacology of cyclophosphamide that will illustrate both the specificity and the sensitivity of the selected ion monitoring technique for pattern matching. This drug is not itself active, but is metabolized in the body to one or more metabolites which exhibit cytotoxic activity. In studies of the metabolism of cyclophosphamide in vitro, in which the drug was incubated with mouse liver microsomes, phosphoramide mustard (scheme **12**) was identied as a metabolite (Colvin *et al.*, 1973). This compound had been synthesized some years previously, and was already known to be active against 20 experimental tumors. Thus it seemed a good candidate for an active human metabolite. It was important to determine if this compound could be identified as a metabolite in patients receiving cyclophosphamide.

The concentration of this metabolite in human serum and urine was expected to be considerably lower than in the supernatant of the in vitro incubation, and therefore a method was needed for the characterization which offered both high sensitivity and also selectivity to minimize interference from endogenous compounds. Selected ion monitoring was used. Phosphoramide mustard must be derivatized for gas chromatography mass spectrometric analysis. Treatment with diazomethane was found to produce a family of suitable derivatives, the mono-, di-, and trimethyl analogs shown in scheme **12**. The electron impact low resolution mass spectrum of the

$(ClCH_2CH_2)_2N-P(=O)(NH_2)OH \longrightarrow (ClCH_2CH_2)_2N-P(=O)(NRR')OCH_3$

m/e 185 and 187	$R = R' = H$
m/e 199 and 201	$R = CH_3, R' = H$
m/e 213 and 215	$R = R' = CH_3$

12

trimethyl derivative of phosphoramide mustard is shown in Fig. 17. Of particular importance is the chlorine-isotope doublet at *m/e* 213 and 215, which occurs with the relative intensity of 3:1. These fragment ions, which contain one chlorine atom, are formed by cleavage in the C–C bond in the side chain (see scheme **12**) analogous to that which occurs in cyclophosphamide (Table IV). These peaks were chosen for selected ion monitoring because they were characteristic of the sample, both in their mass and in their relative intensity, and because they were among the most intense peaks in the spectrum.

The ion records of a synthetic standard treated with diazomethane and injected into the GCMS are shown in Fig. 18. This sample is characterized not only by the masses and relative intensities of the fragments detected, but also by its gas chromatographic retention time. The second tracing in the

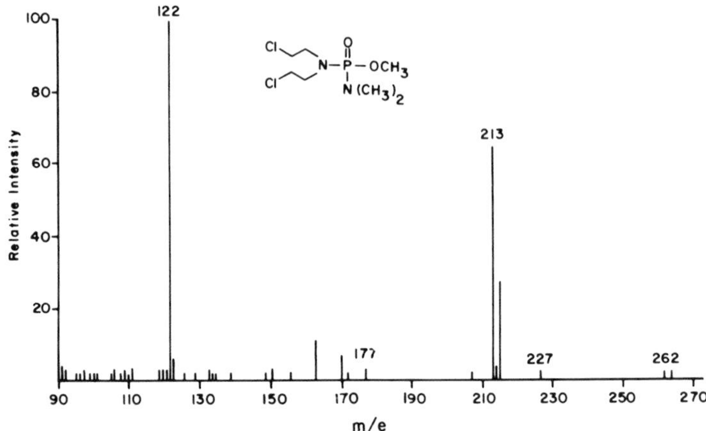

Fig. 17 Electron impact mass spectrum of phosphoramide mustard trimethyl derivative. [Reproduced by permission from C. Fenselau, M. N. Kan, S. Billets, and M. Colvin, *Cancer Res.* **35**, 1453–1457 (1975).]

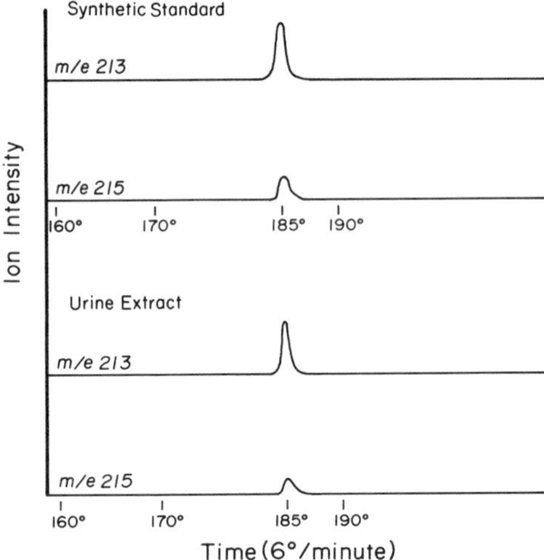

Fig. 18 Selected ion profiles of authentic trimethyl phosphoramide mustard and a urine extract from a patient receiving cyclophosphamide. [Adapted from C. Fenselau, M. N. Kan, S. Billets, and M. Colvin, *Cancer Res.* **35**, 1453–1457 (1975).]

figure shows the same selected ion profiles of XAD-2 resin extract of urine from a patient receiving cyclophosphamide. This extract has also been treated with diazomethane. The derivative of the metabolite was judged to be present in this urine extract based on the chemical history of the sample, on the coincidence of gas chromatographic retention times, on formation of the

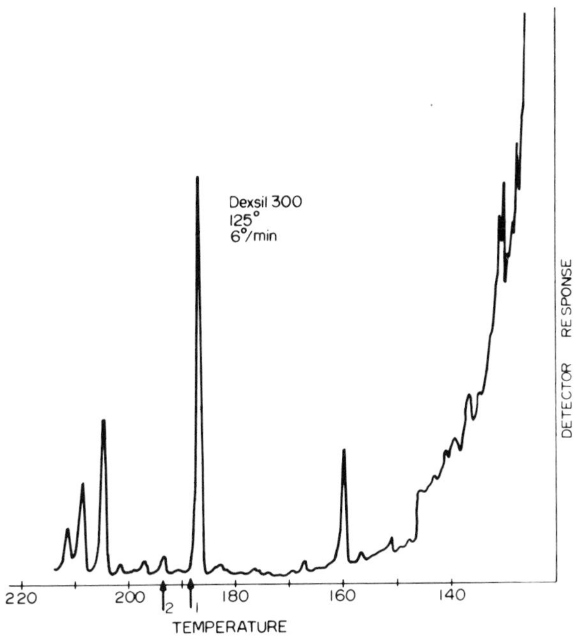

Fig. 19 Gas chromatogram detected by flame ionization of a urine extract from a patient receiving cyclophosphamide. The arrows indicate the points at which di- and trimethyl derivatives of phosphoramide mustard were detected by selected ion monitoring. [Reproduced by permission from C. Fenselau, M. N. Kan, S. Billets, and M. Colvin, *Cancer Res.* **35**, 1453–1457 (1975).]

characteristic ions being monitored and on their relative intensities, which reflect the presence of chlorine isotopes (Fenselau *et al.*, 1975). The dimethyl and monomethyl derivatives were also characterized. Because the sample is derived from physiologic milieu, impurities are present, and some of these do fragment to ions of the masses being monitored and are detected in the selected ion records shown. In order to put the question of contamination into proper perspective, the gas chromatograph of the urine extract detected with a nonspecific flame ionization detector is shown in Fig. 19. The arrows indicate the points at which the di- and trimethyl derivatives of phosphoramide mustard were detected by the selected ion monitor. The trimethyl derivative appears on the trailing edge of one of the major peaks in the chromatogram. A conventional scan of this portion of the chromatogram is shown in Fig. 20. The spectrum is not recognizable as that of a phosphoramide mustard derivative or of a chlorine containing compound of any kind. Rather it appears to be a spectrum of a phthalate ester, one of the plasticizers encountered so often in biological samples. In this case, it is probably a

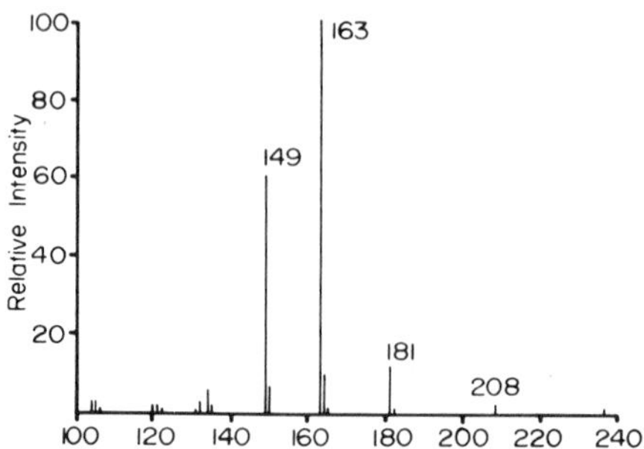

Fig. 20 Mass spectrum scanned at $\uparrow_1$ in Fig. 19. [Reproduced by permission from C. Fenselau, M. N. Kan, S. Billets, and M. Colvin, *Cancer Res.* **35**, 1453–1457 (1975).]

methylalkylphthalate. The most intense peaks generated by the phosphoramide mustard derivative have intensities of less than 1% relative to the base peak at *m/e* 163 contributed by the coeluting plasticizer, and thus are not even plotted in this spectrum which has an intensity scale of 1 to 100. It is not likely that the metabolite would have been identified if the identification

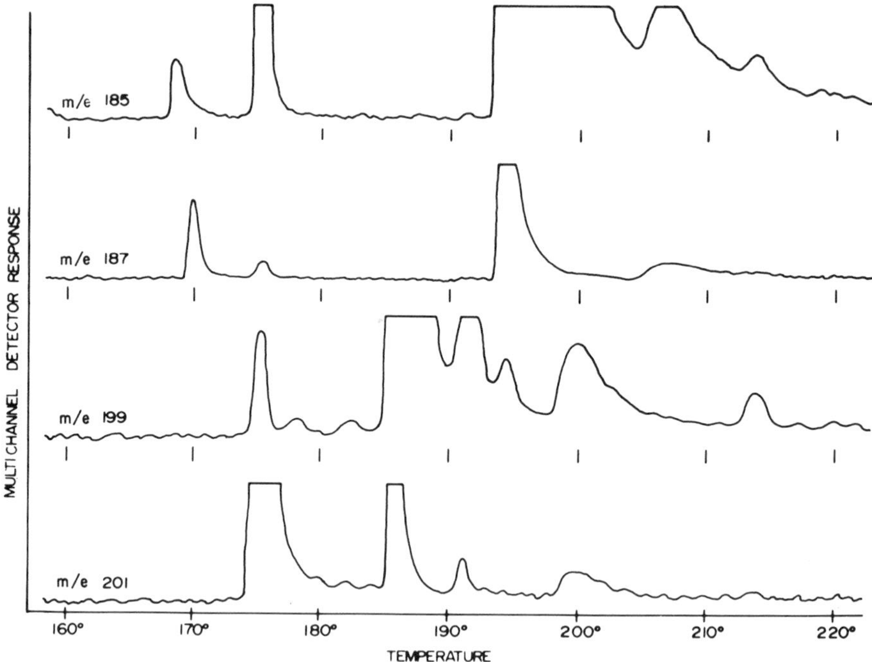

Fig. 21 Selected ion profiles (electron impact) of a plasma extract from a patient receiving cyclophosphamide. [Reproduced by permission from C. Fenselau, M. N. Kan, S. Billets, and M. Colvin, *Cancer Res.* **35**, 1453–1457 (1975).]

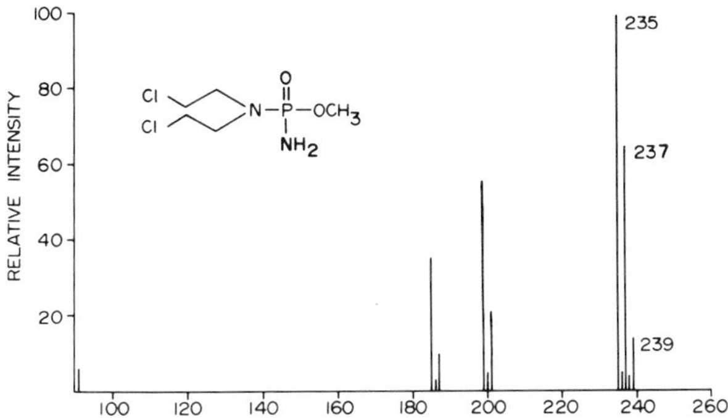

Fig. 22 Chemical ionization mass spectrum of phosphoramide mustard methyl ester. [Reproduced by permission from C. Fenselau, M. N. Kan, S. Billets, and M. Colvin, *Cancer Res.* **35**, 1453–1457 (1975).]

rested on conventional scanned spectra like this one. However, the gross contaminant does not fragment to form ions of the masses being monitored (m/e 213 and 215), and thus it is completely ignored in selected ion monitoring. In this case the mass spectrometer is functioning as a specific GC detector.

Examination of plasma for this active metabolite of cyclophosphamide posed a more challenging problem, because the concentration is lower, lipophillic contamination greater, and available samples much smaller. Thus the more abundant monomethyl and dimethyl derivatives were monitored at the characteristic masses 185, 187, 199, and 201 (see scheme **12**). These selected ion records are shown in Fig. 21. Interference from endogenous compounds is very great, and even relying heavily on retention times, it is difficult to convince oneself that the metabolites have been identified here. In order to achieve greater selectivity in detecting the metabolite, chemical ionization was then substituted for electron impact ionization. The chemical ionization spectrum of the monomethyl derivative of phosphoramide mustard is shown in Fig. 22. It can be seen immediately that the $[M + H]^+$ ions have a much greater abundance in the chemical ionization spectrum than do the M^+ ions formed under electron impact. Three $[M + H]^+$ ions are detected (2 chlorine atoms) with relative intensities of approximately 9:6:1. Concomitantly, the spectrum contains fewer fragment ions.

Not only does chemical ionization simplify the fragmentation of phosphoramide mustard and enhance its molecular ions, but it also minimizes the fragmentation of contaminants in the sample and thus decreases the possibility of their contributing coincident ions to the ion records. As already stated, the higher the mass, the less likely the possibility of contributions from contaminants. Thus the ability to monitor molecular ion species under chemical ionization is a distinct advantage. In Fig. 23 we see the profiles for the three $[M + H]^+$ ions recorded from synthetic phosphoramide mustard

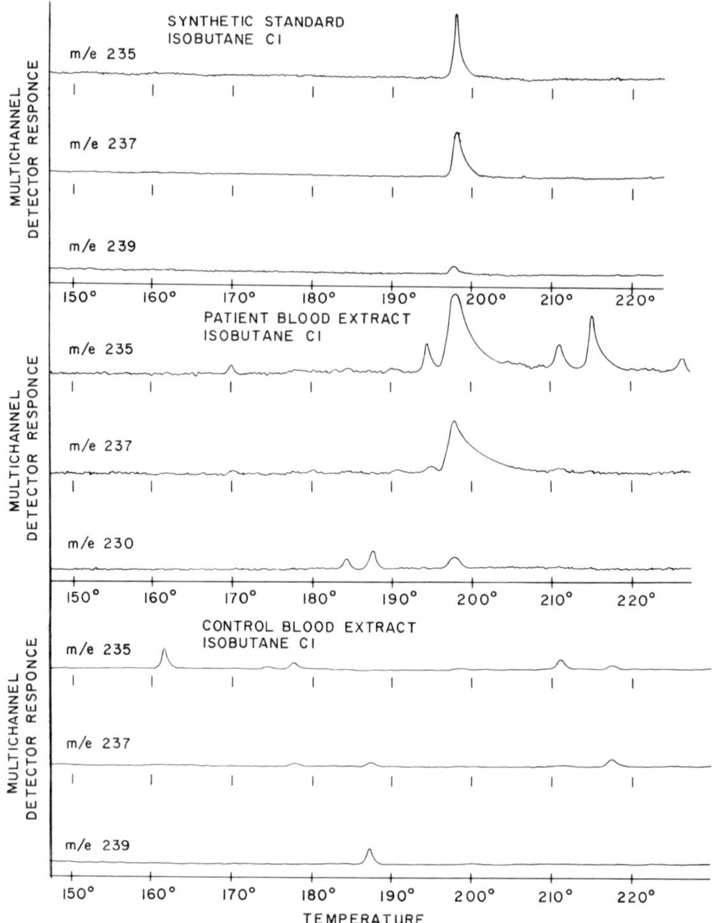

Fig. 23 Selected ion profiles (chemical ionization) of authentic phosphoramide mustard methyl ester and of a plasma extract from a patient receiving cyclophosphamide. [Reproduced by permission from C. Fenselau, M. N. Kan, S. Billets, and M. Colvin, *Cancer Res.* **35**, 1453–1457 (1975).]

derivatized with diazomethane and also those same profiles traced from an XAD-2 resin extract of plasma from a patient receiving cyclophosphamide. The ions are absent in the control analysis, but can be clearly detected in the patient extract, characterized by retention time comparable to that of the synthetic standard, by formation of ions characteristic of the compound, and by the relative intensities of these ions. On the basis of this evidence, then, phosphoramide mustard was reported to be circulating active metabolite of cyclophosphamide in humans (Fensalau *et al.*, 1975). Sensitivity was found to be around 1 ng injected into the GCMS.

2. *High Resolution*

Additional specificity may be attained by carrying out selected ion monitoring in high resolution. Studies of environmental pollutants provide particularly interesting examples. Thus *N*-nitrosamine residues in meat can be detected by high resolution single ion monitoring of the appropriate molecular ions (Stephany *et al.*, 1976). The record of ions of mass 114.08 ± 0.02 Daltons formed from an extract of fat spiked with *N*-nitrospiperidine (molecular weight 114.08 Daltons) is shown in Fig. 24. Resolving power of 1 part in 4000 was reported to provide sensitivity to 0.2 μg/kg.

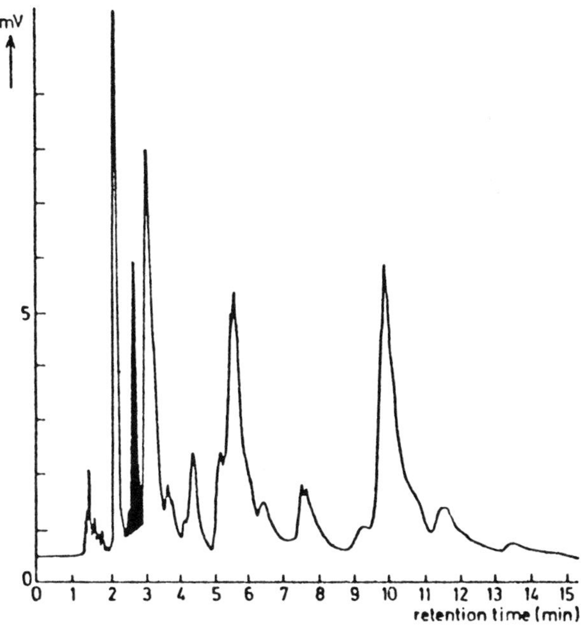

Fig. 24 High resolution single ion profile (*m/e* 114.08 ± 0.02) of an extract of frying fat spiked with *N*-nitrosopiperidine. [Reproduced by permission from R. W. Stephany, J. Freudenthal, E. Egmond, L. G. Gramberg, and P. L. Schuller, *J. Agr. Food Chem.* **24**, 536–539 (1976).]

A method for detection of carcinogenic bis(chloromethyl)ether in air has been developed (Evans *et al.*, 1975) based on single ion monitoring of ions of mass 78.995, $ClC_2H_4O^{+\cdot}$, which contribute the base peak in the spectrum. Sensitivity as low as 0.1 ppb was reported on a 1-liter air sample. Figure 25 shows the resolution of various isobars of nominal mass 79 under the high resolution GCMS conditions used. Single ion monitoring (*m/e* 79) in low resolution afforded somewhat greater sensitivity, but was judged not be sufficiently specific. Multiple ion monitoring of bis(chloromethyl)ether in

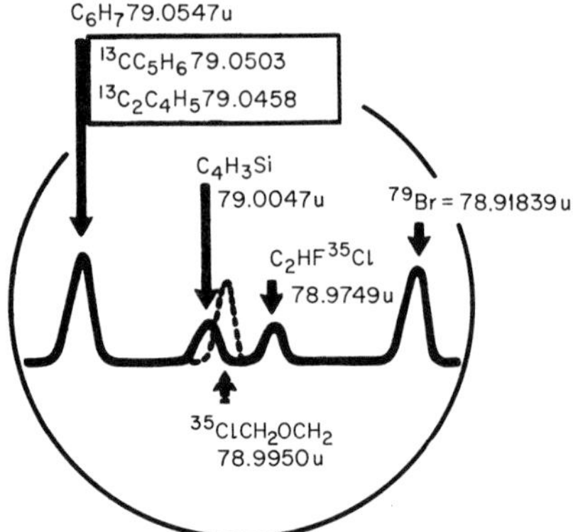

Fig. 25 Typical appearance at 3800 resolving power of the *m/e* 79 region of an air sample being examined for bis(chloromethyl) ether. [Reproduced by permission (K. P. Evans, A. Mathius, N. Mellor, K. Silvester and A. E. Williams, *Anal. Chem.* **47**, 821–824 (1975)). Copyright © by the American Chemical Society.]

low resolution has, also been reported with sensitivities of 0.1 ppb (Ellgehausen, 1975).

In dealing with both nitrosamines from meat and fat, and bis(chloromethyl)ether from air, investigators found that the additional specificity provided by the use of higher resolution was required to distinguish samples unambiguously from background and contaminants. In both cases, the resolution used was below the maximal power of the instruments, reflecting the efforts of investigators to balance the tradeoff between resolution and sensitivity.

D. Mass Chromatograms

Mass chromatograms reconstructed from series of repetitively scanned spectra may be used in much the same manner as selected ion profiles. For example, Schuetzle (1975) has used mass chromatograms reconstructed from repetitively scanned high resolution spectra to fingerprint solid components of atmospheric aerosols introduced into the mass spectrometer on the direct probe.

It has been pointed out elsewhere (Fenselau, 1977) that the longer sampling time permits selected ion monitoring to be done with sensitivity up to 10^4 times greater than that of mass chromatograms reconstructed from computer-controlled repetitive scans. If sensitivity is not a problem, however, mass chromatograms can provide the same selectivity as selected ion monitoring. Mass chromatograms may be obtained in either low resolution or high resolution, although the latter requires instrumentation less commonly available.

E. Limited Mass Range Chromatograms

Several variations of mass spectrometry were used to fingerprint Kepone® as an environmental degradation product of the insecticide mirex in soil 12 years after treatment and also in mirex-treated bait exposed to the elements for five years (Carlson *et al.*, 1975). Acetone and hexane extracts were concentrated for analysis by a GCMS instrument interfaced to a computer for automated data acquisition and manipulation. Kepone® and mirex were fingerprinted in the more concentrated bait extracts by comparison of their scanned methane chemical ionization mass spectra with those of authentic standards. Several hydroxylated degradation products were also recognized by comparing limited mass range chromatograms with those of authentic samples (see Fig. 26). In this technique the intensities of ions in a 6 to 10 Dalton mass range were summed and plotted by the computer from a series

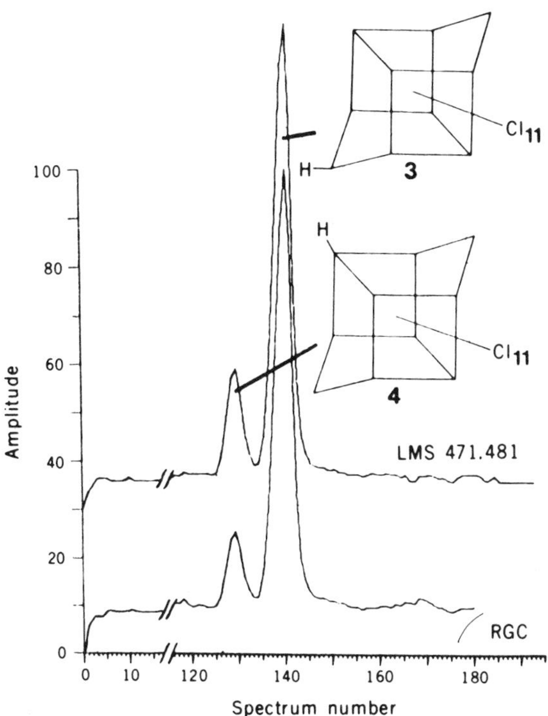

Fig. 26 Computer reconstructed gas chromatogram (RGC) and limited range mass chromatogram (LMS) from a mixture of monohydro derivatives of mirex. [Reproduced by permission from D. A. Carlson, D. D. Konyha, W. B. Wheeler, G. P. Marshall, and R. G. Zaylskie, *Science* **194**, 939–941 (1976).]

of spectra scanned repetitively throughout the gas chromatographic separation. Summing provided better sensitivity than recording profiles of individual ions from these polychlorinated samples, because of the large number of isotopic species. For example, mirex has 12 chlorine atoms and 13 molecular ions. Kepone® was also fingerprinted in the soil extracts, where it was present in much lower amounts (estimated as well above the 500-pg limit of detection, however) by single ion monitoring on an electron impact GCMS, using accurate mass measurements at 454.7 ± 0.1 Daltons in the low resolution mode. This ion, $C_{10}{}^{35}Cl_7{}^{37}Cl_2O^+$, is formed from the molecular ion by loss of 1 chlorine atom.

F. Metastable Ion Spectra

Although most fingerprinting is done with scanned spectra or selected ion profiles, any kind of mass spectal measurements can be used to compare reference and analytical samples. Gallegos (1976) has made the interesting suggestion that the metastable ions (Cooks *et al.*, 1973) that can be detected in the formation of class characteristic ions, e.g., mass-191 ions for terpenes, can be used to identify members of that class present in a mixture without chromatographic separation. The magnetic analyzer in a double-focusing

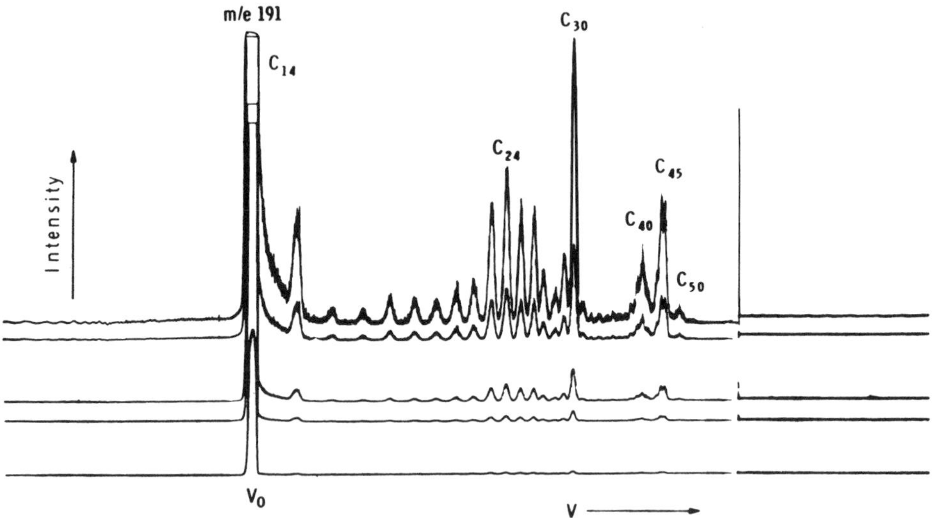

Fig. 27 Metastable scan showing precursors of mass-191 ions in a branched/cyclic hydrocarbon fraction from Green River shale. [Reproduced by permission (E. J. Gallegos, *Anal. Chem.* **48**, 1348–1351 (1976)). Copyright © by the American Chemical Society.]

instrument is adjusted to bring the characteristic daughter ion into the detector, and the accelerating voltage is scanned to transmit precursor ions through a wide mass range and to produce metastable transition spectra such as that shown in Fig. 27. This metastable transition analysis was carried out on an extract of Green River shale characterized as the high molecular weight saturated hydrocarbon fraction. The sample was introduced on a direct probe cooled to $-180°C$. The C_{30}, C_{40}, C_{45}, and C_{50} precursors were separated by 1 or 2 isoprene units, which implied that they were terpenes. Internal ratios could be calculated from these analyses that were consistent with ratios calculated from GCMS data. This kind of analysis can best be used if the decomposition processes are well understood. In favorable situations the bulk of the contaminants in the sample are not detected. Metastable ions may also be recorded by a variety of other instrumental configurations (Cooks *et al.*, 1973; Kruger *et al.*, 1976) and used to fingerprint samples.

G. *Collisionally Induced Dissociation*

Collisionally induced dissociation, or collisional activation, is a technique in which a particular ion species is resolved in the analyzer and subjected to subsequent fragmentation by collision with neutral molecules. Ultimately a spectrum is observed for ions which have been formed from precursor ions of a single m/e value. In addition to providing information on the structure of the precursor ion(s), it has been suggested (Levsen *et al.*, 1974) that this techniques could be useful in fingerprinting isomers with very subtle structural differences.

V. Stable Isotope Analysis

Historically, the most important contribution of mass spectroscopy to science is probably the demonstration of the existence of stable isotopes. In 1913, Thompson reported the separation of the helium isotopes of mass 20 and 22 on his mass spectrograph. In subsequent decades, mass spectroscopy was employed to determine the exact masses and natural abundances of most of the stable isotopes (Nier, 1950).

Isotope abundances of many different elements are currently being measured in applications of mass spectrometry in geology, biology, medicine, archaeology, and cosmochemistry. The isotopes which fall within the "organic" scope of this chapter include those of carbon, oxygen, nitrogen, hydrogen, and sulfur.

A. High Precision Isotope Ratio Measurements

Isotope ratio measurements appear to be characterized by two diverse points of view. On the one hand, the need for highly precise measurements of enrichments near the natural level are met most satisfactorily by dual inlet instruments where the differential ratio of isotopes in an unknown sample is measured against that of isotopes in a standard sample. Relatively large amounts of sample are required, and these measurements are made on gases or gaseous combustion products of solid samples. On the other hand, the requirement of many environmental and biomedical workers for high sensitivity is being met with lowered precision at higher levels of enrichment by selected ion monitoring. This will be discussed later.

Geochemists and cosmochemists usually report the difference between the isotope ratio of the sample and that of the unknown as a delta value. For example, the $\delta^{13}C$ for a differential ratio of $^{13}C^{16}O_2/^{12}C^{16}O_2$ is defined as

$$\delta^{13}C = \frac{(^{13}CO_2/^{12}CO_2)_u - (^{13}CO_2/^{12}CO_2)_r}{(^{13}CO_2/^{12}CO_2)_r} \cdot 10^3$$

where the subscripts u and r denote the unknown and reference samples, respectively. The quantity $\delta^{13}C$ represents the relative difference in isotope ratios expressed in parts per thousand, and can have either a positive or negative value, depending on whether the unknown is richer in ^{13}C than the standard or depleted in ^{13}C relative to the standard. The units for δ are termed "per mil" and are given the symbol ‰. Delta values are also used to characterize enrichments of ^2H, ^{18}O, ^{15}N, ^{34}S, etc.

In the biochemical literature, stable isotope enrichment is usually reported as atom percent or atom % excess. Atom % ^{13}C, for example, is defined from ion intensities of the sample,

$$\text{atom \% } ^{13}C = \frac{(^{13}CO_2)}{(^{12}CO_2) + (^{13}CO_2)} \cdot 100 = \frac{100}{(^{12}CO_2/^{13}CO_2) + 1}$$

and atom % excess ^{13}C as the differences between ratios of unknown and reference,

$$\text{atom \% excess } ^{13}C = \left[\frac{1}{(^{12}CO_2/^{13}CO_2)_u + 1} - \frac{1}{(^{12}CO_2/^{13}CO_2)_r + 1}\right] \cdot 100$$

The calculations for atom % ^{15}N, ^{18}O, and ^2H involve equilibrium constants for isotope species in the diatomic molecules being analyzed, and may vary according to the extent of enrichment. These are well described by Capriolli (1972) among others. Absolute or "one-sided" ratios may be measured with high precision, although accuracy will depend on the chemical and instrumental procedures.

Another characteristic of isotope ratio mass spectrometers is that they are designed to measure very small amounts of one isotope in the presence of large amounts of another, i.e the natural abundance of ^{13}C at levels around 1.1%, of ^{18}O at levels around 0.2%, and ^{15}N at levels of the order of magnitude of 0.4%. Working in that range of enrichment, the ratio $^{13}C^{16}O_2/^{12}C^{16}O_2$ might typically be measured with a precision of 1 part in 10^5 using 500–1000 μg CO_2 (Beckinsale *et al.*, 1973) and with precision as high as 1 part in 10^7 using larger samples. Limitations of this approach include the need for relatively large amounts of sample and for combustion of the sample without isotope fractionation.

1. *Glucose Metabolism*

A striking application of differential isotope ratio measurements involves their use in a noninvasive assay for glucose metabolism. When glucose that is enriched in ^{13}C is administered to a patient, it is ultimately transformed to CO_2 enriched in ^{13}C and exhaled. Carbon dioxide thus exhaled by patients is trapped at appropriate time intervals and analyzed for an increase in ^{13}C content, which reflects metabolism of the enriched glucose.

In one such study, the administration of glucose enriched in ^{13}C led to a marked rise in ^{13}C in expired air, which reached its maximum after 4 hr and then declined. Figure 27 shows the time lag between increased levels of blood glucose (mg %) and increased levels of exhaled ^{13}C ($\delta^{13}C$ per mil) (LaCroix, *et al.*, 1973).

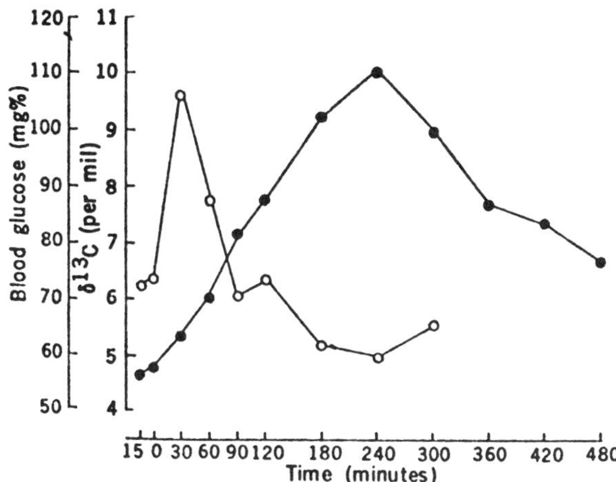

Fig. 28 Changes in blood glucose (○) and expired $\delta\ ^{13}CO_2$ (●) after oral administration of glucose at zero time (mean of six subjects.) [Reproduced by permission from M. LaCroix, F. Mosora, M. Pontus, P. Lefebvre, A. Luyckx, and G. Lopey-Habib, *Science* **181**, 445–446 1973).]

The Belgian workers who published Fig. 28 also reminded the reader that the ^{13}C enrichment in plant-derived glucose varied, according to whether photosynthesis in the plant proceeded by the Calvin pathway or the dicarboxylic acid pathway described by Hatch and Slack (Smith and Epstein, 1971). Low base line levels of ^{13}C enrichment in expired CO_2 in their patient population reflected the fact that beet sugar was generally consumed in West European countries. The enrichment of ^{13}C in glucose derived from cane or maize was more than fourfold higher, and they found that this naturally enriched glucose could be used as tracer material in their beet-eating patients. In dealing with other groups of patients, glucose that has been synthetically enriched in ^{13}C has been used.

2. *Protein Biosynthesis*

The stable isotopes of nitrogen and oxygen are particularly important for tracer work, since no useful radioisotope of either element is available. Shortly after methods were developed for concentrating ^{15}N, Schoenheimer and his colleagues (1938) began to use it in studies of protein and amino acid metabolism, and this still constitutes a major field of research. In a recent report by Halliday and McKeran (1975), for example, the rate of biosynthesis of muscle protein was studied during continuous infusion of lysine labeled with ^{15}N. These authors reported that their measurements on samples of 25- to 100-μg nitrogen gas had short term (replicate) standard deviations of

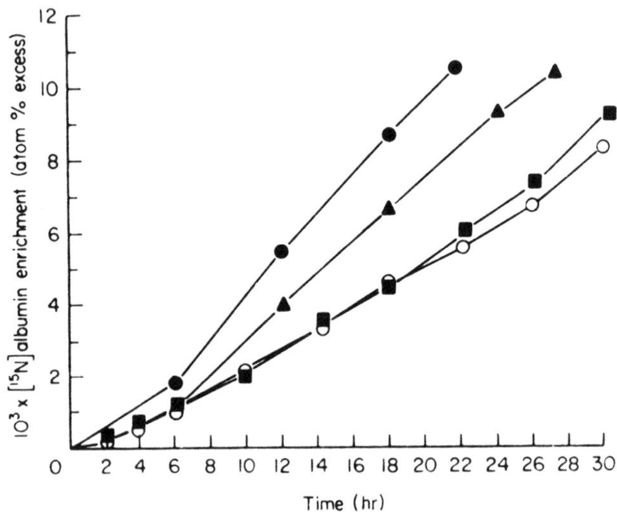

Fig. 29 Increase in the enrichment of ^{15}N in plasma albumin during continuous intravenous infusion with ^{15}N lysine in four subjects. [Reproduced by permission from D. Halliday and R. O. McKeran, *Clin. Sci. Mol. Med.* **49**, 581–590 (1975).]

less than $\pm 0.02\%$ of the enrichment, or ± 0.000073 atom % excess, and a standard deviation of $\pm 0.21\%$ over 14 months. The increases in enrichment of ^{15}N in plasma albumin throughout the infusion in four subjects are shown in Fig. 29. These workers concluded that total muscle protein synthesis accounted for 53% of total body protein turnover and thus occupied an important place in protein metabolism.

3. *Paleoclimates from Fossil Peat*

One of the unique aspects of isotope ratio analysis of combusted samples is the ability to examine isotope enrichment in entire organisms, e.g., plants, bacteria, in parts of organisms, e.g., wood, muscle, or in unfractionated solutions, e.g., urine, soil extracts, in contrast to the analysis of purified compounds. This has been exploited in extensive studies of the variations in abundances of isotopes in different natural sources (Scalan and Morgan, 1970).

The abundance of ^{18}O in rain, for example, has been found to vary with cloud temperature and other climatic conditions (Hintenberger, 1966). Organic matter has been found to be enriched in ^{18}O relative to environmental water (Hardcastle and Friedman, 1974). For example, moss from a cool lake in Tierra del Fuego was found to be enriched in ^{18}O by about 35‰ relative to the lake water it grew in.

Moreover, interesting studies by Ferhi *et al.* (1975) demonstrated that ^{18}O abundance in fossil peat differed according to the climate in which it grew. Isotope ratio measurements for modern moss and fossil peat from warm and cold phases of the late Pleistocene are shown in Table VIII. The modern moss from a cool site and fossil peat from a cold phase have similar $\delta^{18}O$ values, whereas fossil peat from the warm period was relatively depleted in ^{18}O. The authors concluded that the fractionation by plants of the isotopes of oxygen

TABLE VIII

Isotope Composition of Modern Moss and Ancient Peat[a]

Sample[b]	$\delta^{18}O$ (‰)[c]
Fossil peat, cold period	24.65 ± 0.21
Fossil peat, warm period	18.89 ± 0.37
Modern moss, Tierra del Fuego	24.95 ± 0.06

[a] From Ferhi *et al.* (1975).

[b] About 10 mg of organic matter was converted to CO_2 for each determination.

[c] The values shown are means of 2 to 4 determinations.

in water was temperature dependent and that its measurement could be used to retrieve information about paleoclimates in given locales. They were able to calculate a coefficient for the temperature dependent fractionation in peat.

4. *Tracers for Chemical Isolation*

Hayes (Von Unruh *et al.*, 1974) has pointed out that the capability to measure the isotope enrichment in mixtures and solutions allows stable isotopes to be used as tracers through many steps in a sample work-up. The partitioning of the label may be assayed at each step, analogous to the detection of radioisotopes. As little as 0.01% of the material need be consumed in each analysis. Usually isotope ratio measurements on mixtures such as urine cannot be made with the same precision as ratio measurements on cleaner samples (Von Unruh *et al.*, 1974).

B. Isotope Ratios in Large Molecules

Abundance ratios of isotopic species may also be determined in large intact organic molecules without combustion. Some tradeoff in precision and accuracy must be made in return for working with heavier samples which contain many combinations of naturally occuring stable isotopes. For this same reason the minor isotopic components must be enriched to higher levels, in higher molecular weight samples than are required for analysis of small gaseous molecules. This is reflected in Table IX in the column entitled "isotope dilution." In most work with intact organic molecules, molar enrichment is measured and reported directly, and % atom excess has little significance.

Generally speaking, isotope ratios or abundances in large molecules have been measured either by averaging the peak heights in repetitively scanned spectra, or by generating an analog trace of ion intensities through the entire period of ion production. Integrable analog signals may be obtained by selected ion monitoring (see Section IV), by reconstructing mass chromatograms with a computer, and by repetitively scanning through short mass ranges. These methods differ in precision and sensitivity, and this is reflected to some extent in Table IX.

Many of the problems addressed by measuring the isotope enrichment in large molecules require the highest sensitivity. In selected ion monitoring, sensitivity, that is, the minimum sample size, is greatly extended relative to that of the high precision ratio measurements already discussed. However, this is done at the expense of accuracy and precision as is reflected in Table IX. Sensitivity, especially as determined by signal-to-noise ratios, varies with the chemical history of the sample, with the characteristics of its mass spectrum, and with its gas chromatographic properties if a combined GCMS is being

TABLE IX

Examples of Isotope Ratios Measured with Single-Detector Mass Spectrometers (Standard Samples)

Method	Sample and *m/e* Measured	Amount	Precision (%)	Isotope Dilution	Reference
Beamswitching	CO_2 45/44	300 ng	±0.05	1/40,000	Schoeller and Hayes (1975)
Selected ion monitoring	CO_2 45/44	100 μg	±0.3	1/33,000	Klein *et al.* (1972)
Repetitive scanning	Creatinine 114.06226/114.05595	5 μg	±1.1	1/50	Hadden *et al.* (1973)
Selected ion monitoring	Derivatized alanine 120/116	200 ng	±1.78	1/100	Holmes *et al.* (1973)
Selected ion monitoring CI	Derivatized nornitrogen mustard 244/240	10 ng	±5.5	1/10	Jardine *et al.* (1975)
	Phencyclidine 248/243	100 ng	±2.3	2/1	Lin *et al.* (1975)
	N-acetyl procainamide 278/283	2 μg	±1.0	1/15	Dutcher *et al.* (1975)
	Methylstearate 300/299	90 pg	±0.3	1/4	Klein *et al.* (1975)
Selected ion monitoring	Derivatized glucose 304/297	200 ng	±1.62	1/200	Holmes *et al.* (1973)
Selected ion monitoring CI	Phenylbutazone 311.173/309.160	2 μg	±3.0	1/5	Weinkam *et al.* (1977)
Repetitive scanning CI	Phenylbutazone 311.173/309.160	2 μg	±14.3	1/5	Weinkam *et al.* (1977)
Selected ion monitoring	Methyl chenodeoxycholate 373/372	3 μg	±0.4	1/200	Klein *et al.* (1975)
	Derivatized prostaglandin E_2 379/375	1.5 μg	±10	1/250	Hubbard and Watson (1973)
	Derivatized prostaglandin $F_{2\alpha}$ 427/423	20 ng	±2–4	1/50	Watson *et al.* (1973)
	Derivatized uric acid 458/456	20 ng	±1.72	1/33	Walker *et al.* (1975)

used. Nonetheless, the minimum size required for selected ion monitoring is occasionally reported in picograms (10^{-12} g) and less frequently at sub-picogram levels.

Some typical examples from the literature of isotope ratios measured in mass spectrometers with single detectors are presented in Table IX. The entries for CO_2 are similar to those measured on classical double detector instruments. From these high precision measurements on the triatomic molecule, through less precise measurements made on molecules with molecular weights around 400, the tradeoffs in terms of precision, sensitivity, and isotope dilution can be seen. A similar table containing values for a different group of compounds may be found in a review on selected ion monitoring compiled by workers at Vanderbilt University (Falkner *et al.*, 1975).

The use of stable isotope-labeled internal standards in absolute quantitation will be examined in a later section.

C. *Scanned Spectra*

Mass spectral analysis of intact molecules labeled with stable isotopes offers several advantages over analysis of combusted gases—the ability to characterize the exact population of a set of isotope enriched analogs, and to provide information about the location of these labels in the molecule. To be more specific, consider a sample, 50% of the molecules of which contain two ^{16}O and no ^{18}O atoms and 50% of which contain no ^{16}O and two ^{18}O atoms. If this sample is combusted for isotope ratio analysis, the analysis would indicate that all the molecules in the sample contain one atom of ^{18}O and one of ^{16}O. If, on the other hand, the sample is analyzed intact, the duality of the population will be determined directly by the presence of molecular or fragment ions 4 mass units apart of equal intensities. In this latter case, the assessment might be 50% plus or minus $>1\%$, while in the former case, the average incorporation of ^{18}O could be measured with much greater accuracy and precision.

Equally important is the ability to locate stable isotopes within a molecule by examination of the fragments formed in the mass spectrometer. Thus one could envision a molecule labeled with two ^{15}N atoms, which fragments into two pieces. On one hand, each piece might carry one of the ^{15}N labels, or on the other, half of the molecule might carry both ^{15}N atoms and the other half none. These situations could be easily distinguished in a scanned spectrum.

1. *Metabolism of Ethanol*

An interesting example of this kind of analysis comes from the studies of Cronholm and Sjovall on the human metabolism of ethanol (Cronholm and

Sjovall, 1970). It is well known that ethanol is oxidized metabolically to acetaldehyde, accompanied by reduction of the cofactor NAD to NADH (scheme **13**). These authors hypothesized that an overabundance of NADH

$$CH_3CH_2OH + NAD \longrightarrow CH_3CHO + NADH$$

RO + NADH ⟶ + NAD

13

would lead to increased reduction in overall metabolism, which should be reflected in steroid metabolites secreted in the urine. They designed an experiment whereby ethanol labeled with deuterium on the first carbon was fed to human subjects. Subsequently, sterols were isolated from the subjects' urine and examined mass spectrometrically for incorporation of deuterium.

In initiating their work they wisely chose to examine the extent and distribution of deuterium isotope labels in the ethanol which they had purchased. Commercial 1,1-d_2-ethanol was analyzed by converting it to its benzoate derivative. As indicated in Table X, the mass spectrum of ethyl benzoate contained a molecular ion at mass 150 and $M - 15$ peak at m/e 135. The spectrum of the deuterated material contained several molecular ions as indicated. On the basis of the relative intensities of these molecular ions, the investigators estimated that 71% of the sample contained 2 deuterium atoms, 18% contained 3 deuterium labels, 9% contained 1 deuterium atom, and a

TABLE X

Analysis of Deuterium Population and Distribution in the Benzoate Derivative of Labeled Ethanol

	m/e	%
$M^{+\cdot}$	153	18
	152	71
	151	9
	150	~1
$(M - 15)^+$	137	82
	136	17
	135	~1

trace amount contained no deuterium atoms. The loss of 15 mass units, known to comprise the terminal methyl group, allowed the investigators to distinguish the position of the deuterium atoms in ethanol as being on the alpha carbon or the methyl group. Their analysis of the methyl ions (Table X) led them to report that 82% of the molecules contained 2 deuterium atoms in the alpha carbon, 17% of the molecules contained 1 deuterium atom in the alpha carbon, and a small amount of material contained no deuterium atoms in the alpha carbon.

Two hundred milliliters of this labeled ethanol was consumed by a volunteer, and 15 min later a urine sample was colleted. Six sterols were examined from this urine for the presence of deuterium atoms. In a spectrum of the bis(trimethylsilyl) derivative of 3,17-dihydroxyandrostane, the molecular ion normally is observed at mass 436. However, in 20% of the molecules isolated from that 15-min urine collection, the molecular ion appeared at mass 437, revealing the incorporation of one deuterium atom (scheme **14**).

TMSO — OTMS ⟶ TMSO — ⊕OTMS, H ⟶ ⊕OTMS, H (D)

$M^{+\cdot}$ 436 80%
$M^{+\cdot}$ 437 20%

m/e 129 80%
m/e 130 20%

14

A fragment ion of mass 129 is found in the spectrum as well, comprising carbon atoms 15, 16, and 17 from the D ring. Again in the spectrum of the urine sample, 20% of these fragment ions weighed 1 mass unit more and thus carried the deuterium isotope of hydrogen. This observation localizes the deuterium atom in the steroid D ring and the authors' reasonable interpretation was that it arrived there by a reduction of a 17-carbonyl group with deuterated reduced cofactor.

2. *Mechanism of an Alkylation Reaction*

A second example of the use of mass spectrometry to characterize isotope populations and also to locate isotopes within a molecule is drawn from studies of the pharmacology of cyclophosphamide (Colvin *et al.*, 1976). The active metabolite phosphoramide mustard was known to be an alkylating agent that reacted with nucleophiles with the loss of chloride. Two simple mechanisms may be envisaged for this reaction (scheme **15**). In one case, the

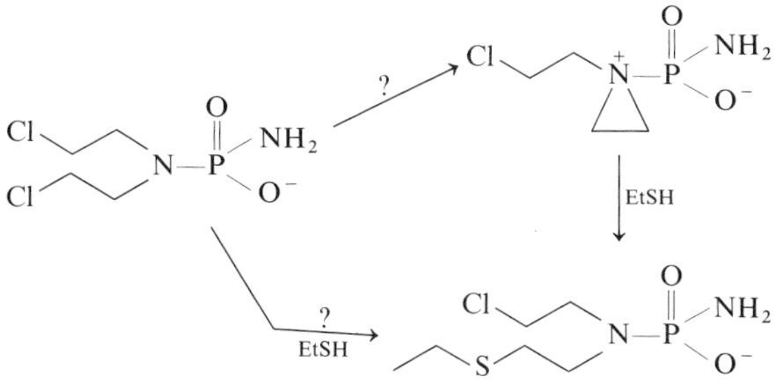

15

chloride ion is displaced directly in an SN_2 mechanism. In the other, an aziridinium ion is formed with anchimeric assistance, which is subsequently opened at either end by attack of the nucleophile. In an experiment designed to distinguish between the two mechanisms, phosphoramide mustard was synthesized carrying two deuterium atoms on the beta carbon atom in each 2-chloroethyl side chain (scheme **16**).

This labeled alkylating agent was then allowed to react with the model nucleophile ethylthiol. The product was characterized as the trimethyl derivative of the bis(ethylthio-ethylene) adduct shown in scheme **16**. In this

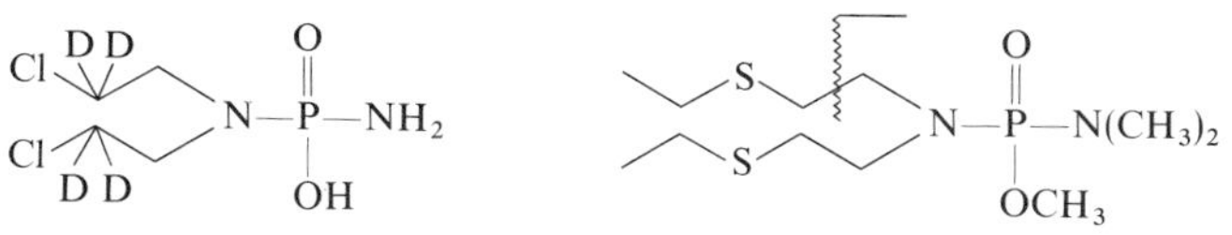

d_4-Phosphoramide mustard

Alkylation product (trimethyl derivative)

16

compound, like the other members of the family (scheme 12), facile cleavage occured between the 2 carbon atoms in the mustard side chains, permitting the deuterium content of each methylene group to be distinguished. Alkylation via SN_2 displacement should lead to a product containing deuterium atoms only in their original beta positions. Alkylation via the symmetrical aziridinium intermediate should lead to adducts carrying deuterium labels in the alpha methylene groups as well as the beta positions. The mass spectrum of the product is shown in Fig. 30. The 1:1 ratio of the peaks at *m/e* 241 and 243 is most reasonably interpreted as confirming that the deuterium labels have been equally distributed between the two positions in each chain, and that alkylation takes place via the aziridinium intermediate.

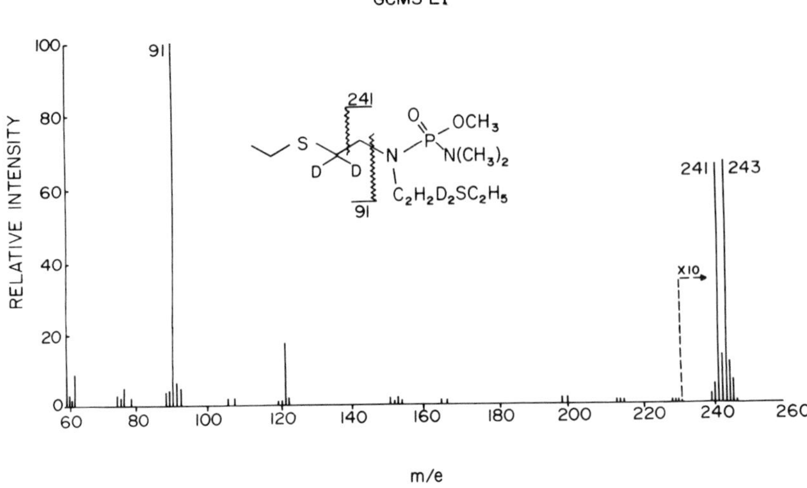

Fig. 30 Mass spectrum of the trimethyl derivative of the product of the reaction of ethanethiol with *N,N*-bis(2,2-d_2-2-chloroethyl) phosphorodiamidic acid (d_4-phosphoramide mustard). [Adapted from M. Colvin, R. B. Brundrett, M. N. Kan, I. Jardine, and C. Fenselau, *Cancer Res.* **36**, 1121–1126 (1976).]

3. *Distinguishing Optical Isomers*

Mass spectrometry has been widely used for many years to assay isotopic populations and molecular locations in mechanistic studies of both chemical reactions and biochemical transformations. As stable isotopes become available more economically and the restrictions on the use of radioisotopes increase, new applications are being devised to exploit the use of stable isotopes as tracer labels. For example, the competitive metabolism of R and S isomers of the chiral psychotomimetic amine 1-(2,5-dimethoxy-4-methylphenyl)-2-aminopropane has been studied by labeling one of the stereoisomers with 6 deuterium atoms (Weinkam *et al.*, 1976: Zweig and Castagnoli, 1977). A mixture of optical isomers, one labeled and one not, was incubated with rabbit liver microsomes. Metabolites were isolated as racemates and characterized without resolution. The relative amount of each optical isomer in a given metabolite product was assayed mass spectrometrically by measuring the relative abundances of compounds carrying no deuterium and 6 deuterium atoms.

4. *Isotope Clusters*

Another approach to employing stable isotopes as tracers has been to use artificial mixtures of isotopically labeled species that will produce mass spectra with unusual clusters of peaks. For example, equimolar mixtures of propoxyphene and propoxyphene labeled with 7 deuterium atoms in the benzyl moiety (see scheme **17**) were coadministered to rats (Due *et al.*, 1976).

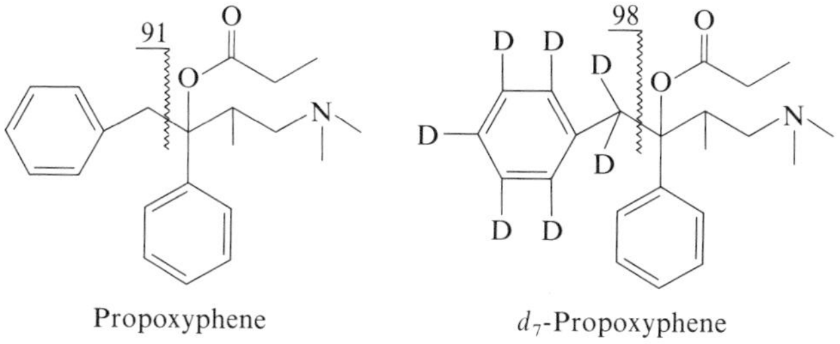

Propoxyphene

d_7-Propoxyphene

17

The benzyl ion $C_7H_7^+$ was readily formed in the electron impact induced fragmentation of propoxyphene and related compounds. The resulting peak at *m/e* 91 shifted to *m/e* 98 when the ion was formed from the deuterated analog. Thus, among the many compounds present in extracts of the rat urine, seven were found to provide mass spectra that contained twin peaks at *m/e* 91 and 98 with roughly 1:1 relative intensities, and on this basis these seven were judged to be drug-related compounds. In an eighth spectrum, twin peaks at *m/e* 107 and 113 were recognized as representing hydroxybenzyl ions formed from a drug metabolite hydroxylated in the benzyl ring. In studying chemical or biochemical transformations of chemicals which contain chlorine or bromine atoms, the naturally occuring halide isotope clusters can be used to find the products of interest.

5. *Monitoring Illegal Drug Supplementation*

Although in many instances deuterium-labeled analogs have markedly different physiologic activity than unlabeled drugs, this usually reflects an isotope effect in cleavage of one of the H/D bonds and can be avoided by using isotopes of heavier nuclides or by placing the deuterium labels away from the site of metabolic transformation. Thus 1,1,1-d_3-methadone has been synthesized and its physiologic activity has been found to be indistinguishable from that of methadone administered to rats (Hsia *et al.*, 1976). The suggestion has been advanced (Hsia *et al.*, 1976) that defined mixtures of d_0- and d_3-methadone could be administered to patients in methadone maintenance programs. Subsequently, isotope ratios could be monitored in the drug in urine or blood to detect illegal supplementation. The method advanced for measuring this ratio involved GCMS with chemical ionization. Repetitive scans made through the mass range of the protonated molecular ions (*m/e* 306–315) generated intensity envelopes of the $M + H$ ion for each compound. The areas under these curves were integrated and compared.

6. *Atmospheric Tracer Studies*

In a futuristic demonstration of isotope analysis by mass spectrometry, an ultrahigh vacuum scanning mass spectrometer has been used to measure methane-21 ($^{13}C^{2}H_4$) at concentrations in air around 1 part in 2×10^{16} by volume, by comparing the intensity of the peak at *m/e* 21 with that at *m/e* 16 from natural methane ($^{12}C^{1}H_4$) concentrated in air at about 1.4 parts in 10^7 by volume (Cowan *et al.*, 1976). Eighty-four grams of methane-21 were released into the atmosphere at Idaho Falls, Idaho. Methane and krypton were concentrated from the atmosphere in cryogenic air samplers at locations throughout the central United States, and separated by chromatographic desorption. The abundance ratios of *m/e* 21:*m/e* 16 were determined from scanned spectra. Precision was poor. For example, the ratio was measured in two commercial methane samples as $(3.7 \pm 1.2) \times 10^{-11}$ and $(2.7 \pm 1.7) \times 10^{-11}$. Nonetheless, the tracer material could be detected above the background as far away as Detroit, Michigan (about 2500 km).

VI. Quantitation

From the viewpoint of the chemist providing the sample, there are two ways in which quantitative measurements may be made with a mass spectrometer. In one, large amounts of sample are employed to provide a steady ion flux for carefully controlled measurements that take as long as several minutes. In the other approach, the high sensitivity and the high specificity of the mass spectrometer are exploited to provide rapid measurements of very small samples, usually introduced through a gas chromatograph.

A. Steady State Sampling

In the history of analytical applications of mass spectrometry, an early period of nearly 25 years may be designated in which the most important application was the identification of isotopes, and eventually, the measurement of their natural abundances (Kiser, 1965; McDowell, 1963). As has been discussed in Section V, relatively large amounts of sample must be combusted in order to provide an even flow of all gaseous isotopic species into the instrument throughout the time (as long as several minutes) required to make highly precise ratio measurements. The high precision measurement of isotope ratios continues to be an active area of mass spectrometry, although currently the measurement of tracer enrichment is emphasized more than the measurement of natural abundances.

A new kind of application was introduced around 1940, when the first commercially produced mass spectrometers became available, sold primarily for quality control in petroleum production (McDowell, 1963; Beynon, 1960). These instruments were developed to permit the relative proportions of components of fuel mixtures to be determined by comparing intensities of selected peaks in conventionally scanned spectra. The instruments were designed for accurate control of sample temperature and pressure, and all instrumental parameters, in order to produce a spectrum that was constant through the several minutes required for the analysis and that was reproducible from day to day and instrument to instrument. Slow scanning and ion detection by a Faraday cup increased precision and accuracy. Several grams of sample were commonly employed.

A third area of quantitative measurements using steady state sampling can be found in the analysis of partial pressures in mixtures of gases. An example of a clinical application is shown in Fig. 31, where the partial pressure of oxygen in canine arterial blood gas was monitored while the gas administered to the animal via a respirator was switched from air to 100% oxygen (Brantigan *et al.*, 1970). The ion current record from the mass spectrometer was calibrated to read partial pressures directly. Real time measurements provided by this kind of mass spectrometer for CO_2, N_2, O_2, as well as for diagnostic gases such as $C^{18}O$, freon, and acetylene, have had a significant impact on the understanding of respiratory physiology, cardiac output, and management of the critically ill patient (Roboz, 1975). Gas mixtures encountered in many industrial projects are also readily analyzed by mass spectroscopy.

Although the mass spectral techniques that require steady state sampling could be used to provide absolute quantitation, this is usually not the issue when large samples are available. Rather, these kinds of mass spectrometric

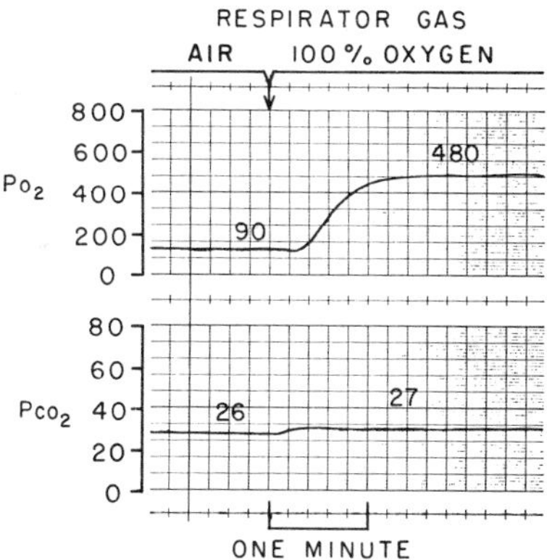

Fig. 31 In vivo demonstration of the rapid response of the mass spectrometric monitoring system to a rapid change of arterial oxygen tension. [Reproduced by permission from J. W. Brantigan, V. L. Gott, M. L. Vestal, G. J. Fergusson, and W. H. Johnston, *J. Appl. Physiol.* **28**, 375–377 (1970).]

measurements provide precision, dynamic range, ease and speed for measurements of relative proportions of mixtures.

B. Dynamic Sampling

In the second approach to quantitation with a mass spectrometer, the objective is usually to measure the absolute amount of a single compound. Usually only a small amount of sample is available, or else it would be assayed by an easier, less sensitive method. The ion flux produced by such small samples is rarely steady, but rises and falls quickly. Thus peak intensities are either integrated through time or averaged. The outstanding feature of mass spectrometry in this kind of assay work is its sensitivity, usually exceeding that of all other techniques except radioisotope labeling. The second strength of mass spectrometry is its great specificity, especially remarkable in such a generally applicable technique. These assays are commonly carried out with combined gas chromatograph mass spectrometers, although the direct probe is also used, particularly for nonvolatile samples.

1. *Integrated Signals*

The vapor pressure of a small sample is likely to rise and fall rather quickly, and one satisfactory approach to such assays is to generate an analog tracing of ion current as a function of time that can be integrated and quantitated. Selected ion monitoring, described earlier, generates ion profiles that resemble conventional chromatograms and can be integrated and analyzed in

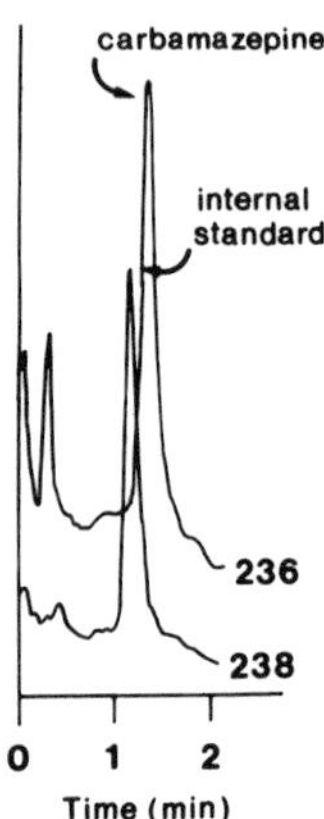

Fig. 32 Selected ion profiles of molecular ions of carbamazepine (*m/e* 236) and dihydrocarbamazepine (*m/e* 238) coinjected into the GCMS. [Reproduced by permission from L. Palmer, L. Bertilsson, P. Collste, and M. Rawlins, *Clin. Pharmacol. Therap.* **14**, 827–832 (1973).]

much the same way. Repetitive scanning through narrow mass ranges can also provide quantitative envelopes of ion current. These recording methods can be used in high or low resolution, with or without internal standards, on samples introduced by the GC or by the direct probe.

(*a*) *Selected Ion Monitoring* Selected ion monitoring has provided the opportunity to assay samples against an internal standard. In Fig. 32, profiles of the molecular ions of carbamazepine and dihydrocarbamazepine are shown, recorded as the mixture elutes from a gas chromatograph. Known amounts of the latter compound were added as internal standards to plasma samples being assayed for carbamazepine (Palmer *et al.*, 1975). A standard curve is shown in Fig. 33, in which measured ratios of the intensities of the molecular ions are shown to correlate linearly with known ratios of the compounds added to plasma. This assay was used to delineate the time course of clearance of carbamazepine from circulation of two volunteers who took the drug (Fig. 34). As will be discussed, stable isotope labeled analogs can also be used as internal standards.

Many assays have been devised which employ single ion monitoring instead of multiple ion monitoring, in order to exploit the optimal sensitivity of the former, and because a single mass can be continuously monitored with any mass spectrometer. In some cases, these assays employed no internal standard. Brooks, for example, has reported that the response of the single ion detector to ions from several estrenols (introduced via the GC) was linear with respect to sample size through the range 50–1000 pg (Brooks and Middleditch, 1971). On the other hand, a nonlinear response has been

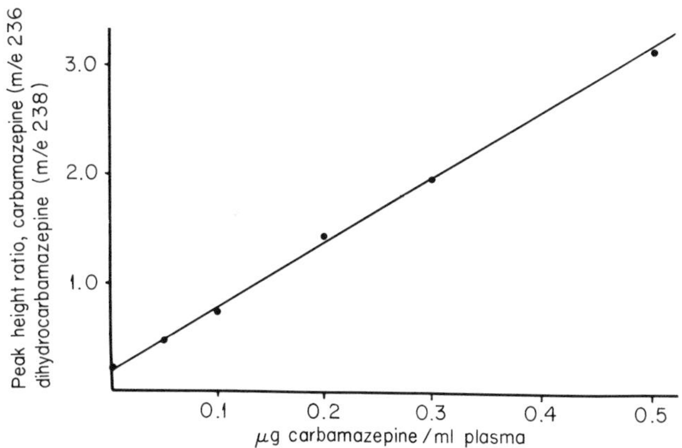

Fig. 33 Standard curve for the quantitative determination of carbamazepine in plasma. [Reproduced by permission from L. Palmer, L. Bertilsson, P. Collste, and M. Rawlins, *Clin. Pharmacol. Therap.* **14**, 827–832 (1973).]

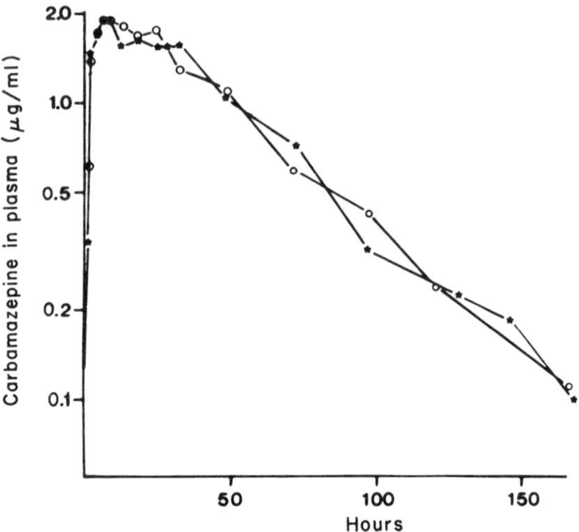

Fig. 34 Concentrations of carbamazepine in plasma of two healthy subjects measured through 7 days following 200-mg oral doses. [Reproduced by permission from L. Palmer, L. Bertilsson, P. Collste, and M. Rawlins, *Clin. Pharmacol. Therap.* **14**, 827–832 (1973).]

reported in single ion detection of diethylstilbestrol (Day *et al.*, 1975). Use of an internal standard might have improved the linearity of this DES assay.

It is possible to use internal standards with single ion monitoring, if the standard elutes from the gas chromatographic inlet system separately from the sample, and if ions of a common mass are formed from both sample and standard. These ions must be sufficiently characteristic and intense to be reliably monitored. For example, Fig. 35a shows the ion current profile monitored at m/e 436.319 as a set of isomeric androstanediols (molecular weight 436.319) was eluted from the GC. Figure 35b shows the same ion current profile of an extract of prostatic tissue to which the 3β, 17α-diol isomer was added as internal standard (Millington *et al.*, 1975). Excellent correspondence was found between sample size (ratio) and signal (ratio), with sensitivity to 20 pg of 5α-androstene-3β, 17β-diol and an accuracy of $\pm 5\%$.

Some workers have reported success with the difficult technique of refocusing the mass spectrometer after one sample has been eluted from the GC in order to monitor a single ion from the next eluted compound.

Selected ion monitoring can also be used effectively in quantitating samples introduced on the direct probe. The direct probe does not offer the same potential for purification as a gas chromatographic inlet system, and compensation must be made in sample purification, in careful probe temperature programing, in the use of "soft ionization" techniques (Section III) and, most effectively, in high resolution mass analysis. Direct probe introduction has

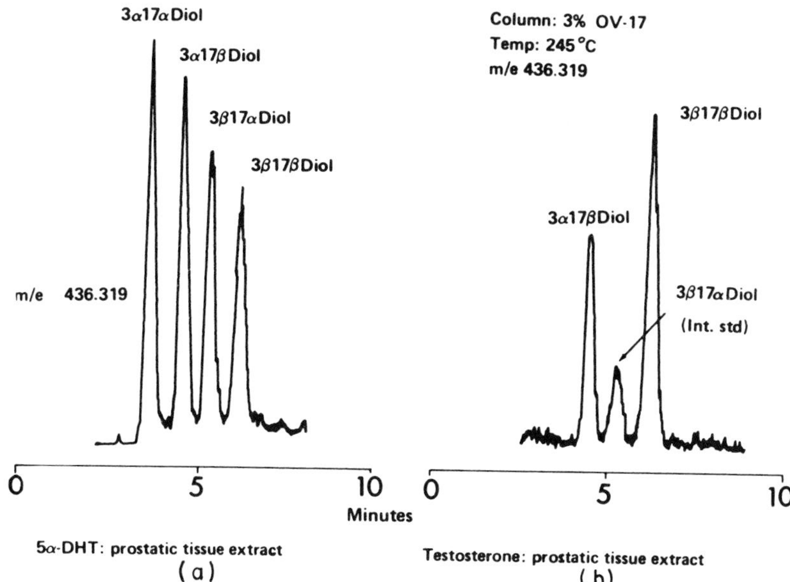

Fig. 35 High resolution single ion profile (*m/e* 436.319) of the molecular ions (a) of a set of isomeric 5α-androstanediol standards and (b) of an extract of prostate tissue. [Reproduced by permission from D. S. Millington, M. E. Buoy, G. Brooks, M. E. Harper, and K. Griffiths, *Biomed. Mass Spectrom.* **2**, 219–224 (1975).]

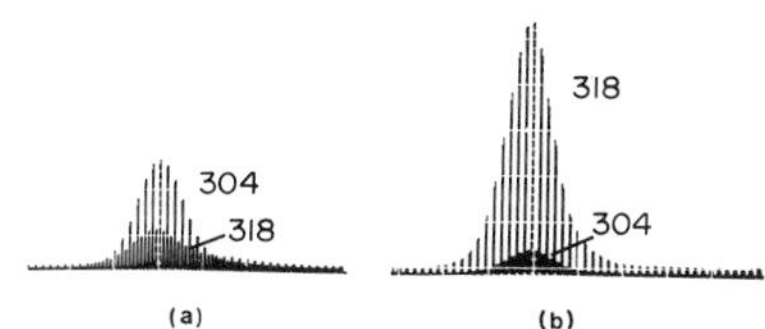

Fig. 36 Molecular ion profiles of dansylated piperidine (*m/e* 318) from (a) the brain of an active mouse and (b) the brain of a dormant mouse, compared with molecular ion profiles from dansylated pyrrolidine (*m/e* 304) added as internal standard. [Adapted from M. Stepita-Klauco, H. Dolezalova, and R. Fairweather, *Science* **183**, 536–537 (1974).]

been useful because not all samples can be vaporized for gas chromatography, and not all mass spectrometers are interfaced to gas chromatographs. Figure 36 shows molecular ion current profiles for exceptionally well purified samples of dansylated piperidine (*m/e* 318) (the 5-dimethylamino-1-naphthalenesulfonyl derivative) isolated from brains of (a) an active mouse and (b) a dormant mouse. Dansylated pyrrolidine (*m/e* 304) has been added as internal standard (Stepita-Klauco *et al.*, 1974).

Two of the problems often encountered in direct probe assays are illustrated in Figs. 37 and 38. Figure 37 presents GCMS ion current profiles

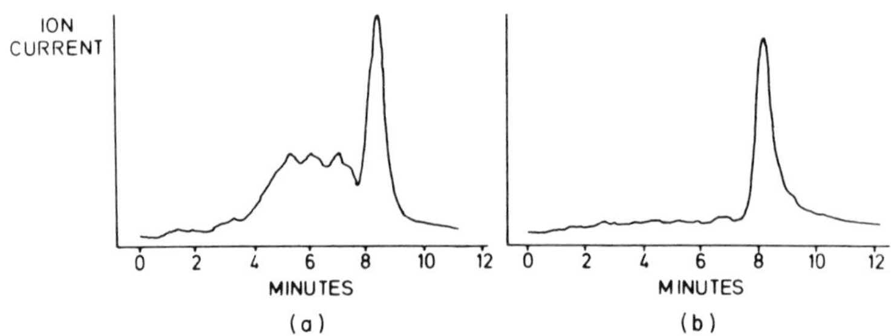

Fig. 37 Single ion profiles recorded at m/e 438 from a purified plasma extract at (a) resolving power of 1000 and (b) resolving power of 3000 (multiplier gain increased to maintain the height of the testosterone peak). [Reproduced by permission of J. R. Chapman and E. Bailey, *J. Chromatogr.* **89**, 215–224 (1974).]

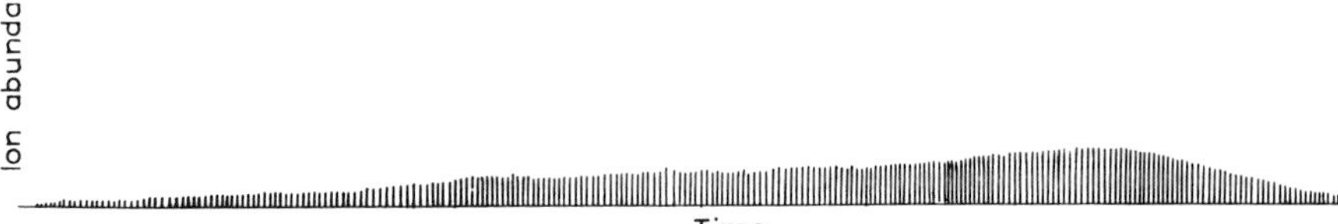

Fig. 38 Record of the ion current at m/e 108 during evaporation off the probe of a urine extract. [Reproduced by permission from A. A. Boulton and J. R. Majer, *J. Chromatogr.* **48**, 322–327 (1970).]

obtained at 1000 resolution and 3000 resolution of a testosterone sample extracted from plasma and purified by thin layer chromatography as though for analysis by direct probe (Chapman and Bailey, 1974). The ion current profile at the lower resolution (Fig. 37a) indicates the presence of a substantial amount of material that would have interfered with direct probe analysis. At medium resolution (Fig. 37b) interference is reduced to a level suitable for direct probe work. Figure 38 shows an example of a compound whose volatility is poor or impaired, resulting in an ion current profile that is difficult to integrate (Boulton and Majer, 1970). The authors suggested that this phenolic amine extracted from urine was complexed as an acetate salt and thus its volatility was impaired. They altered their extraction procedure and obtained a nearly Gaussian ion current profile.

(*b*) *Mass Chromatograms* Ion current profiles reconstructed from computer-acquired repetitive scans can provide assay measurements analogous to selected ion profiles. For example, computer-controlled repetitive scans were used to provide assays of CCl_3F and CCl_2F_2 at parts per trillion concentrations 80-liter samples of air (Tyson, 1975). In this case, profiles of the base peaks in the sample spectra were compared to those from external standards. In general, however, mass chromatograms are used less frequently for assay work than selected ion profiles because of their reduced sensitivity.

(*c*) *Limited Range Repetitive Scanning* Peak heights from repetitive scans made rapidly through a short mass range can be plotted, as shown in Fig. 39, to compose an integrable analog signal. In the example illustrated in Fig. 39a, (Cala *et al*., 1972) a sample of pyrimethamine extracted from chicken liver was introduced into the mass spectrometer via a gas chromatograph. The ion envelope from the sample is compared in size to that generated analogously from an external or bracketing standard (Fig. 39b). If a stable isotope analog is used as an internal standard, the masses of the ions being recorded will fall within a few Daltons of each other and ion profiles for both sample and standard can be recorded in the same series of short range scans.

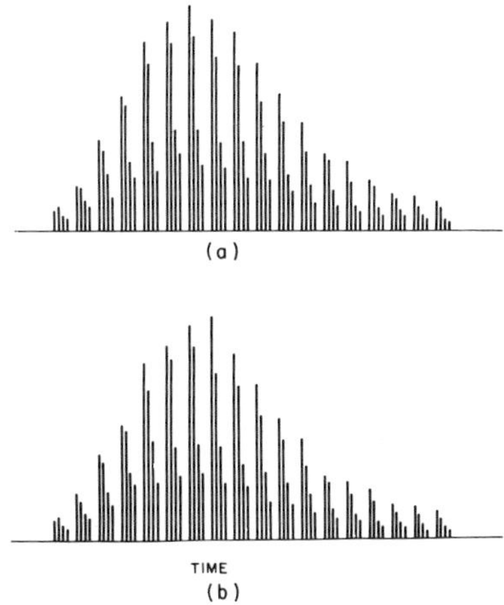

Fig. 39 Envelopes obtained by repetitive scanning through the range *m/e* 247–250 as samples (a) of an extract from chicken liver and (b) pyrimethamine external standard were eluted from the gas chromatograph. [Reproduced by permission from P. C. Cala, N. R. Trenner, R. P. Buhs, G. V. Downing, J. L. Smith, and W. J. A. VandenHeuvel, *J. Agr. Food Chem.* **20**, 337–340 (1972).]

2. *Signal Averaging*

Averaging of repetitive scans has also been used effectively to provide accurate assays at high sensitivity of sample introduced by both the gas chromatograph and the direct probe.

An example may be drawn from the classic analysis by Baughman and Meselson (1973a,b) of levels of 2,3,7,8-tetrachlorodibenzo-*p*-dioxin (TCDD) in animals and fish in areas that had been treated with 2,4,5-T defoliant.

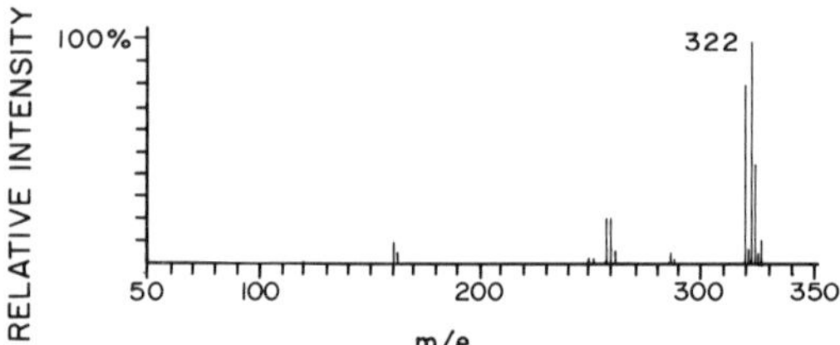

Fig. 40 Low resolution mass spectrum of TCDD.

These workers required an analytical method with sensitivity to 10^{-12} g or 1 pg. They also required good specificity, since purified tissue extracts were to be presented on the direct probe. The low resolution electron impact mass spectrum of TCDD is shown in Fig. 40. The molecular ion peaks at *m/e* 320 and 322 were judged to be most intense and most characteristic signals to use for assaying.

Several approaches were evaluated for measuring these ion currents. Repetitive scanning of the region *m/e* 310–330 at 10,000 resolution yielded ion profiles integrable to a sensitivity of about 100 pg. A selected ion monitoring technique (utilizing peak matching circuitry) allowed the investigators to scan alternately narrow mass ranges around *m/e* 314 (a perfluorotributylamine peak used for accurate mass calculations) and 322 in high resolution.

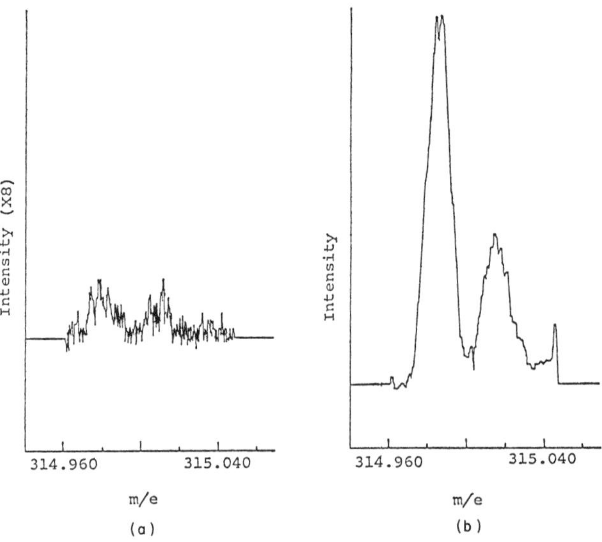

Fig. 41 Effect on signal to noise ratio of averaging 60 scans made at 12,000 resolving power: (a) one scan, (b) 60 scans (1 scan/sec). [Reproduced by permission from R. Baughman and M. Meselson, "Chlorodioxins—Origin and Fate" (E. Blair, ed.), pp. 92–104. American Chemical Society, Washington D.C. (1973).]

This provided sensitivity to 20 pg TCDD. Ultimately the 1-pg sensitivity required for analysis at the parts-per-trillion level was achieved by averaging 7–15 of these repeated narrow range scans. The method has subsequently been adapted to newer instruments in which as many as 400 narrow range scans are averaged over 90 sec (Ryan *et al.*, 1974). The effectiveness of averaging in enhancing the signal-to-noise ratio is illustrated in Fig. 41.

The importance of high resolution in the analysis of TCDD in tissues is illustrated in Fig. 42, in which nominal mass interference at *m/e* 322 from other environmental residues, the polychlorinated biphenyls (PCB) and dichlorophenyl-dichloroethylene (DDE) is shown to be resolved above 10,000. Sample purification techniques have also been developed to help eliminate DDE and PCB from the sample.

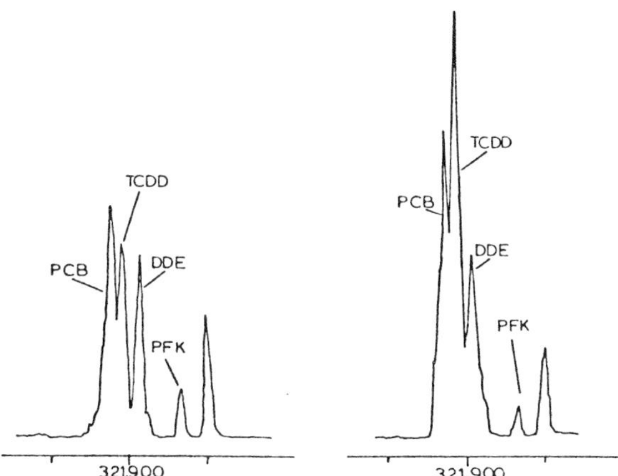

Fig. 42 High resolution time averaged scans of the *m/e* 322 region of an extract of shrew tissue. [Reproduced by permission from J. F. Ryan, F. J. Biros, and R. L. Harless, *Proc. Annu. Conf. Mass Spectrom. Allied Topics, 22nd, Philadelphia, Pennsylvania, March 19–24, 1974.*]

Although assays of TCDD have been carried out primarily through the use of external bracketing standards, TCDD labeled with four ^{37}Cl isotopes has been used as carrier and internal standard in measuring absolute recoveries of TCDD from tissue (Baughman and Meleson, 1973b).

Signal averaging has also been used in conjunction with the introduction of other kinds of samples via the gas chromatograph (Biros, 1970). It is required in analyses by field desorption, where the ion flux is typically small.

3. *Internal Standards*

It is hard to envision an assay that is not better carried out with internal standards than with external or bracketing standards. Even if the response of the detector is reliably linear in relation to the amount of sample, the internal standard offers the great convenience of correcting for sample loss or decomposition during purification and derivatization. Although the recovery efficiency can be measured independently, it is an approximation to assume that it will be constant throughout the preparation of many samples. Homologous internal standards are well accepted in gas chromatographic quantitation and can be used in GCMS assays as well. In addition, the mass spectrometer provides the opportunity to use stable isotope-labeled analogs as internal standards. Both kinds of internal standard have strengths and weaknesses, and these are often hotly debated.

(*a*) *Homologous Internal Standards* Homologous internal standards are easier to obtain and offer the potential advantage (already discussed) that more sensitive single ion monitoring can be used. A number of examples of the successful use of homologous internal standards has already been given (e.g., Figs. 32 and 35). Another interesting case has been found in the phthalates. Figure 43 shows the selected ion profile of ions of mass 149 for an assay of di(ethylhexyl)phthalate from neonatal heart tissue in which dioctylphthalate was used as internal standard (Hillman *et al.*, 1975). Both these plasticizers fragmented to form abundant ions of mass 149. In this case, ions of mass 167 were also monitored to provide further validation of the di(ethylhexyl)phthalate peak. Although single ion monitoring would be expected to be more sensitive, we are reminded once again that the need for sensitivity has to be balanced against the need for specificity.

Homologous samples have also been used for quantitation using the direct probe. For example, naturally occuring thymine has been used as an internal standard to assay the minor base 5-methylcytosine in DNA (Deutsch *et al.*, 1976). The mixture of free bases derived from DNA was presented on the direct probe. Intensities of the molecular ion of 5-methylcytosine, m/e 125.0589, and the $M - 1$ ion from thymine, 125.0339, were recorded by selected ion monitoring (using peak matching circuitry) at 5000 resolution

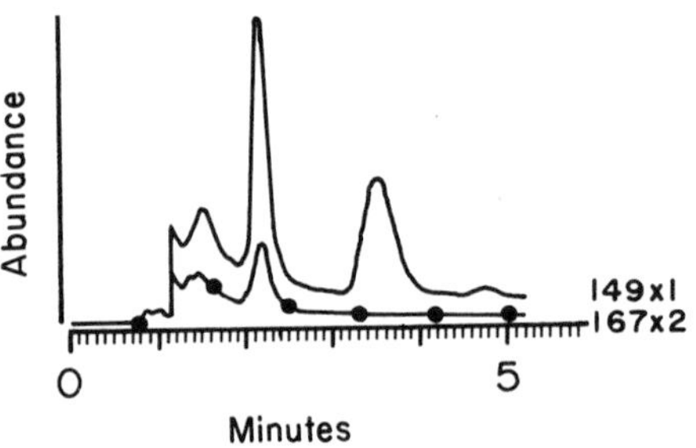

Fig. 43 Selected ion profiles of an extract of neonatal heart tissue and dioctylphthalate added as internal standard. [Adapted from L. S. Hillman, S. L. Goodwin, and W. R. Sherman, *New England J. Med.* **292**, 381–386 (1975).]

and their ratio was calculated. For absolute quantitation of the minor base, the amount of thymine must be determined independently.

(*b*) *Isotope-Labeled Analogs as Internal Standards* Isotope-labeled analogs have offered the major advantage of behaving chemically like the sample. Their partition coefficients are reliably similar, as are their rates of derivatization and decomposition. The possibility of a significant isotope effect in any of these processes can be envisioned only when deuterium is used, and then can be circumvented by thoughtful positioning of the label(s). Although isotope-labeled analogs are sometimes partially resolved from unlabeled samples by gas chromatography, it is easy to find conditions at which the two coelute, providing the advantages of simultaneous measurement to a GCMS assay. As a last advantage, some workers have used isotope-labeled analogs as carriers in reverse isotope dilution techniques for analysis of low level unlabeled samples. Perhaps the principle disadvantage of isotope labeled analogs is that they must usually be synthesized.

Although deuterium has been most commonly employed, ^{15}N, ^{13}C, ^{18}O, ^{37}Cl, and others have been used effectively. Most of the considerations (already discussed in Section V.B. on high-sensitivity low-precision measurements of isotope ratios applies directly to measuring the amount of sample relative to a known amount of stable isotope-labeled analog added as internal standard.

(i) GCMS and chemical ionization

Some of the strengths and weaknesses of stable isotope-labeled internal standards can be illustrated from assays of blood levels of phosphoramide mustard, the cytotoxic metabolite of cyclophosphamide (Jardine *et al.*, 1976, 1978). *N*,*N*-Bis(1,1-d_2-2-chloroethyl)phosphorodiamidic acid was synthesized (scheme **18**) for use as internal standard. Specified amounts of this

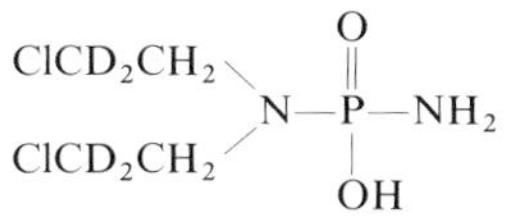

$$(ClCD_2CH_2)_2N{-}P(=O)(OH){-}NH_2$$

18

d_4-analog were added to plasma samples before each was percolated through a column of Amberlite XAD-IV. The absorbed metabolite and the internal standard (along with other compounds) were recovered from the column with methanol, treated with diazomethane, and injected into the GCMS. In an extension of the methodology developed for simple detection of this metabolite (Section IV), a pair of $M + H$ ions were monitored from the trimethylated derivative, simultaneously with two $M + H$ ions from the trimethylated internal standard. Intramolecular peak–height ratios provided a test for interference by ions from other compounds while intermolecular peak–height ratios permitted quantitation of the metabolite.

Figure 44 shows profiles of the four ions that were monitored, recorded as standard samples of phosphoramide mustard and of d_4-phosphoramide mustard (trimethyl derivatives) eluted from the gas chromatograph. It can be seen that there is some small contribution by the labeled standard to ion currents at *m/e* 263 and 265, being monitored as characteristic of the unlabeled sample. In assaying phosphoramide mustard, this problem was overcome by working within ratios between 10:1 and 1:10. Other workers have employed mathematical corrections for cross contributions. Perhaps the most effective approach to minimizing cross contribution is to increase the molecular weight of the labeled standard by 4 or more Daltons. The chlorine isotopes in phosphoramide mustard compromise the mass increment introduced by the four deuterium atoms.

It should be pointed out that the ions being monitored are $M + H$ ions formed by isobutane chemical ionization. In discussing the fingerprinting of this metabolite in blood in Section IV, the value of chemical ionization was emphasized in reducing interference, and permitting the more charac-

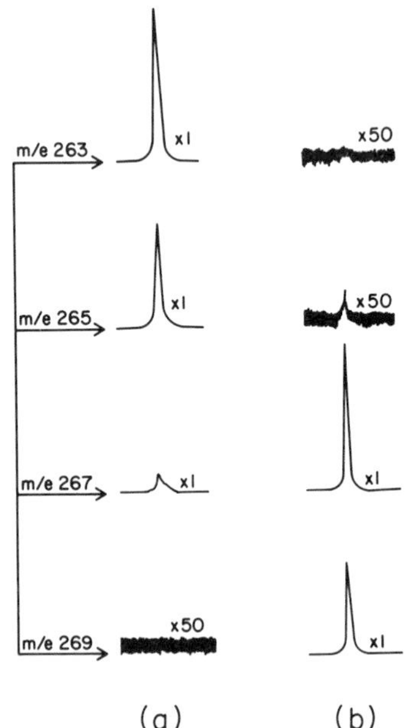

Fig. 44 Selected $M + H$ ion profiles from (a) trimethyl phosphoramide mustard and (b) trimethyl d_4-phosphoramide mustard. [Adapted from I. Jardine, C. Fenselau, M. Appler, M. N. Kan, R. B. Brundrett, and M. Colvin, *Cancer Res.* **38**, 408–415 (1978).]

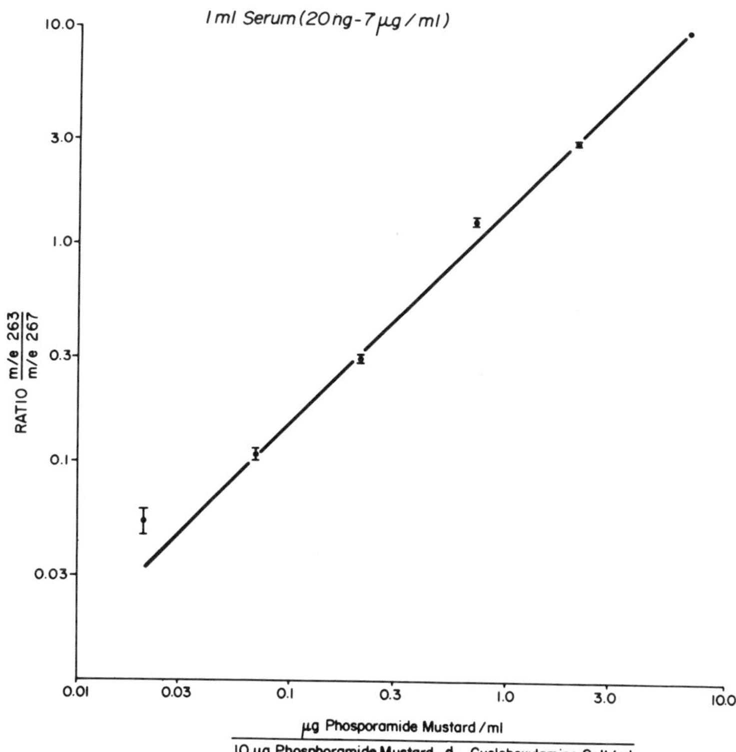

Fig. 45 Calibration curve demonstrating the correlation of peak intensity ratios with the ratio of phosphoramide mustard and internal standard extracted from plasma. [Reproduced by permission from I. Jardine, C. Fenselau, M. Appler, M. N. Kan, R. B. Brundrett, and M. Colvin, *Cancer Res.* **38**, 408–415 (1978).]

teristic molecular ions ($M + \text{H}$) to be monitored at increased intensities. Chemical ionization has been reported to provide enhanced sensitivity and specificity for assays of many other compounds as well, and it seems likely that chemical ionization and selected ion monitoring are an optimal combination.

A standard curve is shown in Fig. 45, in which the ratio of peak areas (or heights in this case) of profiles of the sample ion and the internal standard ion can be seen to correspond linearly to the ratio of the amounts of the two compounds added to plasma. Precision through most of the range is better than $\pm 3\%$. The concentration as a function of time of phosphoramide mustard, cyclophosphamide, and nornitrogen mustard in the plasma of a patient receiving 75 mg/kg cyclophosphamide is shown in Fig. 46.

Variations of this assay were also used to study the decomposition of phosphoramide mustard and its protein binding. In one experiment 11% of

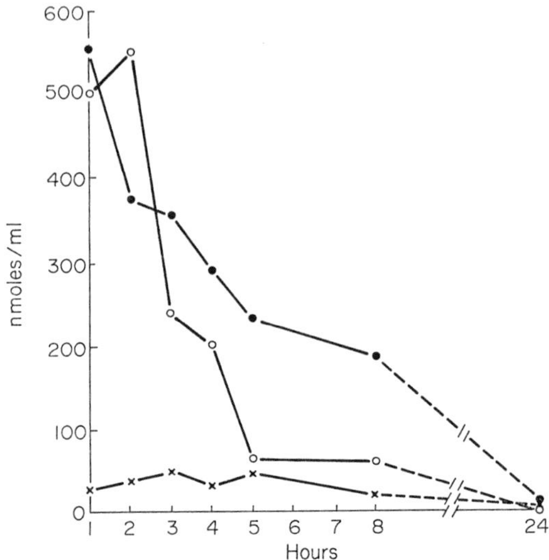

Fig. 46 Concentration as a function of time of cyclophosphamide (●), phosphoramide mustard (×), and nornitrogen mustard (○) in plasma, commencing at the end of a 1-hr infusion of 75 mg/kg cyclophospharmide. [Reproduced by permission from I. Jardine, C. Fenselau, M. Appler, M. N. Kan, R. B. Brundrett, and M. Colvin, *Cancer Res.* **38**, 408–415 (1978).]

a phosphoramide mustard sample was found to decompose spontaneously to *N*,*N*-bis(2-chloroethyl)amine under the conditions used for extraction and derivatization in clinical assays. The use of a stable isotope-labeled analog as an internal standard permits the most accurate and reliable correction for chemical decomposition (as well as extraction efficiency, etc.), because it will undergo decomposition at the same rate. The stable isotopes must be introduced in positions that will not introduce isotope effects into the decomposition reaction.

The drug metabolite was found to be 39% protein-bound in a plasma sample obtained from a patient 1 hr after administration of cyclophosphamide. Protein binding is commonly encountered in drug metabolism studies, and complicates "total drug" assays. A stable isotope-labeled internal standard may be rapidly equilibrated with sample that is reversibly bound to proteins, thus eliminating tedious denaturization procedures.

(ii) Direct probe introduction

Stable isotope-labeled internal standards are also very effective in assays of samples introduced by direct probe. Figure 47 shows ion current profiles of tyramine and d_2-tyramine as bisdansyl derivatives. The molecular ions of the two compounds, *m*/*e* 603.1861 and 605.1987, were recorded by selected

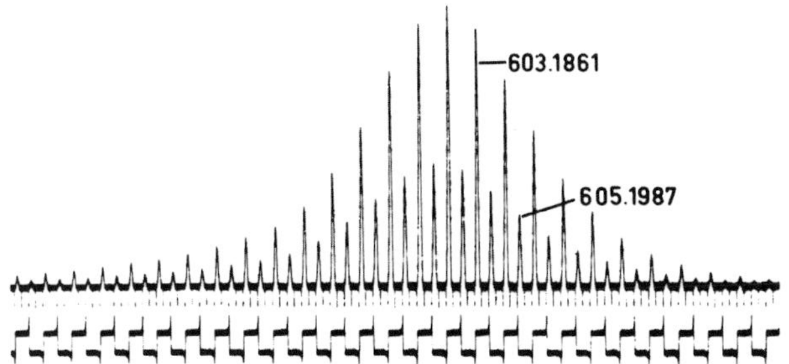

Fig. 47 High resolution selected ion profiles of the molecular ions of bisdansyl *p*-tyramine and bisdansyl d_2-*p*-tyramine extracted from rat kidney. [Reproduced by permission from S. R. Philips, D. A. Durden, and A. A. Boulton, *Can. J. Biochem.* **52**, 366–373 (1974).]

ion monitoring (peak matching circuitry) at 10,000 resolution in a variety of rat tissues (Philips *et al.*, 1974).

Stable isotope-labeled internal standards have also been used in assay methods that employ field desorption of samples introduced on the direct probe.

(iii) Isotope carrier technique

One unique use of stable isotope-labeled analogs is as reversed isotope carrier material. This technique has been applied especially in the analysis of low levels of prostaglandins, their metabolites, and analogs. For example, the structures of the triacetyl derivative of the methyl ester of prostaglandin $F_{2\alpha}$ and of its 3,3,4,4-d_4-analog are shown in scheme 19, along with the major

OAc COOCH$_3$ AcO OAc → COOCH$_3$ $]^{+\cdot}$

m/e 314
$M - (HOAc)_3$

OAc $(CD_2)_2CH_2COOCH_3$ AcO OAc → $(CD_2)_2CH_2COOCH_3$ $]^{+\cdot}$

m/e 318
$M - (HOAc)_3$

19

fragment ions that were monitored (Green *et al.*, 1973). The peaks at m/e 314 and 318 are particularly suitable for assay work because they are intense and fall at the high mass end of the spectrum. When a mixture was introduced by gas chromatography, containing 100 pg of derivatized prostaglandin $F_{2\alpha}$ and 100 ng of the d_4-carrier, the thousandfold excess of carrier material was reflected in the relative size of the ion current profiles. Nonetheless, correspondence of the peak ratio to the ratio of sample to carrier (internal standard) was reported to be linear. Table XI contains results of studies of the assay precision for prostaglandin $F_{2\alpha}$ and some related compounds using the carrier method. This kind of GCMS assay has been consistently used to evaluate radioimmune assays as these have been developed for various members of the prostaglandin family.

TABLE XI

GCMS Assay Precision Determined by Repetitive Analysis of Standard Mixtures[a]

Compound	Injected Deuterated Carrier (ng)	H-form (%)	Mean ± SD (10 Injections) (%)	Total H-form Injected (ng)
$PGF_{2\alpha}$	100	0.1	0.092 ± 25	0.1
	100	0.4	0.390 ± 3.7	0.4
15-Keto-dihydro-$PGF_{2\alpha}$	200	0.1	0.095 ± 18	0.2
	200	0.4	0.382 ± 5.6	0.8
Dihydro-$PGF_{2\alpha}$	500	0.1	0.095 ± 14.3	0.5
	500	0.4	0.392 ± 3.6	2.0
PGE_2	250	0.1	0.098 ± 16	0.25
	250	0.4	0.410 ± 4.8	1.0

[a] From Green *et al.* (1973).

Short range repetitive scanning has also been evaluated for the analysis of prostaglandins in conjunction with d_4-labeled carrier/internal standards. Major ions in the spectra of the tris(trimethylsilyl)methyl ester of prostaglandin $F_{2\alpha}$ and its d_4-analog, introduced together into the GCMS, were measured in a series of repetitive scans (Baczynskyj *et al.*, 1973). The areas of the two envelopes drawn by the repetitive scans were computer calculated and converted to a ratio. Ion envelopes were measured in 200-ng mixtures whose ratios ranged from 1000:1 to 1000:20 d_4:d_0-prostaglandin. Standard deviations between 2 and 10% were reported, based on five determinations at each sample level.

If an isotope-labeled sample is being studied, then of course the unlabeled analog may be used as carrier and internal standard. For example, analyses of deuterated hippuric acid formed by metabolism of labeled 1-ephedrin were made off the direct probe after unlabeled hippuric acid was added in hundredfold excess (Kawai and Baba, 1975).

4. *Sources of Error*

The topic of sources of error in mass spectrometric or GCMS assays is specialized somewhat beyond the level and length of this chapter. Nonetheless, some leading references will be presented in a brief overview.

In general, sources of error in quantitative assays may be assigned to the areas of sample preparation, instrumental measurement, and data analysis.

A key aspect of data analysis is the inverse linear regression by which observed signal size is related to sample size. Most workers argue that the standard curve for such a regression analysis should show a linear correlation between sample size and signal size. However, some approaches have been suggested for correcting or working with nonlinear standard curves (Chapman and Bailey, 1974; Jenden, 1975; Schoeller, 1976). Special statistical treatments have been developed for analyses in which stable isotope-labeled internal standards are used (Chapman and Bailey, 1974; Schoeller, 1976; Pickup and McPherson, 1976; Campbell, 1974). Obviously the considerations of precision and accuracy in isotope ratio measurements discussed in Section V apply here as well.

For the most part, the sources of errors due to instrumental measurements may be minimized by maintaining the instrument in good working order for any application, e.g. the GC column should not be dirty, the MS electron multiplier should not be fatigued (Millard, 1978; Watson, 1976; Falkner *et al.*, 1975). When repetitive scans are being averaged for quantitation, care should be taken to use optimal scan range, scan speed, and scan averaging time. For selected ion monitoring, dwell and cycle times are critical (Millard, 1978). One careful study concluded that the unidirectional sampling pattern (ABCABC . . .) is superior to the bidirection pattern (ABCCBA . . .) in selected ion monitoring (Hayes and Matthews, 1976).

Most workers agree that variation and errors in sample preparation are almost always much larger than variation in the mass spectral or GCMS measurements. This includes sample loss, sample decomposition (e.g., hydrolysis of derivatives), as well as errors in preparing standard solutions, variable GC injection, etc. This demonstrated variability (Millard, 1978; Claeys and Markey, 1977) is the most compelling argument for the use of internal standards. An interesting study by Claeys and Markey (1977) concludes that some homolog standards are better than others, and that stable isotope-labeled analogs can provide the best accuracy and precision for the overall process of sample preparation and instrumental measurement. Claeys and

Markey remind us that accuracy is largely a function of the correct use of the standard curve, while precision is influenced more by chemical and instrumental parameters.

References

Baczynskyj, L., Duchamp, D. J., Zieserl, J. F., and Axen, U. (1973). Computerized quantitation of drugs by gas chromatography–mass spectrometry, *Anal. Chem.* **45**, 479–487.

Bakke, J. E., Feil, V. J., Fjelstul, C. E., and Thacker, E. J. (1972). Metabolism of cyclophosphamide in sheep, *J. Agr. Food Chem.* **20**, 384–388.

Baughman, R., and Meselson, M. (1973a). An analytical method for detecting TCDD (dioxin): Levels of TCDD in samples from Vietnam, *Environ. Health Perspect.* Exp. Issue No. 5, pp. 27–34.

Baughman, R., and Meselson, M. (1973b). An improved analysis for tetrachlorodibenzo-*p*-dioxins, *in* "Chlorodioxins-Origin and Fate" (E. Blair, ed.), pp. 92–104. Amer. Chem. Soc., Washington, D.C.

Beckey, H. D. (1971). "Field Ionization Mass Spectrometry." Pergamon, Oxford.

Beckey, H. D., and Schulten, H. R. (1975). Field desorption mass spectrometry, *Angew. Chem. Int. Ed.* **14**, 403–415.

Beckinsale, R. D., Freeman, N. J., Jackson, M. C., Powell, R. E., and Young, W. A. T. (1973). A 30 cm radius 90° sector double collecting mass spectrometer with a capacitor integrating detector for high precision isotopic analysis of carbon dioxide, *Int. J. Mass Spectrom. Ion Phys.* **12**, 299–308.

Beynon, J. H. (1960). "Mass Spectrometry and Its Applications to Organic Chemistry." Elsevier, Amsterdam.

Biemann, K. (1962). "Mass Spectrometry. Organic Chemical Applications." McGraw-Hill, New York.

Biemann, K., Bommer, P., and Desiderio, D. M. (1964). Element-mapping, a new approach to the interpretation of high resolution mass spectra, *Tetrahedron Lett.* 1725–1731; (1964). Element maps depict complete mass spectra. *Chem. Eng. News* pp. 42–44. (July 6).

Biros, F. J. (1970). Enhancement of mass spectral data by means of a time averaging computer, *Anal. Chem.* **42**, 537–540.

Biller, J. E., and Biemann, K. (1974). Reconstructed mass spectra, a novel approach for the utilization of gas chromatograph–mass spectrometer data, *Anal. Lett.* **7**, 515–528.

Boulton, A. A., and Majer J. R. (1970). Mass spectrometry of crude biological extracts. Absolute quantitative detection of metabolites at the submicrogram level, *J. Chromatogr.* **48**, 322–327.

Brantigan, J. W., Gott, V. L., Vestal, M. L., Fergusson, G. J., and Johnston, W. H. (1970). A nonthrombogenic diffusion membrane for continuous in vivo measurement of blood gases by mass spectrometry, *J. Appl. Physiol.* **28**, 375–377.

Brooks, C. J. W., and Middleditch, B. S. (1971). The mass spectrometer as a gas chromatographic detector, *Clin. Chim. Acta* **34**, 145–157.

Budzikiewicz, H., Djerassi, C., and Williams, D. H. (1967). "Mass Spectrometry of Organic Compounds." Holden-Day, San Francisco, California.

Cala, P. C., Trenner, N. R., Buhs, R. P., Downing, G. V., Smith, J. L., and VandenHuevel, W. J. A. (1972). Gas chromatographic determination of pyrimethamine in tissue, *J. Agr. Food Chem.* **20**, 337–340.

Campbell, I. (1974). Review: Incorporation and dilution values—Their calculation in mass spectrally assayed stable isotope labeling experiments, *Bioorg. Chem.* **3**, 386–397.

Capriolli, R. M. (1972). Use of stable isotopes, *in* "Biochemical Application of Mass Spectrometry" (G. R. Waller, ed.). Wiley (Interscience), New York.

Carlson, D. A., Konyha, D. D., Wheeler, W. B., Marshall, G. P., and Zaylskie, R. G. (1976). Mirex in the environment: Its degradation to kepone and related compounds, *Science* **194**, 939–941.

Chapman, J. R., and Bailey, E. (1974). Determination of plasma testosterone by combined gas chromatography–mass spectrometry, *J. Chromatogr.* **89** 215–224.

Cheer, C. J., Smith, D. H., Djerassi, C., Tursch, B., Braekman, J. C., and Daloze, D. (1976). Applications of artifical intelligence for chemical inference. Chemical studies of marine invertebrates. The computer-assisted identification of (+)-palustrol in the marine organism *Cespitularia sp. aff. subviridis*, *Tetrahedron* **32**, 1807–1810.

Claeys, M., and Markey, S. P. (1977). Variance analysis of error in selected ion monitoring assays using various internal standards. A practical study case, *Biomed. Mass Spectrom.* **4**, 122–128.

Colvin, M., Padgett, C. A., and Fenselau, C. (1973). A biologically active metabolite of cyclophosphamide, *Cancer Res.* **33**, 915–918.

Colvin, M., Brundrett, R. B., Kan, M. N., Jardine, I., and Fenselau, C. (1976). Alkylating properties of phosphoramide mustard, *Cancer Res.* **36**, 1121–1126.

Cooks, R. G., Beynon, J. H., Caprioli, R. M., Lester, G. R., (1973). "Metastable Ions." Elsevier, Amsterdam.

Cornu, A., and Massot, R. (1975). "Compilation of Mass Spectral Data." Heyden, London.

Costello, C. E., Hertz, H. S., Sakai, T., and Biemann, K. (1974). Routine use of a flexible gas chromatograph–mass spectrometer computer system to identify drugs and their metabolites in body fluids of overdose victims, *Clin. Chem.* **20**, 255–265.

Cowan, G. A., Ott, D. G., Turkevich, A., Machta, L., Ferber, G. J., and Daly, N. R. Heavy methanes as atmospheric tracers, *Science* **191**, 1048–1050.

Cronholm, T., and Sjovall, J. (1970). Effect of ethanol metabolism on redox state of steroid sulfates in man, *Eur. J. Biochem.* **13**, 124–131.

Day, E. W., Vanatta, L. E., and Sieck, R. F. (1975). Drug residues in animal tissues. The confirmation of diethylstilbestrol residues in beef liver by gas chromatography–mass spectrometry, *J. AOAC* **58**, 520–524.

Deutsch, J., Razin, A., and Sedat, J. (1976). Analysis of 5-methylcytosine in DNA, *Anal. Biochem.* **72**, 586–592.

Djerassi, C., and Fenselau, C. (1965). Mass spectrometry in structural and stereochemical problems. The hydrogen-transfer reactions in butyl propionate, benzoate and phthalate, *J. Am. Chem. Soc.* **87**, 5756–5762.

Dougherty, R. C., Dalton, J., and Biros, F. J. (1972). Negative chemical ionization mass spectra of polycyclic chlorinated insecticides, *Org. Mass Spectrom.* **6**, 1171–1181.

Dromey, R. G., Stefik, M. J., Rindfleisch, T. C., and Duffield, A. M. (1976). Extraction of mass spectra free of background and neighboring component contributions from gas chromatography/mass spectrometry data, *Anal. Chem.* **48**, 1368–1375.

Due, S. L., Sullivan, H. R., and McMahon, R. E. (1976). Propoxyphene: Pathways of metabolism is man and laboratory animals, *Biomed. Mass Spectrom.* **3**, 217–225.

Duncan, J. H., Colvin, O. M., and Fenselau C. (1973). Mass spectrometric study of the distribution of cyclophosphamide in humans, *Toxicol. Appl. Pharmacol.* **24**, 317–323.

Dutcher, J. S., Strong, J. M., Lee, W., and Atkinson, A. S. (1975). Stable isotope methods for pharmacokinetic studies in man, *Proc. Int. Conf. Stable Isotopes, 2nd* (E. R. Klein and P. D. Klein, eds.). U.S. Energy Res. Develop. Administration.

"Eight Peak Index of Mass Spectra" (1975). Her Majesty's Stationery Office, London.

Ellgehausen, D. (1975). Determination of volatile toxic substances in the air by means of a coupled gas chromatograph–mass spectrometer system, *Anal. Lett.* **8**, 11–23.

Evans, K. P., Mathias, A., Mellor, N., Silvester, K., and Williams, A. E. (1975). Detection and estimation of bis(cloromethyl)ether in air by gas chromatography–high resolution mass spectrometry, *Anal. Chem.* **47**, 821–824.

Falkner, F. C. (1977). Historical account of ion monitoring, *Biomed. Mass Spectrom.* **4**, 66–67.

Falkner, F. C., Sweetman, B. J., and Watson, J. T. (1975). Biomedical applications of selected ion monitoring, *Appl. Spectrosc. Rev.* **10**, 51–116.

Ferhi, A. M., Létolle, R. R., and Lerman, J. C. (1975). Oxygen isotope ratios of organic matter: Analysis of natural compositions, *Proc. Int. Conf. Stable Isotopes, 2nd* (E. R. Klein and P. D. Klein, eds.). U.S. Energy Res. and Develop. Administration.

Fenselau, C. (1972). Applications of mass spectrometry, *in* "Methods in Pharmacology" (C. F. Chignell, ed.), Vol. 2, pp. 401–442. Appleton, New York.

Fenselau, C. (1977). The mass spectrometer as a gas chromatograph detector, *Anal. Chem.* **49**, 563A–570A.

Fenselau C., Kan, M. N., Billets, S., and Colvin, M. (1975). Identification of phosphorodiamidic acid mustard as a human metabolite of cyclophosphamide, *Cancer Res.* **35**, 1453–1457.

Gallegos, E. J. (1976). Analysis of organic mixtures using metastable transition spectra, *Anal. Chem.* **48**, 1348–1351.

Green, K., Granstrom, E., Samuelsson, B., and Axen U. (1973). Methods for quantitative analysis of $PGF_{2\alpha}$, PGE_2, 9α,11α-dihydroxy-15-keto-prost-5-enoic acid and 9α, 11α,15-trihydroxy-prost-5-enoic acid from body fluids using deuterated carriers and gas chromatography–mass spectrometry, *Anal. Biochem.* **54**, 434–453.

Haddon, W. F., Lukens, H. C., Elsken, R. H. (1973). Signal averaging method for high resolution mass spectral measurement of nitrogen-15, *Anal. Chem.* **45**, 682–686.

Halliday, D., and McKeran, R. O. (1975). Measurement of muscle protein synthetic rate from serial muscle biopsies and total body protein turnover in man by continuous intravenous infusion of L-[α-^{15}N]Lysine, *Clin. Sci. Mol. Med.* **49** 581–590.

Hardcastle, K. G., and Friedman, I. (1974). A method for oxygen isotope analyses of organic material, *Geophys. Res. Lett.* **1**, 165–167.

Hayes, J. M., and Matthews, D. E. (1976). Systematic errors in gas chromatography–mass spectrometry isotope ratio measurements, *Anal. Chem.* **48**, 1375–1382.

Hillman, L. S., Goodwin, S. L., and Sherman, W. R. (1975). Identification and measurement of plasticizer in neonatal tissues after umbilical catheters and blood products, *New England J. Med.* **292**, 381–386.

Hintenberger, H. (1966). Precision determinations of isotope abundances, *Adv. Mass Spectrom.* **3**, 517–546, and references therein.

Holland, J. F., Soltmann, B., and Sweeley, C. C. (1976). A model for ionization mechanisms in field desorption mass spectrometry, *Biomed. Mass Spectrom.* **3**, 340–345.

Holmes, W. F., Holland, W. H., Shore, B. L., Bier, D. M., and Sherman, W. R. (1973). Versatile computer generated variable accelerating voltage circuit for magnetically scanned mass spectrometers. Use for assays in the picogram range and for assays of stable isotope tracers, *Anal. Chem.* **45**, 2063–2071.

Holmstedt, B., and Palmer, L. Mass fragmentaography: principles, advantages and future possibilities, *Adv. Biochem. Psychopharmacol.* **7**, 1–14.

Horning, E. C., Carroll, D. I., Dzidic, I., Haegele, K. D., Lin, S., Dortil, C. U., and Stillwell, R. N. (1977). Development and use of analytical systems based on mass spectrometry, *Clin. Chem.* **23**, 13–21.

Hsia, J. C., Tam, J. C. L., Giles, H. G., Leung, C. C., Marcus, H., Marshman, J. A., and LeBlanc, A.E. (1976). Markers for detection of supplementation in narcotic programs. Deuterium labeled methadone, *Science* **193**, 498–500.

Hubbard, W. C., and Watson, J. T. (1976). Determination of 15-keto-13,14-dihydrometabolites of PGE_2 and $PGF_{2\alpha}$ in plasma using high performance liquid chromatography and gas chromatography–mass spectrometry, *Prostaglandins* **12**, 21–35.

Hughes, J., Smith, T. W., Kosterlitz, H. W., Fothergill, L. A., Morgan, B. A., and Morris, H. R. (1975). Identification of two related pentapeptides from the brain with potent opiate agonist activity, *Nature* (*London*) **258**, 577–579.

Jaeger, R. J., and Rubin, R. J. (1970). Plasticizers from plastic devices. Extraction, metabolism and accumulation by biological systems, *Science* **170**, 460–462.

Jardine, I., Kan, M. N., Fenselau, C., Brundrett, R., Colvin, M., Wood, G., Lau, P., and Charlton, R. (1975). Pharmacological studies of chemotherapeutic alkylating agents using chemical ionization and field desorption mass spectrometry, *Proc. Int. Conf. Stable Isotopes, 2nd* (E. R. Klein and P. D. Klein, eds). U.S. Energy Res. Develop. Administration.

Jardine, I., Brundrett, R., Colvin, M., and Fenselau, C. (1976). Approaches to the pharmacokinetics of cyclophosphamide (NSC-26271): Quantitation of metabolites, *Can. Treatment Rep.* **60**, 403–408.

Jardine, I., Fenselau, C., Appler, M., Kan, M. N., Brundrett, R. B., and Colvin, M. (1978). The simultaneous quantitation by gas chromatography chemical ionization mass spectrometry of cyclophosphamide, phosphoramide mustard and nornitrogen mustard in the plasma and urine of patients receiving cyclophosphamide therapy, *Cancer Res.* **38**, 408 415.

Jenden, D. J. (1975). Procedural and statistical considerations in the analysis of multiple ion current measurements to estimate multiple isotopic variants, *In Proc. Int. Conf. Stable Isotopes, 2n*d (E. R. Klein and P. D. Klein, ed.), U. S. Energy Res. Develop. Administration.

Kawai, K., and Baba, S. (1975). Studies on drug metabolism by use of isotopes. Mass spectrometric quantitation of urinary metabolites of deuterated 1-ephedrine in rabbits, *Chem. Pharm. Bull.* **23**, 289–293.

Kiser, R. W. (1965). "Introduction to Mass Spectrometry and Its Applications." Prentice-Hall, Englewood Cliffs, New Jersey.

Klein, P. D., Haumann, J. R., and Eisler, W. J. (1972). Gas chromatograph–mass spectrometer–accelerating voltage alternator system for the measurement of stable isotope ratios in organic molecules, *Anal. Chem.* **44**, 490–493.

Klein, P. D., Haumann, J. R., and Hachey, D. L. (1975). Stable isotope ratiometer–multiple ion detector unit for quantitative and qualitative stable isotope studies by gas chromatography–mass spectrometry, *Clin. Chem.* **21**, 1253–1257.

Kruger, T. L., Litton, J. F., Kondrat, R. W., and Cooks, R. G. (1976). Mixture-analysis by mass-analyzed ion kinetic energy spectrometry, *Anal. Chem.* **48**, 2113–2119.

LaCroix, M., Mosora, F., Pontus, M., Lefebvre, P., Luyckx, A., and Lopey-Habib, G. (1973). glucose naturally labeled with Carbon-13: Use for metabolic studies in man, *Science* **181**, 445–446.

Law, N. C., Aandahl, V., Fales, H. M., and Milne, G. W. A. (1971). Identification of dangerous drugs by mass spectrometry, *Clin. Chim. Acta* **32**, 221–228.

Levsen, K., Wipf, H.-K., and McLafferty, F. W. (1974). Mass spectrometric studies of peptides: applications of metastable ion and collisional activation spectra, *Org. Mass Spectrom.* **8**, 117–128.

Lin, D. C. K., Fentiman, A. F., Foltz, R. L., Forney, R. D., and Sunshine, I. (1975). Quantification of phencyclidine in body fluids by gas chromatography chemical ionization mass spectrometry and identification of two metabolites, *Biomed. Mass Spectrom.* **2**, 206–214.

MacFarlane, R. D., and Torgerson, D. F. (1976). ^{252}Cf-plasma desorption time-of-flight mass spectrometry, *Int. J. Mass. Spectrom. Ion. Phys.* **21**, 81–92.

Maine, J. W., Soltmann, B., Holland, J. F., Young, N. D., Gerber, J. N., and Sweeley, C. C. (1976). Emitter current programmer for field desorption mass spectrometry, *Anal. Chem.* **48**, 427–429.

Masuda, Y., and Hoffman, D. (1969). Quantitative determination of 1-naphthylamine and 2-naphthylamine in cigarette smoke, *Anal. Chem.* **41**, 650–652.

McDowell, C. A. (ed.) (1963). "Mass Spectrometry." McGraw-Hill, New York.

McLafferty, F. W. (1973). "Interpretation of Mass Spectra." Benjamin, New York.

Millard, B. J. (1978). "Quantitative Mass Spectrometry." Heyden, London.

Millington, D. S., Buoy, M. E., Brooks, G., Harper, M. E., and Griffiths, K. (1975). Thin-layer chromatography and high resolution selected ion monitoring for the analysis of C_{19}-steroids in human hyperplastic prostate tissue, *Biomed. Mass Spectrom.* **2**, 219–224.

Munson, B. (1971). Chemical ionization mass spectrometry, *Anal. Chem.* **43**, 28A–43A.

Muxfeldt, H., Shrader, S., Hansen, P., and Brockmann, H. (1968). The structure of pikromycin, *J. Am. Chem. Soc.* **90**, 4748–4749.

Nier, A. D. (1950). A redetermination of the relative abundances of the isotopes of carbon, nitrogen, oxygen, argon and potassium, *Phys. Rev.* **77**, 789–793.

Palmer, L., Bertilsson, L., Collste, P., and Rawlins, M. (1973). Quantitative determination of carbamazepine in plasma by mass fragmentography, *Clin. Pharmacol. Therap.* **14**, 827–832.

Philips, S. R., Durden, D. A., and Boulton, A. A. (1974). Identification and distribution of *p*-tyramine in the rat, *Can. J. Biochem.* **52**, 366–373.

Pickup, J. F., and McPherson, K. (1976). Theoretical considerations in stable isotope dilution mass spectrometry for organic analysis, *Anal. Chem.* **48**, 1885–1890.

Rickards, R. W., Smith, R. M., and Majer, J. (1968). The structure of the macrolide antibiotic picromycin, *Chem. Commun.* 1049–1050.

Roboz, J. (1975). Mass spectrometry in clinical chemistry, *Adv. Clin. Chem.* **17**, 109–191.

Ryan, J. F., Biros, F. J., and Harless, R. L. (1974). Analysis of environmental samples for TCDD residues, *Proc. Annu. Conf. Mass Spectrom. and Allied Topics, 22nd Philadelphia, Pennsylvania*, May 19–24.

Saferstein, R., Chao, J., and Manura J. (1974). Identification of drugs by chemical ionization mass spectroscopy, *J. Forensic Sci.* **19**, 463–485.

Saunders, R. A., Griffith, J. R., and Saalfeld, F. E. (1974). Identification of some organic smog components based on rain water analysis, *Biomed. Mass Spectrom.* **1**, 192–194.

Scalan, R. S., and Morgan, T. D. (1970). Isotope ratio mass spectrometer instrumentation and application to organic matter contained in recent sediments, *Int. J. Mass Spectrom. Ion. Phys.* **4**, 267–281.

Schoeller, D. A. (1976). A review of the statistical considerations involved in the treatment of isotope dilution calibration data, *Biomed. Mass Spectrom.* **3**, 265–271.

Schoeller, D. A., and Hayes, J. M. (1975). Computer controlled ion counting isotope ratio mass spectrometer, *Anal. Chem.* **47**, 408–415.

Schoenheimer, R., Rittenberg, D., Foster, G. L., Keston, A. S., and Ratner, S. (1938). The application of the nitrogen isotope ^{15}N for the study of protein metabolism, *Science* **88**, 599–600.

Schuetzle, D. (1975). Analysis of complex mixtures by computer controlled high resolution mass spectrometry. Application to atmospheric aerosol composition, *Biomed. Mass Spectrom.* **2**, 288–298.

Smith, B. N., and Epstein, S. (1971). Two categories of $^{13}C/^{12}C$ ratios for higher plants, *Plant Physiol.* **47**, 380–384.

Smith, R. G., Daterman, G. E., and Daves, G. D. (1975). Douglas-fir tussock moth: Sex pheromone identification and synthesis, *Science* **188**, 63–64.

Spiteller, G. (1966). "Massenspektrometrische Strukturanalyse Organischer Verbindungen." Verlag Chemie, GmbH, Weinheim.

Stenhagen, E., Abrahamsson, S., and McLafferty, F. W. (1974). Registry of Mass Spectral Data." Wiley, New York.

Stephany, R. W., Freudenthal, J., Egmond, E., Gramberg, L. G., and Schuller, P. L. (1976). Mass spectrometric quantification of traces of volatile *N*-nitrosamines in meat products, *J. Agr. Food. Chem.* **24**, 536–539.

Stepita-Klauco, M., Dolezalova, H., and Fairweather, R. (1974). Piperdine increase in the brain of dormant mice, *Science* **183**, 536–537.

Struck, R. F., Kirk, M. C., Mellett, L. M., El Dareer, S., and Hill, D. L. (1971). Urinary metabolites of the antitumor agent cyclophosphamide, *Mol. Pharmacol.* **7**, 519–529.

Sweeley, C. C., Elliott, W. H., Fries, J., and Ryhage, R. (1966). Mass spectrometric determination of unresolved components in gas chromatographic effluents, *Anal. Chem.* **38**, 1549–1553.

Thompson, J. J. (1913). "Rays of Positive Electricity and Their Application to Chemical Analysis." Longmans Green, London.

Tyson, B. J. (1975). Chlorinated hydrocarbons in the atmosphere—analysis at the parts-per-trillion level by GC–MS, *Anal. Lett.* **8**, 807–813.

Von Unruh, G. E., Hauber, D. J., Schoeller, D. A., Hayes, J. M. (1974). Detection limits of carbon-13 labeled drugs and metabolites, *Biomed. Mass Spectrom.* **1**, 345–349.

Walker, R. W., VandenHueval, W. J. A., Wolf, F. J., Noll, R. M., and Duggan, D. E. (1977). A stable isotope gas–liquid chromatographic–mass spectrometric assay for determining uric acid body pool size, *Anal. Biochem.* **77**, 235–242.

Watson, J. T. (1976). "Introduction to Mass Spectrometry: Biomedical, Environmental and Forensic Applications." Raven Press, New York.

Watson, J. T., Pelster, D. R., Sweetman, B. J., Frolich, J. C., and Oates, J. A. (1973). Display-oriented data system for multiple ion detection with gas chromatography–mass spectrometry in quantifying biomedically important compounds, *Anal. Chem.* **45**, 2071–2078.

Weinkam, R. J., Gal, J., Callery, P., and Castagnoli, N. (1976). Application of chemical ionization mass spectrometry to the study of stereoselective in vitro metabolism of 1-(2,5-dimethoxy-4-methylphenyl)-2-aminopropane, *Anal. Chem.* **48**, 203–209.

Weinkam, R. J., Rowland, M., and Meffin, P. J. (1977). Determination of phenylbutazone, tolbutamide and metabolites in plasma and urine using chemical ionization mass spectrometry, *Biomed. Mass Spectrom.* **4**, 42–47.

Williams D. H., and Howe, I. (1972). "Principles of Organic Mass Spectrometry." McGraw-Hill, New York.

Zweig, J. S., and Castagnoli, N. (1977). On the in vitro O-demethylation of the psychotomimetic amine 1-(2,5-dimethoxy-4-methylphenyl)-2-aminopropane, *J. Med. Chem.* **20**, 414–421.

Atomic Fluorescence and Atomic Absorption Spectroscopy

Thomas J. Vickers

Department of Chemistry
Florida State University
Tallahassee, Florida

ISBN 0-12-430801-5

I. Introduction

Atomic spectra are derived from changes in the electronic configurations of the outer electrons of atoms. Since there are no vibrational levels associated with atoms, these electronic transitions give rise to sharp "line" spectra rather than the "band" spectra found with molecules. The energy changes are such that for most elements the principal lines fall in the near ultraviolet or visible portion of the electromagnetic spectrum. For analytical spectroscopy the lines of greatest interest are the resonance lines of the elements, that is, those lines which arise from transitions in which one of the energy states is the ground state of the atom.

Three types of experiments can be distinguished in atomic spectroscopy—emission, absorption, and fluorescence. Analytical techniques based on the first of these have been discussed elsewhere in this volume by Keliher. The present chapter is devoted primarily to analytical atomic absorption and atomic fluorescence spectroscopy, but must necessarily consider some aspects of emission spectroscopy which bear on absorption and fluorescence measurements. The three kinds of experiments are shown in schematic fashion in Fig. 1.

In the emission experiment, excited atoms are produced principally by collisions of the ground state atoms with energetic species—electrons, other atoms, or molecules—in a high temperature gas, such as that produced by a flame, arc, or spark. Excitation is nonselective. If the system approaches equilibrium, the relative populations of the various energy state are described by a single temperature through the Boltzmann equation

$$n_u/n = (g_u/Z)\exp(-E_{exc}/kT) \tag{1}$$

and the relative intensities of the spectral lines are described by an equation of the form

$$I_e = h\nu l A_{ul} n(g_u/Z)\exp(-E_{exc}/kT) \tag{2}$$

where I_e is the intensity of emission, h Planck's constant, ν the frequency of the emitted radiation, l the thickness of the emitting volume in the direction

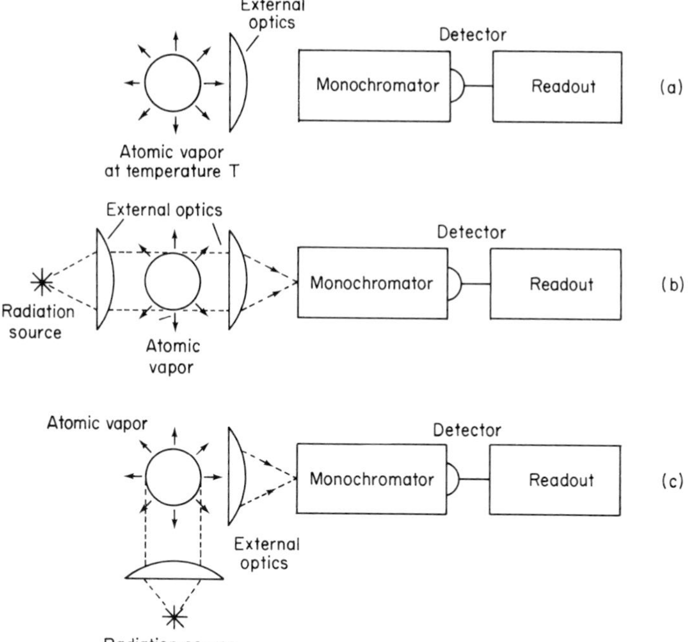

Fig. 1 (a) Emission, (b) absorption, and (c) fluorescence experiments.

of observation, A_{ul} the Einstein transition probability for spontaneous emission for a transition between an upper state u and lower state l (Fig. 2), n the total number of atoms per unit volume, g_u the statistical weight of the upper state, Z the partition function for the atom, E_{exc} the excitation energy required for the upper state, k the Boltzmann constant, and T the absolute temperature.

In the absorption experiment, radiation from a high temperature excitation source (usually an electrical discharge) is passed through the atomic vapor, and the attenuation of the source intensity due to absorption of photons by the atoms is measured. For absorption to occur the photon energy must match the energy difference between the lower energy state in which an atom finds itself and a higher energy state. Since the ground state

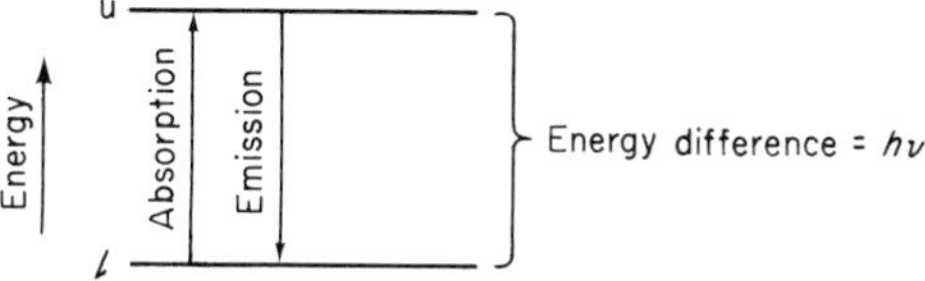

Fig. 2 Transitions between two energy levels, u and l. The transitions are accompanied by absorption or emission of photons of energy $h\nu$.

is almost invariably the state of highest population, resonance absorption is the most important process. Some fraction of the absorbed radiation will be spontaneously reemitted at the same wavelength when the atom returns to the lower energy state, but since this radiation is emitted isotropically, the fraction collected by the optical system is so small as to be insignificant in the measurement. Thermal emission, that is, the spontaneous emission resulting from the collisional excitation already described, poses a potentially more serious problem. For most elements, significant collisional excitation is an inescapable consequence of producing the atomic gas. This difficulty can be overcome by modulating the radiation from the external source. Standard signal handling techniques can then be used in the readout electronics to distinguish between the ac signal due to the source radiation and the approximately dc signal due to the thermal radiation of the atomic vapor.

In the fluorescence experiment, radiation from a high temperature external source is passed through the atomic vapor as in the absorption experiment, but the detector is so arranged that it cannot receive the primary source radiation. Some fraction of the atoms excited by absorption ultimately return to the ground state by spontaneous emission of photons. This reemitted radiation is the fluorescence signal. Thus atomic fluorescence differs from atomic emission only in the mode of excitation. Fluorescence can be distinguished from the thermal emission of the atomic vapor by modulating the source radiation as already described for the absorption experiment.

II. Dependence of Absorption on Atomic Concentration

A. Some Quantities Relating to Absorption

In deriving the quantitative relationship between absorption and atomic concentration, it is convenient to consider a uniform volume of material containing a concentration of absorbing atoms n. The spectral radiant energy density, which is given the symbol u_ν, is the radiant energy per unit of frequency in a unit volume of the sample, and the spectral radiant energy density in a small frequency interval $d\nu$ is given by $u_\nu \, d\nu$. If a resonance line of the atoms lies in this small frequency interval, then radiant energy is absorbed, causing the transition of some atoms from the ground to an excited state (Fig. 2). The total energy absorbed per second in a unit volume is evidently proportional to the concentration of atoms in the lower state, n_l, to the spectral radiant energy density, to the energy of the photons absorbed, $h\nu$, and to the probability of absorption, B_{lu}:

$$\phi_a = n_l h\nu B_{lu} u_\nu \, d\nu \tag{3}$$

B_{lu} is the Einstein transition probability for absorption for a transition from a lower state l to an upper state u. The product $B_{lu}u_\nu\,d\nu$ has units of $\sec^{-1}$.

The lower state atomic concentration is related to the total atomic concentration by the Boltzmann equation (1). If, as is the usual case, the lower state is the ground state, then the concentration is given by

$$n_0 = n(g_0/Z) \tag{4}$$

The partition function is

$$Z = g_0 + g_1 \exp(-E_1/kT) + g_2 \exp(-E_2/kT) + \cdots \tag{5}$$

where the subscripts 1, 2, ... indicate various excited states of the atoms. For the atoms of many elements the excitation energies of the various excited states are sufficiently large that to a good approximation $Z = g_0$ and $n_0 = n$.

In analytical atomic spectroscopy the absorption oscillator strength f_{lu}, a unitless quantity, is more often used than B_{lu} to describe the strength of a transition. The two quantities are related by the expression

$$f_{lu} = (mh\nu/\pi e^2)B_{lu} \tag{6}$$

where m is the mass and e the charge of an electron. The total energy absorbed per second in a unit volume is thus given by

$$\phi_a = n_l(\pi e^2/m)f_{lu}u_\nu\,d\nu \tag{7}$$

Transition probabilities and oscillator strengths for emission are found in the tabulation by Corliss and Bozman (1962). The appropriate quantities for absorption can be obtained from the relationships

$$g_u f_{ul} = g_l f_{lu} \tag{8}$$

$$g_u A_{ul} = (8\pi h\nu^3/c^3)g_l B_{lu} \tag{9}$$

where g_u and g_l are the statistical weights of the upper and lower states and f_{ul} is the emission oscillator strength.

B. Atomic Absorption Coefficient

Consider an experiment in which a parallel light beam with spectral radiant density $u_\nu\,d\nu$ is falling on a layer of atomic gas of unit area and of thickness dl. The total number of lower state atoms in this volume is thus $n_l\,dl$, and, from Eq. (3), the total energy absorbed in the given volume per unit time is

$$\phi_a = n_l\,dl\,h\nu B_{lu}\,d\nu \tag{10}$$

An equation for ϕ_a can also be obtained by a different approach. The radiant energy falling on the layer per unit time is given by $cu_\nu\,d\nu$, where c

is the velocity of light. The number of photons arriving per unit time is thus $cu_\nu \, d\nu/h\nu$. If the effective cross section for absorption for one atom is defined as κ_{lu}, then in the considered unit area the total absorbing area will be $\kappa_{lu} n_l \, dl$, and the number of photons absorbed will be $\kappa_{lu} n_l \, dl \, cu_\nu \, d\nu/h\nu$. The total energy absorbed is the product of the number of photons and their energy:

$$\phi_a = \kappa_{lu} n_l \, dl \, cu_\nu \, d\nu \tag{11}$$

An expression for the effective cross section can be obtained by equating expressions (10) and (11):

$$\kappa_{lu} = (h\nu/c) B_{lu} \tag{12}$$

and from Eq. (6),

$$\kappa_{lu} = (\pi e^2/mc) f_{lu} \tag{13}$$

The atomic absorption coefficient k_{lu} is related to the absorption cross section by the expression

$$k_{lu} = n_l \kappa_{lu} \tag{14}$$

The cross section κ_{lu} has units of area; k_{lu} has units of reciprocal length.

This derivation of the atomic absorption coefficient ignores any effect due to stimulated emission. Stimulated emission is the reverse of the absorption process. It depends on the radiant energy density and the concentration of excited state atoms. Photons produced by stimulated emission travel in the same direction as the stimulating radiation and will therefore reduce the apparent absorption. Taking into account stimulated emission, the absorption cross section is given by

$$\kappa_{lu} = \frac{h\nu}{c}\left(1 - \frac{n_u g_l}{n_l g_u}\right) B_{lu} \tag{15}$$

For analytical atomic spectroscopy the lower state of the transition is usually the ground state ($n_l = n_0$), and the ratio n_u/n_l is so small that the second term within the parentheses of Eq. (15) is negligibly small compared to 1 so that the expression reduces to that of Eq. (12).

C. *Line Profiles*

The spectral line arising from atomic transitions between u and l, as shown in Fig. 2, is not infinitely narrow but covers a definite although relatively narrow frequency interval. The atomic absorption coefficient already derived refers to the total number of photons absorbed of all frequencies within the spectral line and is thus an integrated atomic absorption coefficient. We can also imagine a monochromatic atomic absorption coefficient which is the absorption coefficient in a unit frequency interval. This absorption coefficient is called the spectral atomic absorption coefficient and is given the symbol k_ν. If k_ν is plotted against frequency, we obtain the absorption line profile, as indicated in Fig. 3. The integrated atomic absorption coefficient, henceforth called k_L, is related to the spectral absorption coefficient by

$$k_L = k_{lu} = \int_0^\infty k_\nu \, d\nu = (\pi e^2/mc) n_l f_{lu} \tag{16}$$

The integration is intended to be over the frequency interval of the spectral line; extending the boundaries from zero to infinity assumes the absence of other spectral lines. An important term in describing the profile of a line is the line width, which is usually taken to mean the full width at half height (illustrated in Fig. 3).

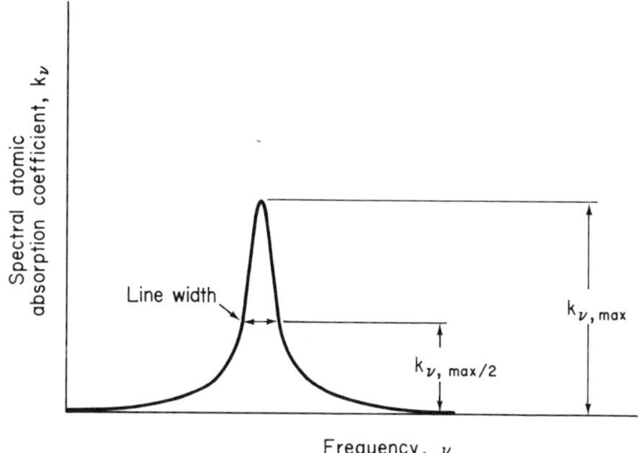

Fig. 3 Profile of an absorption line.

Several factors may contribute to the width of a spectral line. The most fundamental limitation on the narrowness of a spectral line is set by Heisenberg uncertainty. The uncertainty principle requires an uncertainty ΔE in measurement of the energy of an atomic state (u or l in Fig. 2) given approximately by

$$\Delta E = h/2\pi\tau \tag{17}$$

where τ is the lifetime of the state. The resulting spread of frequencies in the spectral line is said to be due to natural broadening, and the line width arising from this cause is called the natural line width. In analytical spectroscopy the contribution of natural broadening to the line width is negligibly small. The most important contributors to the width of the spectral lines are collisional broadening and Doppler broadening.

Collisional broadening occurs when the absorbing or emitting atoms undergo collisions during the natural lifetime of the state. The uncertainty in energy of a state from this cause is given approximately by

$$\Delta E = h/2\pi\tau_c \tag{18}$$

where τ_c is the mean time between successive collisions.

Doppler broadening arises from the random thermal motion of the absorbing or emitting atoms with respect to the observer. The Doppler effect is that phenomenon in which the apparent frequency of radiation is increased if the source is moving toward the observer and decreased if the source is moving away from the observer. Atoms in random motion have radial velocities relative to the observer of all possible values between the limits set by those moving directly toward and those directly away from the observer. The line width of a purely Doppler broadened line, if a Maxwellian distribution of velocities is assumed, is given by

$$\Delta\nu_{D,} = 2\nu_0(2kT \ln 2/m_a c^2)^{1/2} \tag{19}$$

where ν_0 is the frequency at the line center and m_a is the mass of the absorbing or emitting atom.

Doppler broadening produces a line profile described by a Gaussian function. Collisional broadening produces a line profile described by a Lorentzian function. The Lorentzian function falls off much less steeply

TABLE I

Calculated Line Widths for Several Elements in an Air–Acetylene Flame[a]

Line (Å)	Doppler Width (Å)	Collision[b] Width (Å)	Total[b] Width (Å)
Li 6707.80	0.090	0.069–0.165	0.131–0.205
Na 5889.95	0.044	0.035–0.084	0.065–0.103
Cs 8521.10	0.026	0.054–0.131	0.065–0.136
Mg 2852.13	0.021	0.008–0.019	0.025–0.032
Al 3092.71	0.021	0.009–0.022	0.026–0.035
Zn 2138.56	0.009	0.004–0.009	0.011–0.015
Hg 2536.52	0.006	0.005–0.011	0.009–0.014

[a] Parsons *et al.* (1975).

[b] A range of values is given because of uncertainty in the measured collision cross-section values used in calculating the collisional line width.

than the Gaussian. In many cases of interest in analytical atomic spectroscopy, the contributions of Doppler and collision broadening to the line profile are comparable in size, and the line profile is neither Gaussian nor Lorentzian but must be described by a more complex function known as a Voigt profile (Poesner, 1959).

Table I lists the calculated line widths for several elements. Calculated values for several flame conditions are available for most elements (Parsons *et al.*, 1966, 1975).

D. *Beer's Law*

The Beer's law relationship for atomic absorption spectroscopy assumes the form

$$\phi_{\nu,t} = \phi_{\nu,0} \exp(-k_\nu l) \tag{20}$$

where $\phi_{\nu,0}$ and $\phi_{\nu,t}$ are the spectral radiant fluxes before and after passing through a uniform absorbing layer of thickness l, and k_ν is the spectral atomic absorption coefficient. Recall from Eq. (14) that the atomic concentration n is contained within k_ν.

The monochromatic transmittance is given by

$$\tau(\nu) = \phi_{\nu,t}/\phi_{\nu,0} \tag{21}$$

and the absorbance is

$$A(\nu) = -\log \tau(\nu) = 2.303 k_\nu l \tag{22}$$

A third term frequently used in analytical atomic absorption spectroscopy is the absorption factor, which is the fraction of radiation absorbed:

$$\alpha(\nu) = \frac{\phi_{\nu,0} - \phi_{\nu,t}}{\phi_{\nu,0}} = \frac{\phi_{\nu,a}}{\phi_{\nu,0}} = 1 - \tau(\nu) \tag{23}$$

In practice we do not measure the monochromatic absorption but rather an absorption integrated over some finite band of frequencies. The integrated absorption factor is

$$\alpha = \int_0^\infty \phi_{\nu,a}\, d\nu \bigg/ \int_0^\infty \phi_{\nu,0}\, d\nu = \phi_a \bigg/ \int_0^\infty \phi_{\nu,0}\, d\nu \tag{24}$$

where

$$\phi_a = \int_0^\infty \phi_{\nu,0}[1 - \exp(-k_\nu l)]\, d\nu \tag{25}$$

If a continuum excitation source is used, it can be assumed that $\phi_{\nu,0}$ is constant over the range of frequencies passed by the monochromator, and thus

$$\phi_a = \phi_c \int_0^\infty [1 - \exp(-k_\nu l)]\, d\nu \tag{26}$$

The only case that will interest us is the one for which $k_\nu l$ is sufficiently small that we can make use of the approximation in which we represent the exponential term as a series and ignore all terms in which $k_\nu l$ is raised to the second or higher power. Thus

$$1 - \exp(-k_\nu l) = 1 - [1 - k_\nu l + (k_\nu l)^2/2! - (k_\nu l)^3/3! + \cdots] = k_\nu l \tag{27}$$

and

$$\phi_a = \phi_c \int_0^\infty k_\nu l\, d\nu \tag{28}$$

In the usual case the monochromator bandpass is several times the width of the absorption line, and hence the integral is just the integrated absorption coefficient of Eq. (16):

$$\phi_a = \phi_c(\pi e^2/mc)\, n_l f_{lu} l \tag{29}$$

The denominator of Eq. (24) is easily evaluated for a continuum source:

$$\int_0^\infty \phi_{\nu,0}\, d\nu = \phi_c s \tag{30}$$

where s is the half-intensity spectral bandwidth of the monochromator. Thus the absorption factor is given by

$$\alpha = (\pi e^2/mcs) n_l f_{lu} l \tag{31}$$

When a line excitation source is used, a more complex expression is required to describe the relationship between absorption and concentration. If we assume the source emission line width is small compared to the absorption line width, that is, k_ν does not vary appreciably over the range of frequencies covered by the emission line, then Eq. (25) can be written as

$$\phi_a = k_{\nu_0} l \int_0^\infty \phi_{\nu,0}\, d\nu = \phi_L k_{\nu_0} l \tag{32}$$

where the approximation of Eq. (27) has been used, k_{ν_0} is the spectral absorption coefficient at the line center, and ϕ_L is the integrated intensity of the source line.

It is obvious from Eq. (16) that the integrated absorption coefficient does not depend on the factors which affect the width of the line. It is, therefore, inescapable that the peak absorption coefficient k_{ν_0} depends on the width of the line: as $\Delta\nu$ increases, k_{ν_0} decreases. For a purely Doppler broadened line (Gaussian distribution), the peak height is related to the area by the expression

$$k_{\nu_0} = [2(\ln 2)^{1/2}/\pi^{1/2}\, \Delta\nu_D] k_L \tag{33}$$

For a line source the denominator in Eq. (24) is simply ϕ_L, and the final equation for the absorption factor with a narrow line source and a purely Doppler broadened absorption line is

$$\alpha = [2(\ln 2)^{1/2} e^2 \pi^{1/2}/mc\,\Delta\nu_D] n_l f_{lu} l \tag{34}$$

When the source line is not small compared to the absorption line width, then the profiles of both the emission and absorption lines affect the absorption.

III. Dependence of Fluorescence on Atomic Concentration

A. Types of Atomic Fluorescence

A fluorescence line can occur at the same wavelength as the exciting radiation, or it can be of longer or (rarely) shorter wavelength than the exciting radiation. Basically there are two types of atomic fluorescence: resonance fluorescence and nonresonance fluorescence.

Resonance fluorescence occurs when atoms absorb and reemit radiation at the same wavelength. Resonance transitions (i.e., between the ground state and some excited state, as shown in Fig. 4a) account for most of the analytically useful examples of resonance fluorescence, since the transition probabilities for resonance transitions are usually much greater than those for other

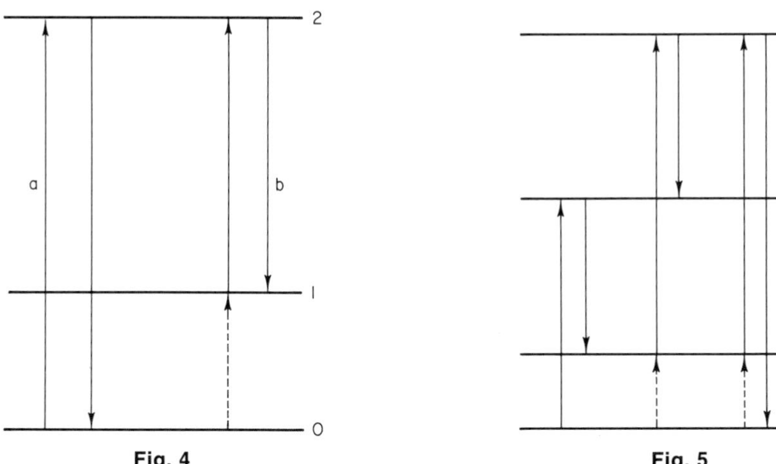

Fig. 4 Types of resonance fluorescence: (a) resonance fluorescence originating from a resonance transition; (b) resonance fluorescence originating from a metastable state. 0, ground state of atom; 1, 2, excited states. —radiational processes; ---, a nonradiational process.

Fig. 5 Direct line fluorescence transitions. Note that for direct line fluorescence the absorption and fluorescence transitions have a common upper state but differ in their lower states.

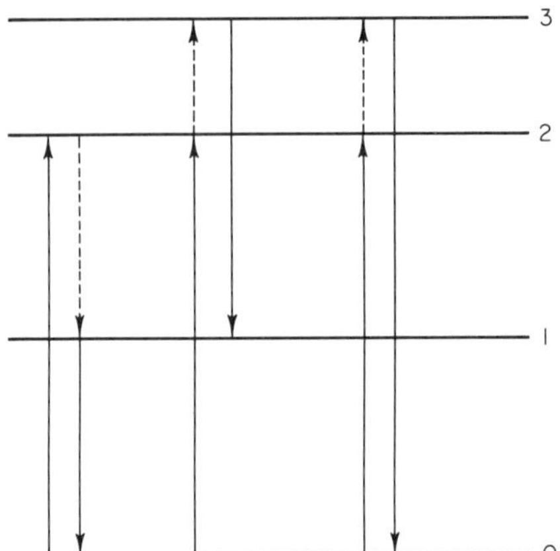

Fig. 6 Stepwise-line fluorescence. Note that for stepwise-line fluorescence the absorption and fluorescence transitions have different upper states.

transitions. The strong fluorescence lines of Zn and Cd at, respectively, 2138.56 and 2288.02 Å are examples of this type of resonance fluorescence.

If an atom has low-lying metastable energy levels which can be thermally populated, resonance fluorescence may originate from the metastable levels, as indicated in Fig. 4b. Indium, gallium, and lead provide examples of elements for which this type of resonance fluorescence is possible.

Nonresonance fluorescence occurs when the exciting line and the observed fluorescence line are of different wavelengths. The two main categories which are usually distinguished are direct-line and stepwise-line fluorescence. Examples of each are given in Figs. 5 and 6.

B. Relationship between Fluorescence and Atomic Concentration

The fluorescence radiant flux ϕ_f from an atomic vapor is given by

$$\phi_f = \phi_a Y_p f_s \tag{35}$$

where ϕ_a is the radiant flux absorbed, as defined in Eq. (25), Y_p the power efficiency of the fluorescence (i.e., the radiant flux reemitted per primary radiant flux absorbed), and f_s a factor which accounts for the fluorescence radiance lost within the gas cell due to reabsorption.

The power efficiency Y_p is closely related to the quantum efficiency Y which is the ratio of the number of photons re-emitted per unit time to the number of photons absorbed per unit time. The relationship between Y_p and Y is thus

$$Y_p = Y(\nu_f/\nu_a) \tag{36}$$

where ν_f and ν_a are the frequencies of the fluoresced and absorbed radiation, respectively.

Consider the term diagram of Fig. 4. The quantum efficiency of the fluorescence transition $2 \to 1$ is the rate of the transition $2 \to 1$ divided by the total rate at which the state 2 is depopulated. Three processes exist for depopulating state 2: radiational transitions $2 \to 1$ and $2 \to 0$ and non-radiational deactivation (quenching) by collisional processes. Thus the quantum efficiency for the transition $2 \to 1$ is given by

$$Y_{2\to1} = A_{2\to1}/(A_{2\to1} + A_{2\to0} + r_2) \tag{37}$$

where $A_{2\to1}$ and $A_{2\to0}$ are the Einstein transition probabilities for spontaneous emission for the transitions indicated by the subscripts, and r_2 is the total psuedo first-order rate constant for quenching from state 2. If the absorption transition $0 \to 2$ is the only process by which state 2 is populated, then the power efficiency is given by

$$Y_{p,\,2\to1} = Y_{2\to1}(\nu_{2\to1}/\nu_{2\to0}) \tag{38}$$

For the fluorescence transition $2 \to 0$ the quantum efficiency is

$$Y_{2\to0} = A_{2\to0}/(A_{2\to1} + A_{2\to0} + r_2) \tag{39}$$

and the power efficiency is equal to the quantum efficiency since the frequencies of fluorescence and absorption are the same (resonance fluorescence).

The geometry of the atomic vapor cell affects the expression for fluorescence intensity much more than the expression for atomic absorption. The reason for the additional complexity can be seen in Fig. 7. Only the radiation absorbed in the cell defined by the intersection of the source and observed paths is effective in producing observable fluorescence, and the intensity of source radiation reaching this cell is affected by the prior absorption occurring along the path length b. The intensity of fluorescence reaching the detector further depends on reabsorption occurring along the path length b'. Of course the excitation and observation paths need not be at right angles,

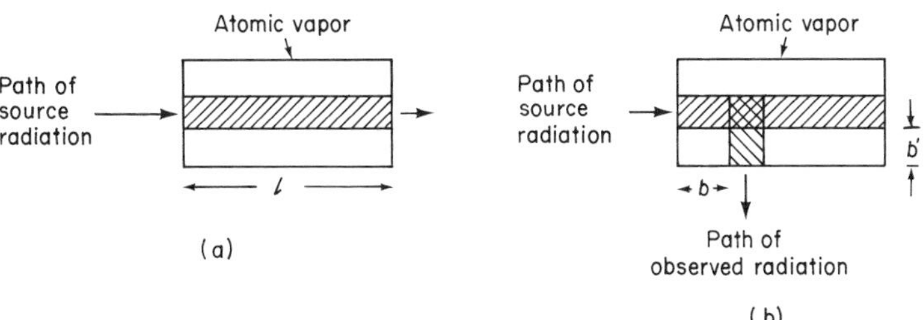

Fig. 7 Cell geometry for (a) atomic absorption and (b) atomic fluorescence measurements.

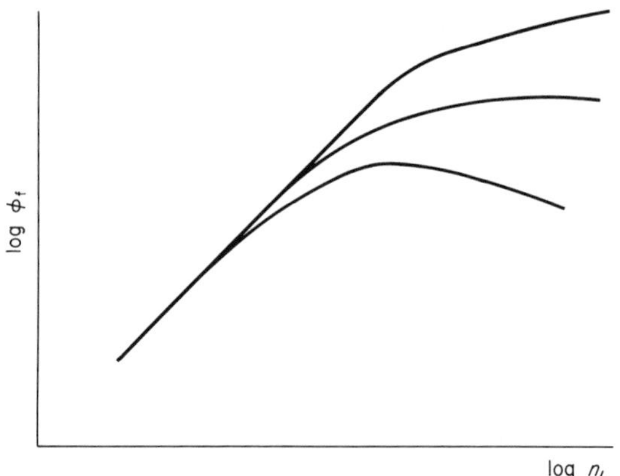

Fig. 8 Some possible growth curves for atomic fluorescence.

and the exact expression relating the fluorescence to atomic concentration will depend on the chosen cell geometry.

Several authors (Hooymayers, 1968; Svoboda *et al.*, 1972) have derived expressions for the intensity of fluorescence in which full account is taken of prior absorption and reabsorption. For our purposes it is satisfactory to sketch only the broad outline of the discussion.

At low atomic concentrations, the situation of greatest interest for analytical purposes, the fluorescence radiant flux is linearly dependent on the atomic concentration. In this range of concentrations, $f_s = 1$ and ϕ_a is given by the expressions derived in the previous section [Eqs. (29) and (32)]. At higher concentrations, f_s is less than 1 and shows a complex dependence on atomic concentration. Some examples of atomic fluorescence growth curves (log–log plots of fluorescence radiant flux versus atomic concentration) are shown in Fig. 8.

IV. Production of Atoms

We have derived expressions relating absorption and fluorescence to the concentration of atoms. In this section we shall be concerned with practical means of producing the atomic vapor.

A method for producing an atomic vapor should satisfy several requirements:

(1) The concentration of free atoms should be as large as possible.

(2) The proportionality between the atomic concentration and the concentration in the sample should not vary.

(3) The method should be simple and reproducible.

No method so far examined is completely satisfactory, and further improvement in the methods of atom production is one of the major goals of current research in analytical atomic spectroscopy.

In the discussion which follows the term atomization is understood to mean the production of atoms. For the more colloquial sense of this term, that is, the reduction of a liquid to a fine spray, we shall use nebulization.

A. Chemical Flames

By far the most popular method of atomization for analytical atomic absorption and atomic fluorescence spectrometry is the use of chemical flames. This is partly due to the adaptation of the method from the older emission flame spectrometry and partly due to the elegant simplicity of the flame in atomizing solution samples.

Figure 9 presents in schematic form the processes which must occur in producing atoms of some metallic element M from a solution of the salt MX. The solution is broken up into fine droplets, which may be injected directly into the flame or may pass through a nebulization chamber in which the larger droplets are separated and discarded and only the smaller droplets are carried into the flame. Solid aerosol particles are produced as solvent evaporates from the droplets. When a nebulization chamber is used, some solvent evaporation surely occurs even before the droplets enter the flame. As the aerosol enters the flame, the remaining solvent is removed and the solid particles are vaporized. Processes at the gas–solid interface are complex and not yet well understood, but one possible further path is shown in Fig. 9: production of molecular species, which may then dissociate to give free atoms. The atoms produced undergo further reactions such as ionization or association with flame gas products to give other molecular species.

A variety of gas mixtures have been used to produce flames for analytical atomic spectroscopy. Some of these and their more important properties are contained in Table II. Flames are customarily classified as premixed or diffusion, depending on how the fuel and oxidant are mixed prior to combustion.

The flame produced on a Bunsen burner is of the premixed type: the fuel and oxidant are mixed before issuing from the burner tube, the gas flow is largely nonturbulent, and the flame exhibits distinct zones of different properties. The rate of combustion in such a flame is described by the burning velocity of the mixture. When a combustible mixture is ignited at some point in the mixture, the rate at which the flame front propagates through the mixture is the burning velocity. If we ignite a combustible mixture that is issuing from a tube, a stable flame will be produced only if the rate of flow of

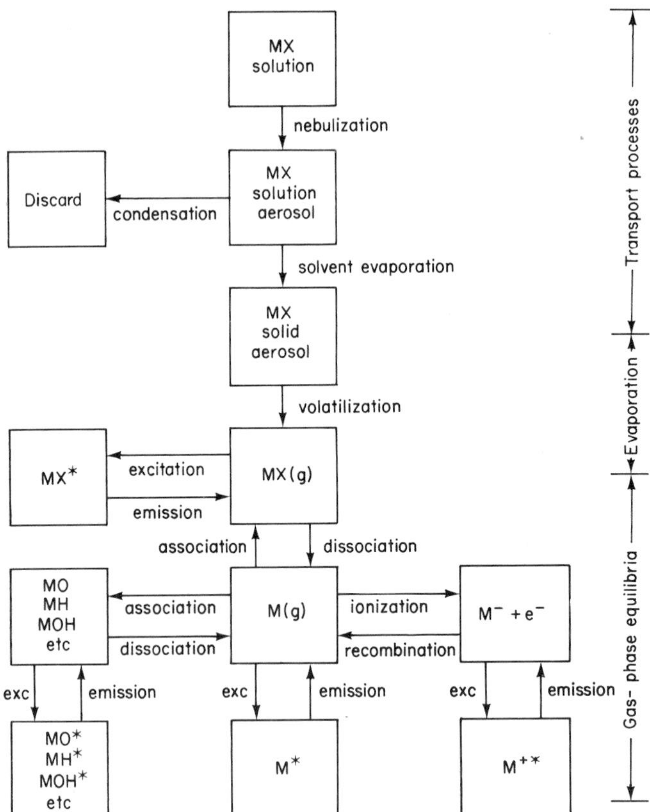

Fig. 9 Flame atomization processes.

TABLE II

Properties of Some Fuel–Oxidant Mixtures

Mixture	Ignition Temp. (°C)	Burning Velocity (cm sec^{-1})	Approximate Max Flame Temp. (°C)
H_2–air	530	440	2045
H_2–O_2	450	3680	2660
H_2–N_2O	—	390	2650
C_2H_2–air	350	160	2125
C_2H_2–O_2	335	2480	3100
C_2H_2–N_2O	400	285	2955

the mixture along some front just equals the velocity with which the flame front propagates in the opposite direction.

In the premixed flames used for analytical spectroscopy, it is usually possible to distinguish the four zones shown in Fig. 10. In the preheating zone the combustion mixture is heated to the ignition temperature. The primary reaction zone is characteristically a rather thin blue cone surrounding the preheating zone. The complex processes of combustion occur in this reaction zone. The region is rich in ions and radical species that have not had sufficient time (not enough collisions) to reach their equilibrium concentration. The reaction zone usually exhibits relatively intense continuum and band emission which, together with its small size, limits its usefulness for analytical atomic spectroscopy. The interconal zone and the secondary combustion zone are the regions of greatest analytical utility. In the secondary combustion zone, excess fuel and products of the primary combustion are burned to stable molecular species by air entrained from the atmosphere. In flames produced from a stoichiometric mixture of fuel and oxidant, that is, where the amounts of each are in the ratio of their coefficients in the balanced chemical equation that best describes the primary combustion reaction, the interconal zone is very small; but in fuel-rich hydrocarbon flames, this zone increases greatly in size and may extend to a height several centimeters beyond the primary reaction zone. It is within this zone that the chemical environment is most conducive to the production of free atoms from those elements that form monoxide molecules with large energies of dissociation.

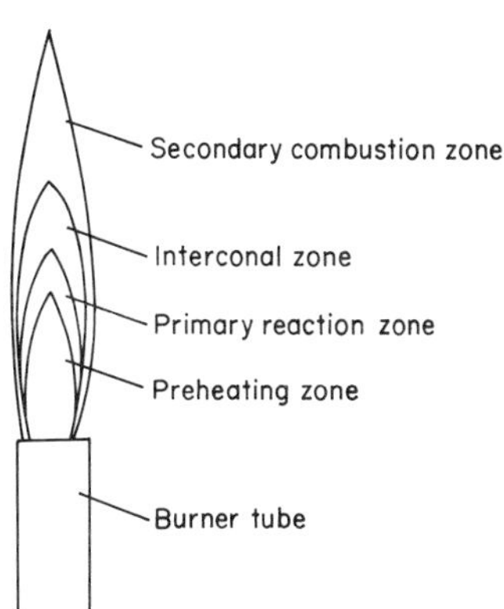

Fig. 10 Structure of a premixed flame.

The distribution of atoms within the flame depends on the gas mixture and the chemical properties of the element. Atomic distributions (Rann and Hambly, 1965) for three elements for a fuel-rich and a fuel-lean flame are shown in Fig. 11. Sodium is representative of the elements that form readily vaporized and easily dissociated compounds. There are rather broad regions

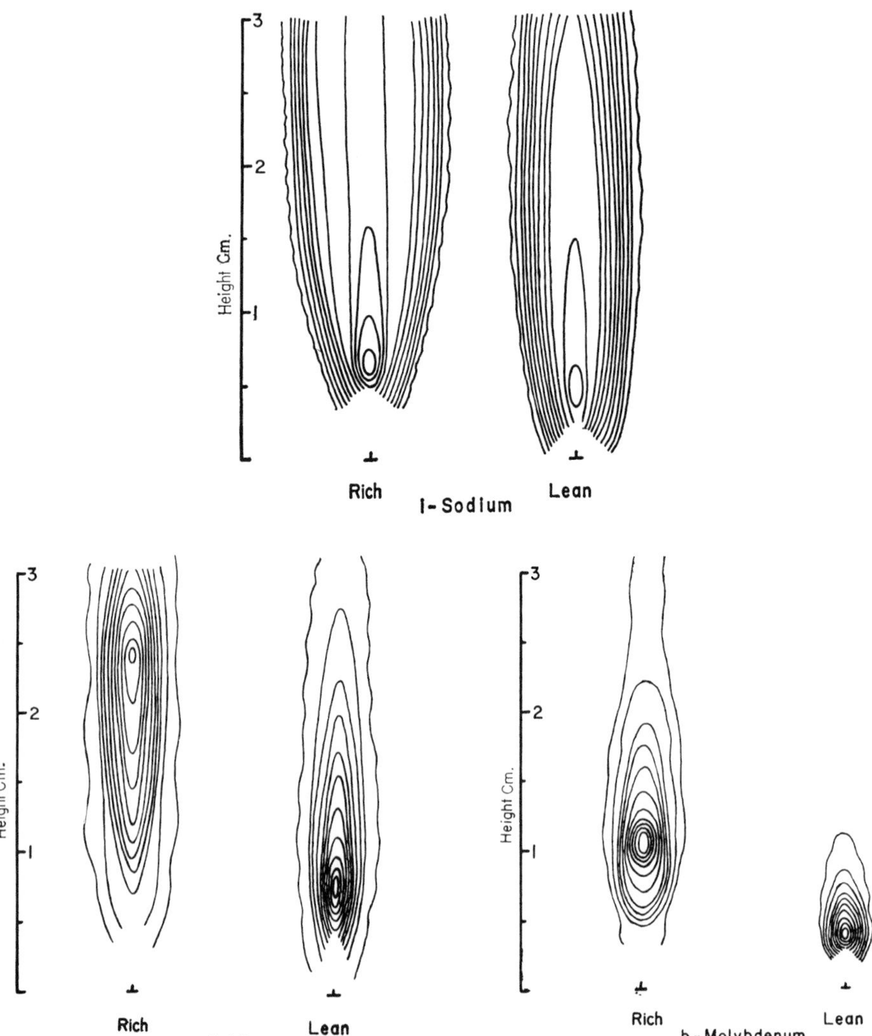

Fig. 11 Distributions of atoms in 10-cm slot-type premix acetylene–air flame. View is end on. Contours are drawn at intervals of 0.1 absorbance units, with maximum absorbance at center. [Reprinted with permission (Rann, C. S., and Hambly, A. N. (1965). *Anal. Chem.* **37**, 880, 881). Copyright © by the American Chemical Society.]

of constant atomic concentration. The maximum concentration occurs low in the flame and the atomic concentration is not strongly affected by the change from lean to rich conditions. Calcium represents an intermediate case in which the evaporation proceeds more slowly, and compound formation becomes important in the cooler and less fuel-rich regions of the flame.

Thus the maximum atomic concentration is rather sharply localized, occurs at some distance from the burner top, and its position is strongly altered by the change from lean to rich conditions. Molybdenum represents those elements which form a stable monoxide in the flame. The atomic concentration at a given point is strongly dependent on the flame gas composition, and fuel-rich conditions are essential to useful measurements.

A burner for premixed flames is shown in schematic fashion in Fig. 12. The sample is aspirated and nebulized by the high velocity flow of the oxidant gas across the capillary tip. The larger droplets collect on the baffles and walls of the chamber and are discarded. The remaining aerosol and the fuel and oxidant are thoroughly mixed in the chamber as they travel to the exit orifice. The exit orifice may be a narrow slot, up to 15 cm long, or an array of small-diameter holes. The slot-type burner is readily constructed and provides a long path suitable for emission or absorption measurements. A burner with a square or circular array of holes is better suited for atomic fluorescence measurements.

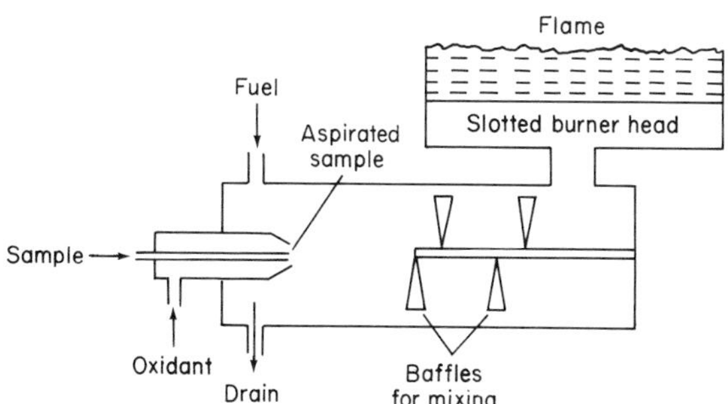

Fig. 12 A premix burner.

By preventing entrainment of the surrounding atmosphere, the secondary combustion zone of premixed flames can be shifted away from the primary reaction zone, thus extending the interconal zone. With flame separation the interconal zone can be observed free of the rather strong background emission arising from the combustion processes occurring in the secondary combustion zone. For measurements in the interconal zone of hydrocarbon flames, flame separation reduces the intensity of the continuum emission by two or more orders of magnitude and the OH band emission is almost completely suppressed. Extension of the interconal zone has the further favorable effect that the whole primary light beam can pass through the region in which conditions for atomization are optimal so that the density of free atoms, particularly for elements forming stable monoxides, is higher.

The separation of zones can be achieved by means of a quartz tube placed above the burner head or by a laminar stream of inert gas, generally nitrogen or argon, surrounding the flame. "Mechanical" separation with a tube has the advantage that no cold gases dilute the interconal flame gases, but when hydrocarbon flames are used, carbon deposition on the tube is a problem. The use of an inert gas shield is simpler and more often employed despite the sometimes substantial temperature drop in the interconal gases due to dilution with the cold shield gas. A burner for producing an inert gas separated flame (West and Cresser, 1973) is shown in Fig. 13.

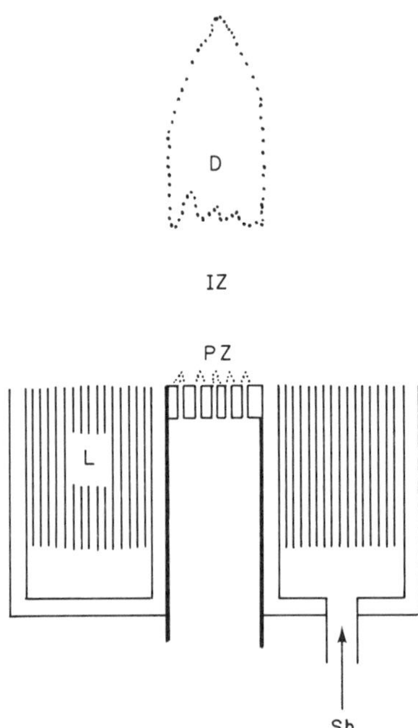

Fig. 13 A flame separator (PZ, primary reaction zone; IZ, interconal zone; D, diffusion flame; L, capillaries or corrugated metal strip for laminar gas flow; Sh, sheath gas inlet). [West, T. S., and Cresser, M. S. (1973). *Appl. Spectrosc. Rev.* 7, 79.]

Diffusion flames differ from premixed flames in several important respects. In a diffusion flame, the pure fuel gas flows from a tube, either into the open atmosphere, where it burns with the oxygen of the air, or into a stream of oxidant gas flowing from an adjacent or concentric orifice. When the rate at which a gas issues from an orifice exceeds a certain critical velocity, it becomes a turbulent jet that entrains the surrounding gas and broadens to form a cone. The flame in such a jet consists of irregular vortexes of fuel and oxidant mixture forming random patches of combustion waves throughout the jet, but no cohesive flame front. Such a turbulent diffusion flame is typical of the diffusion flames employed in flame spectrometry.

A burner for producing a turbulent diffusion flame is shown in Fig. 14. It consists of three concentric chambers—a central capillary, an inner jacket, and an outer jacket. The flow of oxidant over the capillary tip causes a pressure drop at the tip and aspiration of the solution through the capillary. The velocity of the oxidant stream at the capillary tip is sufficiently high that the solution stream is broken into fine droplets as it issues from the capillary. In this arrangement all the solution is sprayed into the flame, and such burners are commonly known as "total consumption" burners. The name should not be understood to mean that all the sample material is atomized in the flame. A rather large range of droplet sizes is produced with this arrangement, and some pass through the flame incompletely vaporized.

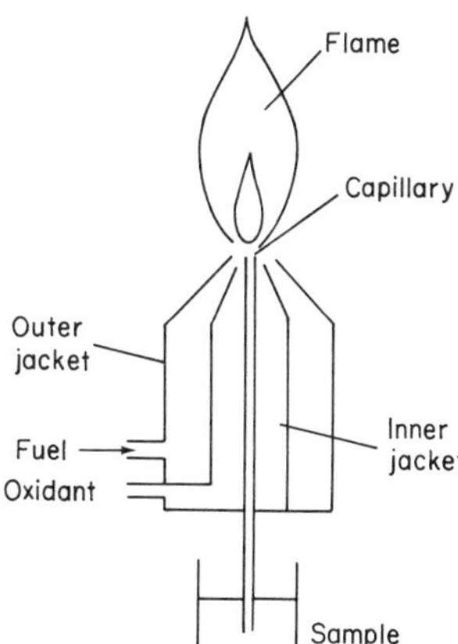

Fig. 14 A total consumption burner.

Turbulent diffusion flames are noisy and provide only a short path for absorption measurements. They achieved widespread use in flame spectrometry because high temperature, oxygen-supported flames could be safely produced in this fashion. Because of the high burning velocity of mixtures in which pure oxygen is the oxidant, it is difficult and hazardous to produce a premixed flame from such a mixture. The development of nitrous oxide-supported flames eliminated this advantage of turbulent diffusion flames. Use of nitrous oxide as the oxidant provides flames of high temperature and low burning velocity which can be produced safely on premixed burners.

At the beginning of this section on the production of atoms we suggested several requirements to be met by any method for atomization. It should be evident that chemical flames, despite the considerable complexity of their chemistry, provide a simple and reproducible method of atom production. Further, the considerable success of atomic absorption and atomic fluorescence flame spectrometry demonstrates that an adequate atom population can be produced in chemical flames. One should recognize, however, that the sample material is enormously diluted in the process of flame atomization. At least several liters per minute, measured at room temperature, of fuel and oxidant are required with conventional burners for stable flame production, and as the gas mixture burns and is raised to the final flame temperature there is a manyfold increase in the gas volume. The remaining requirement in atom production, that is, that the proportionality between the atomic concentration and the concentration in the sample should not vary, requires some additional consideration.

The concentration of an element in a sample cannot be directly determined by measuring the absorbance or fluorescence intensity produced. The response of the system must be calibrated by running one or, usually, more samples of known concentration (standards). An interference occurs if the proportionality between the response and the solution concentration is different for the standards and other samples. It is convenient to discuss possible interferences in the categories suggested by Fig. 9: transport interferences, evaporation interferences, and gas phase equilibria interferences.

To transport interferences belong all factors affecting the amount of sample entering the flame. The most important factors are viscosity (influencing the sample aspiration rate), surface tension (influencing the size of the aerosol droplets), and the vapor pressure of the solvent (influencing solvent evaporation rate and condensation losses). Because transport effects affect all elements equally, they constitute interferences not specific to particular elements.

Evaporation interferences arise from alterations in the rate of evaporation of solid aerosol particles in the flame. These may be either specific to particular elements or nonspecific. A specific interference of this type occurs when a chemical reaction between the analyte and a concomitant in the solution leads to the formation of a compound that vaporizes at a different rate. An example of this type of behavior is found in the effect of phosphate anion on calcium. In the presence of phosphate, a relatively involatile Ca–P–O compound is formed, and a marked depression of the Ca atomic concentration occurs in the lower regions of the flame. A nonspecific evaporation interference can occur when the analyte occurs in a solution of high total solids. After solvent evaporation, the test element may be embedded in a large salt particle from which it cannot be vaporized quickly.

Gas phase equilibria interferences arise from shifts in the dissociation and ionization equilibria. These equilibria may be considered in much the same fashion as solution equilibria. Thus for the ionization reaction,

$$\mathrm{M} \rightarrow \mathrm{M}^{+} + \mathrm{e}$$

the equilibrium constant customarily takes the form

$$K_{\mathrm{i}} = [(p_{\mathrm{M}^{+}})(p_{\mathrm{e}})]/p_{\mathrm{M}} \tag{40}$$

where p_{M^+}, p_e, and p_M are the partial pressures of ions, electrons, and atoms, respectively. The partial pressure of a species may be related to its concentration by the ideal gas law. For example, the partial pressure of M, p_M, is related to the concentration of M in atoms per cubic centimeter of flame gases, n_M, by

$$p_M = n_M kT \tag{41}$$

From Eq. (40) it can be seen that if the partial pressure of electrons in the flame is increased—for example, by the addition of another easily ionized element to the flame—that the partial pressure of M (and n_M) is increased. This type of effect can be observed for Na in the presence of K.

The detection and elimination of interferences depends largely on the ingenuity and experience of the analyst, but we can note some common approaches that have proved useful.

One very general approach is to prepare standards that imitate the sample composition. This method is reliable but laborious. If the overall sample composition is not known or varies from sample to sample, then the standard additions approach may be a better choice. It is usually possible, by either the imitation or standard additions approach to ensure that physical properties of samples and standards are sufficiently similar that transport interferences are avoided.

Interferences may sometimes be eliminated by addition of a relatively large concentration of the interferant species to standards and samples so that the natural variation is "buffered" out. Thus, it is customary to add a large excess of an easily ionized species to repress ionization. The buffer approach has sometimes been adopted to eliminate evaporation interferences, such as the effect of phosphate on calcium, but it is less useful in this case because of the rather considerable depression of the signal which results. A better approach in such cases is the use of "releasing agents," which prevent the depression by setting up a competitive equilibrium in the aerosol droplet. For example, La is sometimes used as a releasing agent for Ca since it effectively "competes" with Ca for the phosphate anion, and the involatile Ca–P–O compound is avoided.

The choice of flame, flame region, and flame gas composition may also have a profound effect on the efficiency of atom production. Higher temperature flames reduce the effect of slow evaporation and reduce compound formation in the flame. The effect of slow evaporation steps may also be reduced by making observations higher in the flame, but compound formation may be greater in this region because of greater entrainment of ambient air and cooling of the flame. The flame gas composition affects the efficiency of atomization by its effect of gas phase equilibria. Many elements form stable monoxides in flames, and a substantial gain in atomic concentration can sometimes be obtained by increasing the amount of fuel in the flame mixture, thereby reducing the free oxygen concentration in the flame and shifting the equilibrium reaction

$$MO \rightarrow M + O$$

to favor formation of M.

Conventional measurements with a flame atomizer require 0.5 ml or more of sample for each reading. For many applications, and especially with biological materials, trace element determination must be carried out with limited amounts of sample material. Recognition of this need has led to extensive development of microsampling techniques for atomic absorption and atomic fluorescence spectrometry. Electrothermal atomizers are described in the next section. Several methods of microsampling with flame atomizers deserve mention here.

Figure 15 illustrates the principal features of the "Delves cup" technique (Delves, 1970). In this technique, solution residues are vaporized from nickel crucibles into an absorption tube placed in a long path flame. Both the crucibles and absorption tube are heated by the flame, and the atoms are retained in the optical path by the tube for considerably longer than with a conventional burner.

The Delves technique appears to have achieved wide acceptance for the determination of lead in blood (Hicks *et al*., 1973; Olsen and Jatlow, 1972). Some similar approaches suited to atomic fluorescence spectrometry have also been described (Sychra and Kolihova, 1973; Grime and Vickers, 1974).

Another approach to atomic absorption flame spectroscopy with microsamples makes use of a tantalum sampling boat. The boat, as first described by Kahn *et al*. (1968), is about 5 cm long and 1–2 mm wide and deep. A simple slide device allows the boat to be loaded about 10 cm away from the flame, then pushed close to the flame to dry the sample, and finally placed within the long path flame for sample atomization.

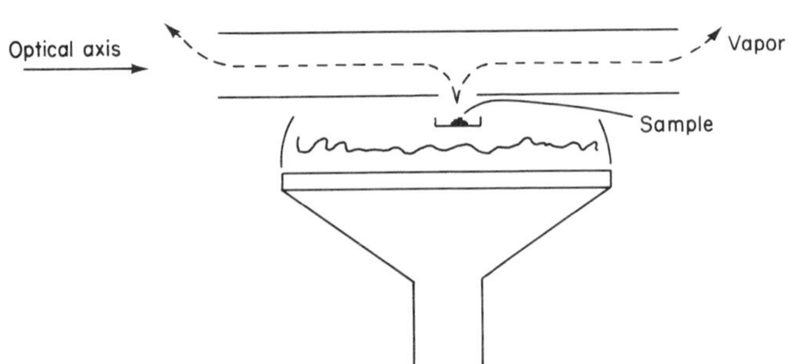

Fig. 15 Arrangement for "Delves cup" atomic absorption flame spectrometry. [Delves, H. T. (1970). *Analyst* **95**, 431.]

In a third approach (Grime and Vickers, 1975; McCullough and Vickers, 1976), suitable to both absorption and fluorescence measurements, a small volume of sample (typically 10 μl) is placed on a tantalum filament which can be heated by passage of electrical current. By controlling the current, the sample is successively dried, ashed, and volatilized from the filament. The volatilized material is swept by a gas stream into the mixing chamber of a premixed burner and is atomized in the flame. Both air and nitrous oxide-supported flames have been used with this microsampling device.

B. *Electrothermal Atomizers*

The first furnace atomizer for analytical atomic absorption spectroscopy was described by L'vov (1961) in 1961. An improved version of the L'vov furnace is shown in Fig. 16 (L'vov, 1969). The sample (typically 0.1 mg or less) is placed on the electrode 1, which can be introduced in the hole in the graphite tube 2. The tube is 5–10 cm long with an internal diameter of about 3 mm. The tube is heated by current from a step-down transformer 4. The sample electrode is heated by current from the second step-down transformer 5, which is switched on only after the electrode has been inserted. The furnace tube is lined with pyrolytic graphite to inhibit vapor diffusion through the walls of the tube. The whole system is contained in an argon-filled chamber. Samples may be in solution or powder form. If solutions are to be analyzed, the electrode heads are treated with polystyrene dissolved in benzene to prevent the sample from soaking into the electrode. The theory and practice of electrothermal atomization has been discussed in detail in a book by L'vov (1970).

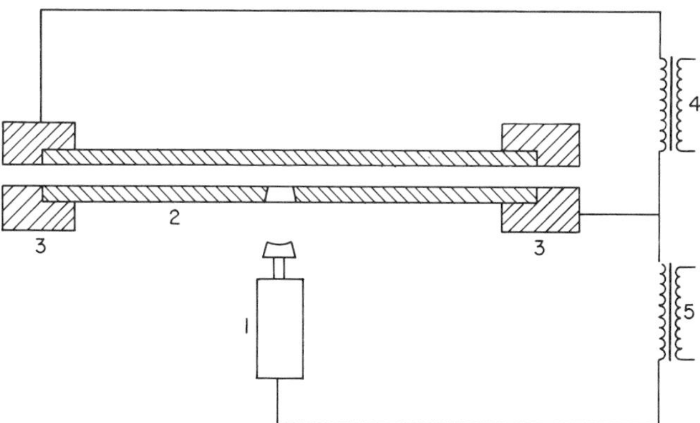

Fig. 16 L'vov furnace. 1, sample electrode; 2, graphite tube; 3, support block; 4, 5, step-down transformers. [L'vov, B. V. (1969). *Spectrochim. Acta* **24B**, 53.]

Massmann (1968) described graphite furnace devices for both atomic absorption and atomic fluorescence spectroscopy. The fluorescence device is shown in Fig. 17. With this device sample volumes of 5 to 50 μl are placed in the graphite cup. The cup has an internal diameter of 6.5 mm and is 40 mm high. Two vertical slots are cut into the wall on opposite sides of the cup to allow the fluorescence radiation to be observed. The exciting radiation enters the cup from above. The holder serves to make the electrical connections and surrounds the cup with an argon atmosphere. A current of up to 400 A is used to heat the device.

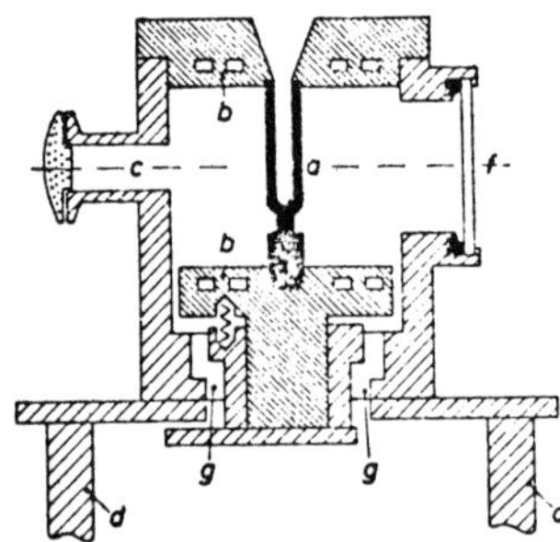

Fig. 17 Massmann furnace for atomic fluorescence measurements. 1, hollow cathode lamp source; 2, quartz window; 3, argon inlet; 4, graphite cup; 5, heating. [Massmann, H. (1968). *Spectrochim. Acta* **23B**, 218.]

The Perkin-Elmer Corporation produced a commercial version of the Massmann furnace for atomic absorption spectroscopy, a schematic diagram of which is shown in Fig. 18 (Manning and Fernandez, 1970). The atomizer is a graphite tube 50 mm long and 10 mm in diameter through which the sample beam passes. A flow of argon directed into an outer chamber enters through several small holes and leaves through the open ends of the tube. The entire assembly is held inside a water-cooled metal cylinder. The sample is loaded into the center of the tube. After placement of the sample, the furnace is heated by a three-stage electrical program to effect drying of the sample, ashing of organic matter, and finally atomization of the sample.

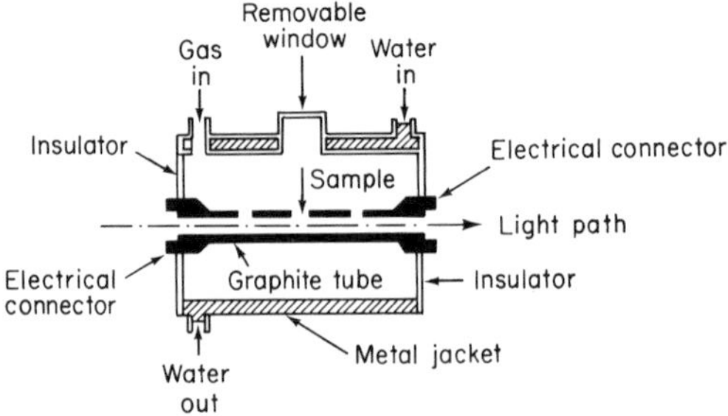

Fig. 18 Perkin-Elmer heated graphite atomizer. [Manning, D. C., and Fernandez, F. (1970). *At. Abs. Newsletter* **9**, 65.]

Varian-Techtron Pty. Ltd. produced a commercial atomizer based on the carbon filament atomizer first described by West and Williams (1969). The original carbon filament was a graphite rod 2 mm in diameter and about 20 mm long. This was supported by water-cooled stainless steel supports which also served as electrical connections. The rod was heated by passage of about 100 A of current. To prevent burning, the first device was placed in an argon atmosphere in a glass chamber with optical windows. Later versions, such as that shown in Fig. 19, eliminated the chamber and protected the rod by arranging a laminar flow of an inert gas around the rod.

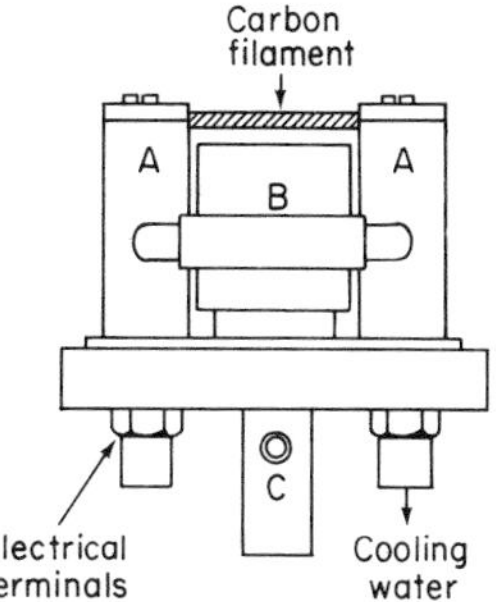

Fig. 19 Varian carbon rod atomizer. A, water-cooled electrodes; B, laminar flow shield gas head; C, inlet for shield gas. [West, T. S., and Cresser, M. S. (1973). *Appl. Spectrosc. Rev* **7**, 79.]

With the carbon rod atomizer a small volume of sample solution (about 1 μl) is placed on a depression in the center of the rod, the solvent is evaporated, organic matter is ashed, and then the rod temperature is raised to 2000–2500°C within about 5 sec. A disadvantage of the rod is the rapidity with which the atomic vapor leaves the hot surface and enters the relatively cool surrounding atmosphere where condensation begins. In order to raise the temperature of the surrounding atmosphere, hydrogen is sometimes used as the shield gas so that the rod is surrounded by a hydrogen diffusion flame.

In addition to those already briefly described, a number of other atomizers have been described in recent publications. These include other furnace designs, tantalum filaments, metal loops, and cathodic sputtering chambers.

The electrothermal atomizers generally provide very good absolute detection limits (of the order of 10^{-12} g) and are capable of handling very small samples. They have provided a satisfactory solution for a host of analysis problems for which the absolute detection limits were inadequate with flame atomizers, but they offer no panacea. Decreased precision, compared to flame atomizers, is frequently the price that must be paid for smaller sample size and pulsed atomization, and atomization with an electrothermal atomizer is susceptible to a variety of solid state matrix interferences not experienced with flame atomizers. For the present it seems likely that, despite all their faults, flame atomizers will not be displaced for the vast majority of samples.

V. Radiation Sources

For analytical atomic absorption spectroscopy, narrow line sources are generally preferred to continuous sources because they provide a higher degree of selectivity and better sensitivity. Both advantages are attributable to the relative sizes of the monochromator spectral bandwidth and the atomic line widths. For the small to medium size monochromators typically employed in analysis, the minimum spectral bandwidth (a few tenths of an angstrom) is several times larger than the emission or absorption line widths.

With a continuous source, the selectivity of the method is established by the spectral bandwidth passed by the monochromator. With a narrow line source the selectivity is established primarily by the source emission line width.

The sensitivity advantage of line sources can be seen in the equations for absorption with a line and continuous source [Eqs. (34) and (31)]. The ratio of the absorptions (line-to-continuous) is

$$\frac{\alpha_L}{\alpha_C} = \frac{2(\ln 2)^{1/2}}{\pi^{1/2}} \frac{s}{\Delta\nu_D} \simeq \frac{s}{\Delta\nu_D} \tag{42}$$

Equation (42) overstates the advantage of the line source since the assumptions of Eq. (34) are seldom met in practice. Nevertheless, with the usual equipment, the monochromator spectral bandwidth s is considerably larger than the emission and absorption line widths, and the sensitivity of measurement is considerably better when a line source is used.

When sensitivity is not the most important concern, a continuous source offers the advantage that only a single source is required, whereas with line sources each element requires its own source. For theoretical studies in which absolute atomic concentrations are to be measured, a continuous source offers a further advantage in that the measured absorption is independent of the line width. With a line source the absolute atomic concentration can be calculated from the measured absorption only if the profiles of both the emission and absorption lines are known.

For atomic fluorescence measurements the source requirements are somewhat different than for atomic absorption measurements, since the radiation source is not viewed by the detector. From Eq. (35) it can be seen that the intensity of fluorescence is proportional to the source radiant flux at the absorption line of interest. In practice, line sources are often more intense than continuous sources at the specific wavelengths of interest, and most atomic fluorescence measurements are carried out with line sources.

In the discussion which follows, various kinds of line and continuous sources will be described, and their advantages and disadvantages for atomic absorption and atomic fluorescence spectroscopy will be considered.

A. Hollow Cathode Lamps

Figure 20 shows the type of hollow cathode lamp which is used for almost all analytical atomic absorption measurements. Hollow cathode lamps provide narrow spectral lines of moderate intensity and thus satisfy the requirements for atomic absorption measurements very well. They are less satisfactory for atomic fluorescence measurements.

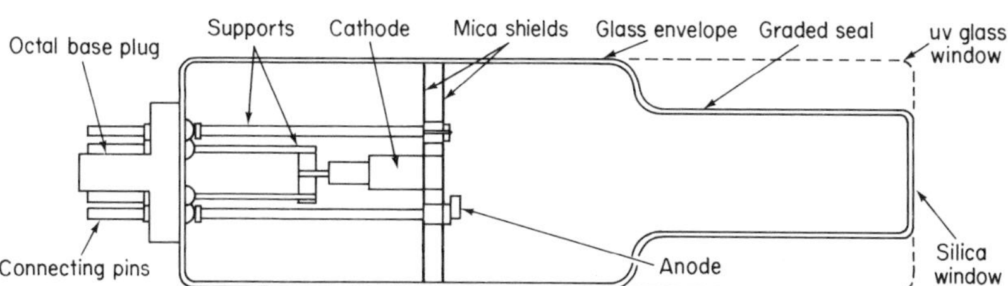

Fig. 20 Diagram of hollow cathode lamp for atomic absorption spectroscopy.

The tube is evacuated and filled at a relatively low pressure (10 Torr or less) with a high purity monatomic gas, usually neon or argon. The hollow cylindrical cathode gives rise to a particular type of low pressure discharge. The positive column, usually a prominent feature of a glow discharge, virtually disappears, and the inside of the cathode is filled with the negative glow. Between the negative glow and the cathode lies the Crookes dark space (Fig. 21), and nearly all the applied voltage falls across this region.

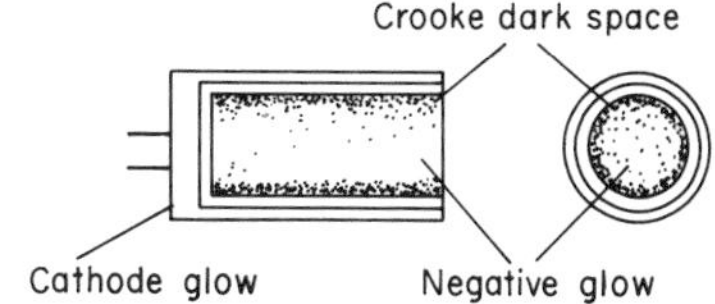

Fig. 21 Diagram of discharge in hollow cathode.

The current is carried by metal and gas ions and electrons. Positive ions, strongly accelerated through the high electrical gradient of the Crookes dark space, bombard the cathode, removing material by sputtering. The sputtered material enters the negative glow region where atoms are excited by collisions with gas ions and electrons which have been slowed enough by previous collisions to excite atomic spectra.

The negative glow region is known to have a low potential gradient in spite of the high electron and ion densities, and broadening due to the electric field (Stark effect) is therefore small. The low gas pressure and relatively low metal vapor density keep collisional broadening small. When low lamp currents are used, the temperature of the negative glow plasma is low and Doppler broadening is also minimized. Thus the narrow line requirement of atomic absorption spectroscopy is fairly well satisfied. As the lamp current is increased, both the temperature (and thus the number of collisions) and the rate of sputtering increase, resulting in broader lines. For many elements this effect is easily demonstrated by a decrease in the measured absorption for a fixed atomic concentration as the lamp current is increased.

The choice of monatomic gas affects the excitation conditions, sputtering rate and lamp life. Neon seems to be the gas of choice for most lamps constructed today. Argon is sometimes preferred when neon spectral lines may interfere with the lines of the cathode element.

Cathode construction differs considerably for the various metals. Where the metal is easily machined and not particularly expensive, it is usual to manufacture the whole cathode from the metal. When the metal is expensive, for example, gold or platinum, the metal is usually inserted as a thin shield in a carrier cathode of some less expensive metal, the spectral lines of which will not interfere with the element of interest. When the melting point of the metal is low, it may be necessary to hold the metal in a carrier cathode which will retain the element of interest even in the molten state.

Some lamps are designed to emit the lines of several elements. Successful multielement lamps can be made only for carefully selected combinations of elements. Such lamps are available for up to five elements. The intensities of the lines are generally lower than from single-element lamps. The chief advantage of the multielement lamps appears to be rapid change from one element to another.

Hollow cathode lamps operated in the low current mode suited for atomic absorption spectroscopy generally provide poor sensitivities when used for atomic fluorescence spectroscopy. Some hollow cathode lamps can, however, prove useful for atomic fluorescence when operated in a high current pulsed mode. Pulsed hollow cathode lamps seem likely to receive increased attention in the future and have already proven useful in the development of atomic fluorescence systems for multielement analysis (Mitchell and Johansson, 1970, 1971; Cordos and Malmstadt, 1972a,b, 1973; Omenetto *et al.*, 1973a).

B. *Electrodeless Discharge Lamps*

Radio- and microwave-frequency electromagnetic fields are very efficient in generating and accelerating electrons and thereby maintaining a gaseous glow discharge without electrodes in the plasma. Good radiation sources for analytical atomic fluorescence spectroscopy have been obtained from microwave powered electrodeless discharges in sealed tubes, such as those in

Fig. 22, containing a monatomic gas at low pressure and some volatile pure metal or metal salt. Both radio-frequency and microwave-frequency electrodeless discharge lamps have been used as radiation sources in analytical atomic absorption spectroscopy for a few of the more volatile elements for which satisfactory hollow cathode lamps are difficult to prepare.

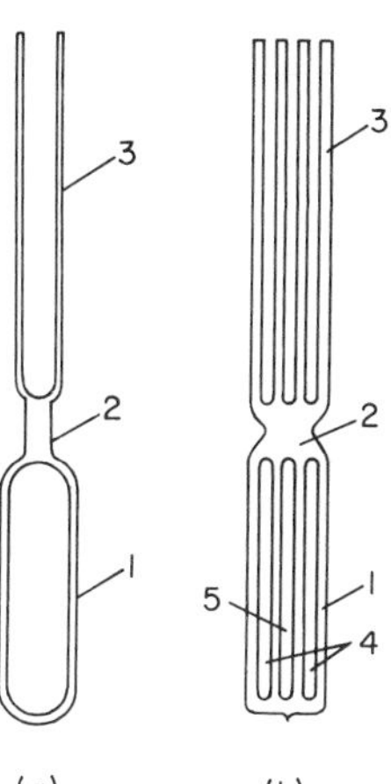

Fig. 22 Typical electrodeless discharge lamp: (a) conventional 9-mm-bore lamp; (b) vacuum jacketed lamp. 1, bulb; 2, seal; 3, stem; 4, outer cavity; 5, inner cavity.

For atomic fluorescence studies medical diathermy units operating at 2450 MHz at powers up to 150 W have been widely used to operate electrodeless discharge lamps. Microwave energy from the generator is led through a coaxial output cable and coupled to the lamp by means of an antenna or resonant cavity. The purpose of these devices is to efficiently transfer power from the generator into the lamp. Sufficient energy must be absorbed in the lamp to effect vaporization, excitation, and ionization to sustain the discharge. The success of the electrodeless discharge lamp depends to a large extent on the proper design and choice of the coupling device.

Antennas were developed primarily for medical use with the diathermy units. One such antenna, type A, is shown in Fig. 23. One limitation of this device in operating electrodeless discharge lamps is its lack of thermal energy for maintaining an adequate vapor pressure of the metallic species in the discharge. This limitation can be overcome by insulating the lamp, as shown in Fig. 23, with quartz wool or a vacuum jacket.

A resonant cavity transfers microwave power to a discharge by matching its own resonance frequency to that of the microwave source. When the cavity is properly "tuned," the power reflected from the cavity is at a minimum. Several cavities which have been used for atomic fluorescence measurements are shown in Fig. 24.

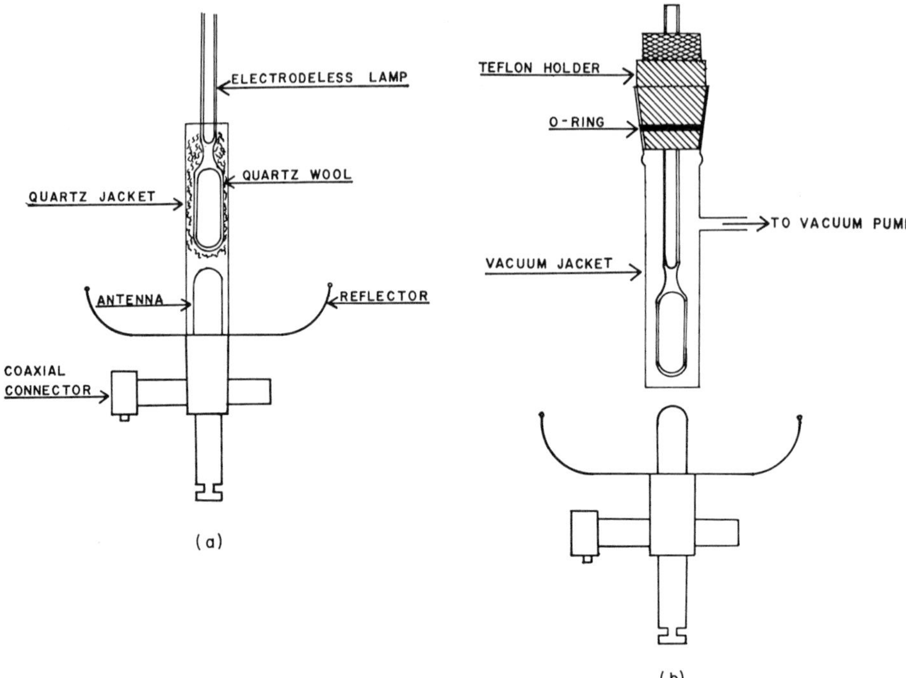

Fig. 23 Type A antenna with (a) quartz wool insulated lamp and (b) vacuum jacket insulated lamp. [Reprinted with permission (Zacha, K. E., Bratzel, M. P., Jr., Winefordner, J. D., and Mansfield, J. M. Jr. (1968). *Anal. Chem.* **40**, 1735). Copyright © by the American Chemical Society.]

The $\frac{1}{4}$-wave, or Evenson, cavity is the most efficient in coupling power into the discharge but requires the greatest care in tuning. Tuning is a tedious procedure since the two tuning stubs are mutually dependent and successive readjustment of each must be used to attain maximum efficiency. Moreover, the tuning is easily spoiled as the vapor and solid material shift in the lamp during operation.

The $\frac{3}{4}$-wave, or Broida, cavity is somewhat less efficient but much easier to use than the $\frac{1}{4}$-wave cavity. The lamp is almost totally immersed in the microwave field so that the coupling efficiency is seldom disturbed by movement of material within the lamp.

The tapered rectangular cavity is less often used than the other devices. It is less efficient than the other two cavities in coupling energy into the discharge, but since it requires no tuning, it is easy to operate.

With all the coupling devices described the discharge is customarily initiated by inducing ionization in the lamp with a simple Tesla coil vacuum tester.

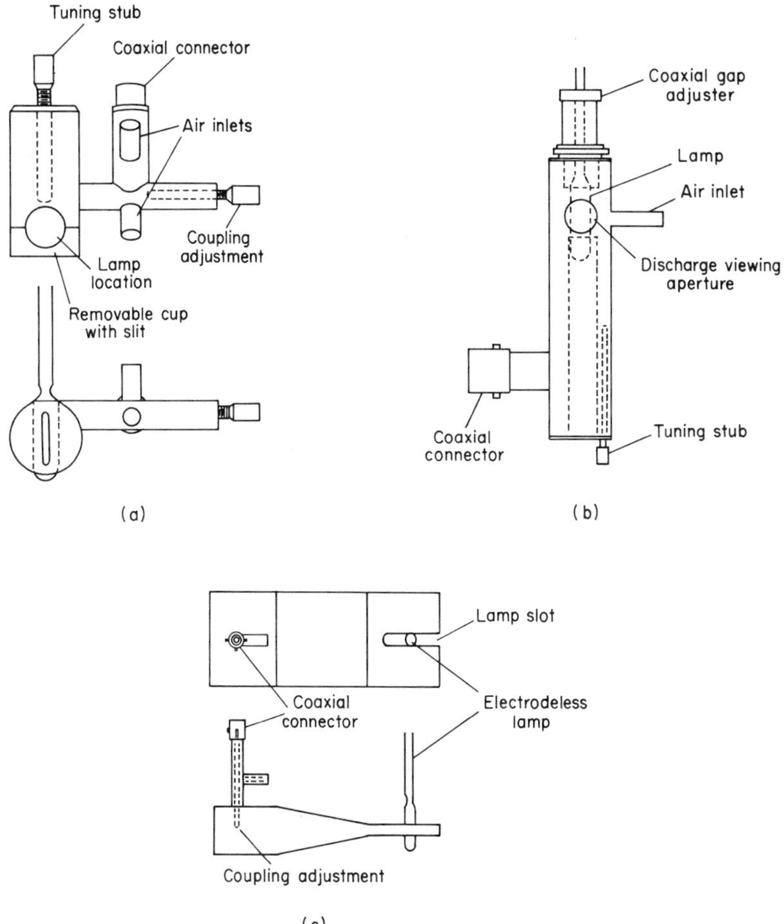

Fig. 24 Cavities used with electrodeless discharge lamps for atomic fluorescence spectroscopy. (a) $\frac{1}{4}$-wave cavity, (b) $\frac{3}{4}$-wave cavity, (c) tapered rectangular cavity.

A number of reports have been published describing the preparation of lamps for atomic fluorescence spectroscopy. As yet there are few clear-cut rules for the successful preparation of lamps for the various elements. It seems that to achieve good stability and intensity that the material placed in the lamp should have a vapor pressure of about 1 Torr at 200 to 400°C. It is this property which determines whether the metal itself or some more volatile compound, usually the chloride or iodide, is used.

Pure element lamps have been prepared mostly for the softer metals, such as zinc, cadmium, mercury, and selenium. In preparing metal halide lamps, the compounds are often formed in the lamp by a direct reaction of the metal with the appropriate halogen. Metal halide lamps have been prepared for a large number of elements, such as aluminum, titanium, bismuth, tellurium, tin, and lead.

Most of the lamps for atomic fluorescence studies have been prepared by the users. However, lamps are at present commercially available from several sources.

For atomic absorption spectroscopy, electrodeless discharge lamps are used chiefly for arsenic and selenium. Hollow cathode lamps for these elements tend to be low intensity, noisy, and short-lived, presumably because of the volatility of the elements. Moreover, the difficulty is compounded because the resonance lines of the elements are below 2000 Å, and in this wavelength range the transmission properties of the optical system and detector sensitivity are generally poor. Electrodeless lamps are readily prepared for these elements and reportedly provide a substantial improvement in intensity, stability and lifetime. At least one manufacturer of atomic absorption equipment offers a complete electrodeless discharge lamp and power supply system as an accessory to the standard atomic absorption equipment.

C. Metal Vapor Discharge Lamps

Lamps of this type were employed in early atomic fluorescence studies and also found use in atomic absorption measurements of the alkali metals before hollow cathode lamps were developed for these elements. They are frequently referred to by the name of the manufacturer (Wotan, Osram, or Philips lamps).

Figure 25 indicates typical construction features. The inner tube is usually made from silica and contains the metal, a monatomic gas at low pressure (1–10 Torr), and two tungsten filaments. The inner tube is thermally insulated by an evacuated jacket. The outer tube is usually made of glass, but silica must be used if ultraviolet lines are to be transmitted. Lamps of this type

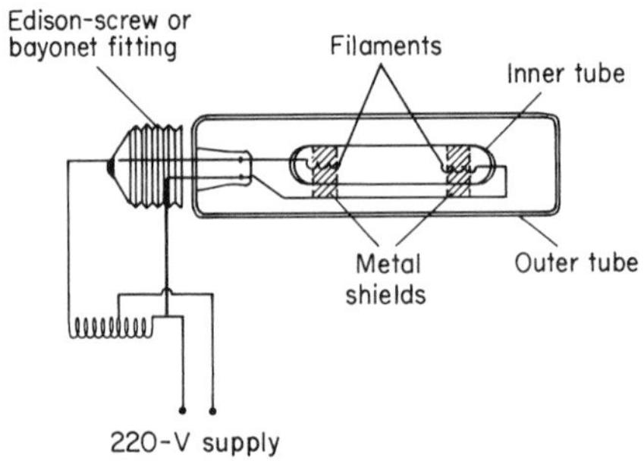

Fig. 25 Diagram of metal vapor discharge lamp.

can be made only for the more volatile elements (Cd, Zn, Hg, Ga, In, Tl, and the alkali metals except Li).

Metal vapor discharge lamps are usually operated from a simple auto-transformer arrangement which provides a high open circuit voltage for initial breakdown and limits the current to about 1 A in operation. For atomic fluorescence and atomic absorption measurements the lamps are frequently run at less than the maximum current to minimize self-absorption and self-reversal of the spectral lines. Even with this precaution the spectral output is less than satisfactory for both fluorescence and absorption measurements, and these lamps are seldom used today for practical analysis.

Mercury vapor arc lamps of a slightly different construction (pen lamps and germicidal lamps) have some utility in low cost apparatus designed specifically for the atomic fluorescence determination of mercury (Muscat *et al.*, 1972). These lamps consist of double bore quartz tubing with the two channels joined at one end and both electrodes at the other end so that the discharge occurs in the form of a long narrow *U*. Very low argon or neon pressure and low operating currents result in relatively cool lamps and in the emission of narrow spectral lines. Approximately 90% of the total output is concentrated at the 2537 Å mercury line, providing an intense source for excitation of fluorescence.

D. Continuous Sources

Tungsten filament, quartz halide, hydrogen- or deuterium-arc, and xenon-arc lamps are all satisfactory sources for atomic absorption measurements (Fassel *et al.*, 1966). Emission from the first two is largely limited to the visible, while hydrogen- and deuterium-arc lamps emit continuous radiation principally in the ultraviolet. Xenon-arc lamps provide strong continuous emission throughout the near ultraviolet (from about 250 nm) and visible regions of the spectrum.

For atomic fluorescence measurements only the xenon-arc lamps seem capable of providing enough energy for reasonable sensitivity. The Eimac short arc high pressure xenon-arc lamp has proven to be particularly useful for atomic fluorescence studies (Chuang and Winefordner, 1975). The Eimac lamp is constructed with an integral elliptical reflector so that virtually all radiation from the point source is transferred to the atomic vapor.

*E. Tunable Lasers**

Sources of this type are not now widely used for atomic absorption or atomic fluorescence measurements, but they seem to offer great promise for

* Soffer and McFarland (1967).

the future if their cost can be reduced and their ease of operation improved. Most of the work so far reported on analytical atomic fluorescence spectroscopy with tunable lasers has been carried out in Winefordner's group at the University of Florida (Fraser and Winefordner, 1971, 1972; Omenetto *et al.*, 1973b,c) with a dye laser pumped by a nitrogen laser. A box diagram of their system is shown in Fig. 26. The dye laser output consists of pulses 2–5 nsec long at up to 25-Hz repetition rate with up to 50-kW peak power. The spectral bandwidth of the laser output is of the order of 0.1 to 1.0 nm and can be adjusted over the range 360–650 nm by changing the angle of the grating in the dye laser cavity and/or by changing the dye cell.

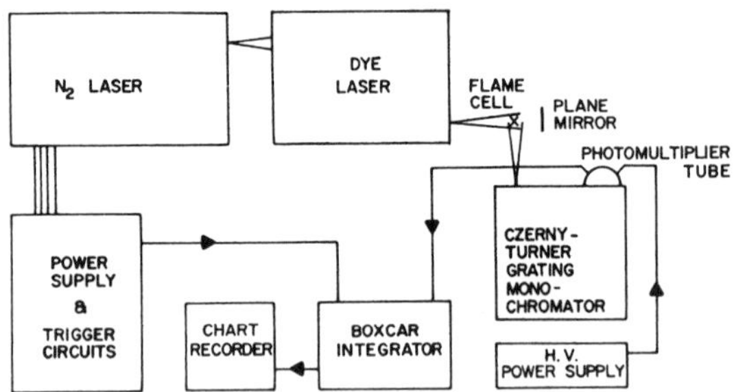

Fig. 26 Block diagram of experimental system for laser excited atomic fluorescence flame spectrometry. [Reprinted with permission (Fraser, L. M., and Winefordner, J. D. (1971). *Anal. Chem.* **43**, 1693). Copyright © by the American Chemical Society.]

The main advantages of the tunable laser sources are very high total power output, tuning capability over a wide wavelength region, and continuous scanning over a small wavelength region around the nominal wavelength chosen. The high cost of these excitation devices limits their use at present.

Because of the high radiant power available with the laser source, a saturation effect can be seen that is not observed with other excitation sources. Equation (14) shows no dependence on source intensity for the atomic absorption coefficient, but when the source radiant flux is large, the absorption coefficient becomes dependent on the source intensity. Saturation occurs when the populations of the upper and lower levels of the absorption transition are equal; beyond this point the radiant flux absorbed and the fluorescence intensity are completely independent of the source radiant flux. Although complete saturation is not achieved, a nonlinear dependence between fluorescence intensity and source intensity can be observed with laser excitation sources.

There are analytical advantages to operating at near saturation conditions. Near saturation, source instability does not affect the stability of the fluorescence signal. Moreover, near saturation, the fluorescence signal should be largely unaffected by collisional quenching. The linear relationship (for dilute atomic vapors) between fluorescence intensity and atomic concentration remains.

VI. Practical Atomic Absorption Spectrophotometry

A. Typical Equipment

The basic equipment for atomic absorption spectrophotometry is shown in Fig. 27. A wide range of equipment is commercially available (Veillon, 1972). This section will describe only typical features.

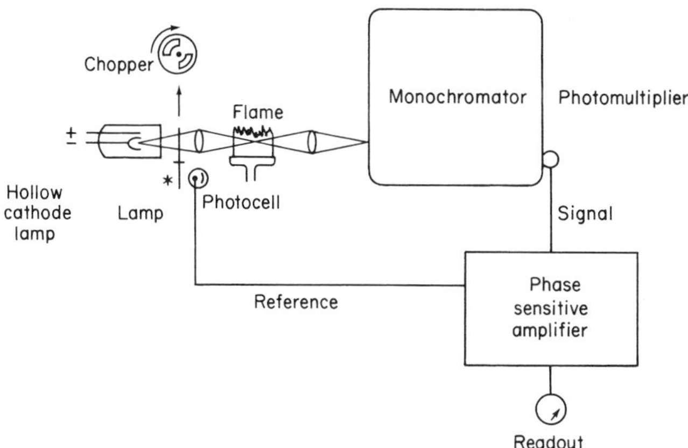

Fig. 27 Basic components for single-beam atomic absorption spectrophotometer.

The radiation source is almost invariably a sealed hollow cathode lamp (usually single element). A separate source is required for each element. The radiation is always modulated, either by mechanically chopping the radiation, as shown in Fig. 27, or by electronic modulation of the current to the lamp.

Premixed flames are usually employed. The burners are designed with slots 5–10 cm long to burn nitrous oxide–acetylene or air–acetylene flames.

The monochromator selects the narrow wavelength range of radiation which falls on the detector. Thus the monochromator limits the flame emission which reaches the detector and, in the ideal situation, isolates a single source line from the emission spectrum of the hollow cathode lamp. This process is shown in schematic fashion in Fig. 28.

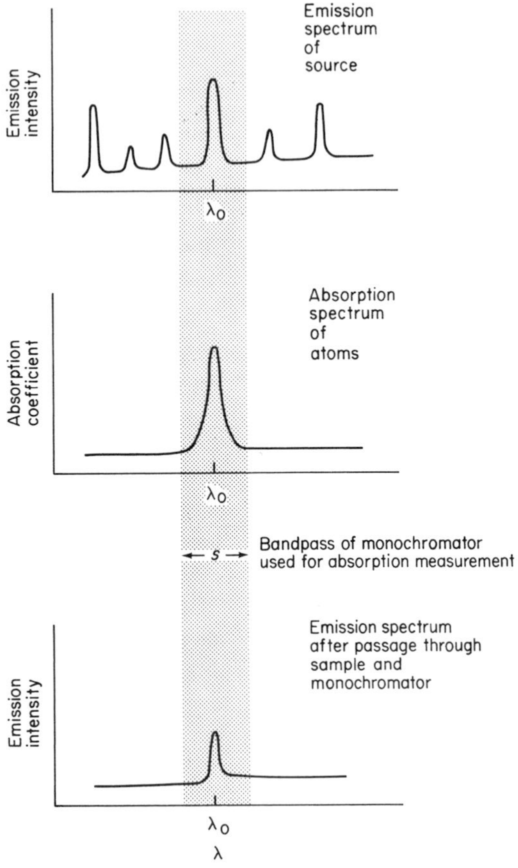

Fig. 28 Isolation of a source line by the monochromator and absorption of the line by atoms in the flame.

Monochromators in atomic absorption spectrophotometers are comparable to those used in other "table-top" ultraviolet–visible region spectrophotometers. The most important terms used in describing their properties are dispersion and aperture. The optical arrangement may also be of interest.

Dispersion is a measure of the ability of the device to separate two wavelengths of radiation. This is most often described as the wavelength interval per unit length in the plane of the exit slit of the monochromator, a quantity properly called the reciprocal linear dispersion. The smaller the reciprocal linear dispersion, the greater the ability of the monochromator to isolate a given wavelength of radiation. The monochromators used for atomic absorption spectrophotometry typically have reciprocal linear dispersions in the range 10–50 Å/mm. The theoretical spectral bandwidth of a monochromator is readily calculated by multiplying the reciprocal linear dispersion by the slit width (assuming the usual arrangement of a symmetrical optical system with equal entrance and exit slits). Spectral bandwidths used for practical atomic absorption measurements usually fall in the range 0.5–50 Å.

Aperture is a measure of the light gathering ability of the instrument, that is, the solid angle of light that can be accepted by the instrument. The aperture of an instrument (monochromator and external optics) is set by the component in the system with the smallest solid angle of acceptance. The aperture is described by the *f*-number of the system. For a circular lens the *f*-number is the ratio of focal length to diameter. Thus a lens with a focal length of 100 mm and a usable diameter of 10 mm is an *f*/10 lens. Note that the solid angle of acceptance increases as the focal-length-to-diameter ratio decreases. Thus, the smaller the *f*-number, the greater the light gathering ability of the system.

In atomic absorption spectroscopy, high light gathering ability is not of prime importance because the hollow cathode lamp is a relatively high brightness source which emits usable radiation through a rather narrow solid angle. The optical systems of atomic absorption spectrophotometers are typically around *f*/8.

Optical arrangements of instruments differ primarily in the optics external to the monochromator. Most instruments employ symmetrical grating monochromators with one of the three mounting arrangements shown in Fig. 29. Atomic absorption spectrophotometers can be single beam, as shown in Fig. 27, double beam, or dual double beam.

In a double-beam instrument, radiation from the source is divided into two beams, only one of which (the sample beam) passes through the flame. Beyond the flame the two light paths are recombined so that the sample and reference beams follow identical paths through the monochromator to the detector. In the usual arrangements the beams are chopped so that the sample and reference beams alternate in reaching the detector, and the ratio of the two signals is taken by the electronic system. Thus, the double-beam system minimizes the effects of variations in lamp intensity, detector sensitivity, optics, and electronics.

In the dual double-beam instrument, illustrated in Fig. 30, two external sources are used simultaneously in a double-beam arrangement. Since the lines of interest in the two channels are usually of different wavelengths, two monochromator–detector systems or one monochromator–detector system and one filter–detector system are employed. The additional channel provides several possibilities: an internal standard element can be used to correct for fluctuations in the atomization process, two elements can be determined simultaneously in a single sample, or the second channel can be used to correct for nonspecific absorption or scatter in the flame.

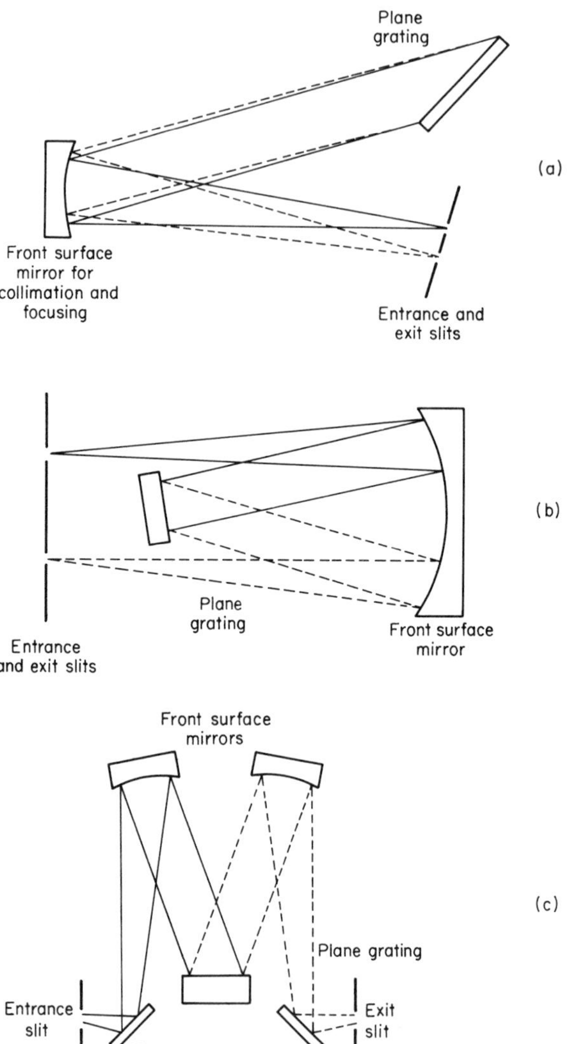

Fig. 29 Typical optical arrangements for "table-top" grating monochromators. (a) Littrow, (b) Ebert, (c) Czerny–Turner with in-line slits.

Photomultiplier detectors are always used. Frequently these are of the multialkali type to provide adequate sensitivity over the wavelength range extending from the arsenic line at 1937 Å to the cesium line at 8521 Å.

Phase sensitive detection (lock-in amplification) is frequently employed. The electronics system is usually designed so that the output signal is linear in the fraction of radiation absorbed or absorbance.

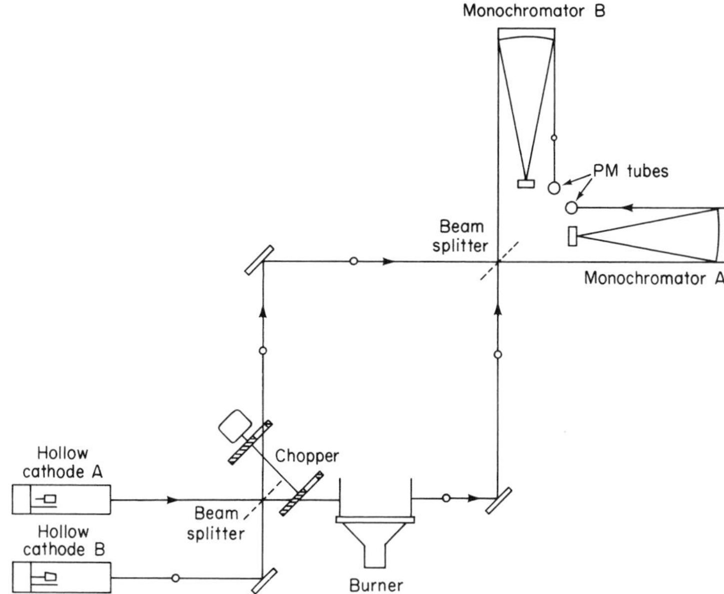

Fig. 30 Optical diagram of a dual double-beam atomic absorption spectrophotometer.

The major manufacturers of atomic absorption equipment offer a variety of accessories for their instruments, the most important of which are electrothermal atomizer systems as replacements for the flame atomizers.

B. Choice of Analytical Line and Possible Spectral Interferences

In analytical atomic absorption spectroscopy the primary function of the monochromator is to isolate a single source line from the emission spectrum of the hollow cathode lamp. The lamp spectrum may contain, in addition to the atomic lines of the element of interest, ionic lines of the element, atomic and ionic lines of the fill gas, and perhaps emission features due to impurities in the fill gas or cathode material. A more complex spectrum is, of course, produced when a multielement cathode is used. Thus a large number of lines may be present in the emission spectrum, but, because the lines are narrow and often relatively far apart, for many elements there is no difficulty in isolating a single line with the moderate resolution monochromators typically employed in atomic absorption equipment. In this section we will consider the factors governing the choice of a measurement line, and the types of spectral interferences which are occasionally encountered.

Any radiation present within the monochromator bandwidth other than the analytical line has an unfavorable effect on the measurement. This statement can be justified by considering the expressions for the absorbance. If only the analysis line is present, the absorbance is given by

$$A = -\log(\phi_t/\phi_0) \tag{43}$$

where ϕ_t and ϕ_0 are the radiant fluxes transmitted and incident, respectively. If additional radiation is present within the monochromator bandwidth, the absorbance expression is

$$A = -\log[(\phi_{t,1} + \phi_{t,2})/(\phi_{0,1} + \phi_{0,2})] \tag{44}$$

where the subscripts 1 and 2 indicate the primary line and the unwanted radiation.

If the unwanted radiation cannot be absorbed by the analysis element (e.g., an ion line or fill gas line), then at every concentration of the element of interest the measured absorbance will be less than the absorbance given by Eq. (43), and as the atomic concentration of the analysis element increases the absorbance will approach a finite value given by

$$A = -\log[\phi_{t,2}/(\phi_{0,1} + \phi_{0,2})] \tag{45}$$

It is evident, therefore, that the effect of the unwanted radiation is to decrease the sensitivity of the measurement and to cause bending of the analytical curve (a plot of absorbance versus concentration).

If the unwanted radiation corresponds to a line which can be absorbed by the analysis element, but for which the absorption coefficient is smaller than for line 1, then the situation is much the same. At every concentration the absorbance will be less than that for the stronger absorption transition taken alone, and, since as the concentration increases the relative sizes of $\phi_{t,1}$ and $\phi_{t,2}$ change, bending of the analytical curve will occur.

If the unwanted radiation is a spectral line, it can sometimes be eliminated by decreasing the slit width (and thus the spectral bandwidth) of the monochromator. The best slit width can be determined by measuring the absorbance as a function of slit width. As long as only a single line is passed, the absorbance should be constant, but as soon as another line appears within the spectral bandwidth, the absorbance begins to decrease. The best slit width is the largest value for which no decrease in absorbance is found. This value provides the largest radiant flux, which permits the lowest gain setting, and therefore provides the largest signal-to-noise ratio.

The relative intensities of atomic and ionic lines depends on choice of fill gas and other details of construction of the lamp, and interfering fill gas lines can be eliminated by use of a different fill gas, but these changes are beyond the reach of the experimenter who is using a commercial sealed hollow cathode lamps. If the unwanted radiation cannot be eliminated even at the smallest available slit setting, then the only alternatives to accepting the decreased sensitivity and analytical curve bending are choice of a different analysis line or purchase of a lamp of a different design.

Some older hollow cathode lamps exhibit appreciable continuous emission due to hydrogen impurity. Complete elimination of the unwanted radiation is obviously impossible in this situation, but improvement is obtained by decreasing the monochromator slit width.

A more serious type of spectral interference occurs when the analysis line emitted by the source is attenuated in the flame by some species other than atoms of the element of interest. Such attenuation can be due to an actual overlap of the source line profile by the absorption line of another element, or it can be due to molecular absorption or scattering of the source radiation by particles in the flame.

Parsons *et al.* (1975) have tabulated a number of predicted and observed direct spectral overlaps in atomic absorption spectroscopy. There are two possible remedies for an interference of this type: select a different analysis line or separate the interfering element from the analysis element prior to measurement.

Molecular absorption and scattering of the source radiation have been observed in flames and, to a greater extent, in electrothermal atomizers. The occurrence of this type of interference can be detected by measuring the apparent absorption of a blank solution which contains the major constituents of the sample but none of the analysis element. When the preparation of a blank solution is difficult, an alternative procedure is to measure the apparent absorption at a wavelength offset from but near to the absorption line of interest. With the usual atomic absorption equipment, this can be accomplished by finding a nearby nonresonance line in the spectrum of the source lamp, by substituting another source lamp with a strong nearby line of an element not present in the sample, or by substituting a continuous source providing good emission intensity near the wavelength of the analysis line. Correction for molecular absorption and scatter may be made by subtracting the background "absorbance" measured by one of the methods already described from the absorbance of the sample solution at the resonance line wavelength. Background correction is important to the successful use of electrothermal atomizers, and devices for automatic background correction, described in the next section, have been developed.

Equation (34) suggests that the line chosen for analysis should be the one for which $n_l f_{lu}$ has the largest value. For most elements this line will be the resonance line (i.e., corresponding to a transition from the ground state) with the largest oscillator strength. The actual choice may differ from this for several reasons.

If the ground state is a multiplet with only slightly differing energies, the number of atoms in a state of slightly higher energy and much larger statistical weight may be larger than the number of atoms in the ground state. The Boltzmann equation [Eq. (1)] predicts this, since for small energies the exponential term approaches 1 and the ratio of the populations is then approximately equal to the ratio of the statistical weights. Aluminum provides an example. The aluminum line at 3082 Å corresponds to a transition, with an oscillator strength $f = 0.22$, from the ground state, which has a statistical weight $g_0 = 2$. The 3093 Å line is more sensitive and is the line of choice for atomic absorption measurements. It has an oscillator strength of 0.23 and corresponds to a transition from a state which is 0.13 eV above the ground state but has a statistical weight $g_1 = 4$.

Other factors affecting the choice of an analysis line are spectral interferences, the relative intensities of source lines, optical system transmission characteristics, and detector spectral sensitivity. Spectral interferences have already been described. The remaining factors relate to the size of the signal produced by the system and hence to the signal-to-noise ratio that can be achieved. Some elements have their strongest lines so deep in the ultraviolet that they are entirely inaccessible. The mercury resonance line at 1850 Å is a good example of this. For other lines the transparency of the flame or the poor sensitivity of the detector at that wavelength may be a consideration.

C. *Automatic Correction for Background Absorption and Scatter*

Methods of automatic background correction make use of a continuous source, usually a deuterium arc lamp, in an arrangement such as that shown in Fig. 31. A reflective half-sector chopper is used to alternately pass radiation from the hollow cathode lamp and the deuterium arc lamp through the atomizer. The electronic system is that of a double-beam spectrophotometer and provides the ratio of the signals due to the two beams. A variable attenuator is placed in the beam of the deuterium arc lamp. With no sample in the atomizer the attenuator is set to make the signals due to the two beams equal. When a sample is atomized, the ratio of the signals in the two beams is the transmittance of the sample corrected for background absorption and scatter. That this is so can be seen from the following considerations.

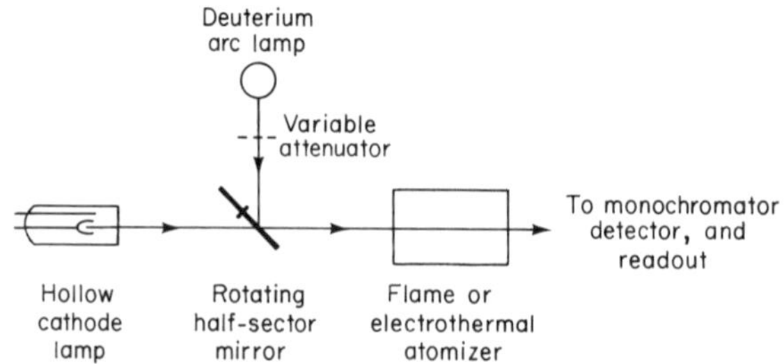

Fig. 31 Schematic arrangement of a deuterium background corrector.

The absorbance of the sample occurring with the line source A_L is the sum of the absorbance due to atomic absorption A and the background "absorbance" due to molecular absorption and scatter, A_b:

$$A_L = A + A_b \tag{46}$$

The apparent absorbance occurring with the continuous source, A_C, is to a good approximation just the background "absorbance":

$$A_C = A_b \tag{47}$$

As shown in Eq. (31), the contribution of atomic absorption to the absorbance with a continuous source can be made negligibly small if the monochromator spectral bandwidth is large. The background "absorbance," however, is a broad-band phenomenon and is independent of the monochromator bandwidth. Spectral bandwidths of several angstroms are customarily employed when using an automatic background corrector. Thus, the desired quantity, the background corrected absorbance A, is given by

$$A = A_L - A_C \tag{48}$$

From the definitions of absorbance and transmittance it follows that

$$A = \log(\phi_{L,0}/\phi_{L,t}) - \log(\phi_{C,0}/\phi_{C,t}) \tag{49}$$

$$= \log(\phi_{L,0}\phi_{C,t}/\phi_{L,t}\phi_{C,0}) \tag{50}$$

where $\phi_{L,0}$, $\phi_{L,t}$, $\phi_{C,0}$, and $\phi_{C,t}$ are the incident and transmitted (0 and t subscripts) radiant fluxes for the line and continuous (L and C subscripts) sources. Since the two beam signals are initially equal,

$$\phi_{L,0} = \phi_{C,0} \tag{51}$$

and

$$A = \log(\phi_{C,t}/\phi_{L,t}) \tag{52}$$

From Eq. (52) it is apparent that the ratio $\phi_{L,t}/\phi_{C,t}$ is the background corrected transmittance of the sample.

D. *Scale Expansion*

Scale expansion facilities are useful on atomic absorption equipment for accurate readings on low concentration samples. Scale expansion is simply an electronic multiplication of the output signal by a known factor. As such it does not improve the inherent signal-to-noise ratio of the information since both signal and noise are multiplied by the same factor. It is advantageous only when the accuracy of the reading is degraded by the coarseness of the graduations on the readout device. For many atomic absorption measurements the stability of the lamp output and the atomization process is such that the noise corresponds to less than 1% absorption, and scale expansion is desirable for small signals.

In single-beam spectrometers with transmittance readout, scale expansion is achieved by increasing the overall gain by the required factor, then backing off the total signal with the zero control until the reading is again 100% transmittance with no sample in the atomizer. For double-beam instruments with readout linear in fraction absorbed or absorbance, scale expansion is easily implemented with a single control. The readout signal is simply multiplied by the required factor by controlling the gain on the final amplifier stage.

E. Sample Preparation

Many aqueous solutions can be nebulized without prior treatment. Others, which cannot be nebulized directly because of high total solids or other problems, can be nebulized after a simple dilution, provided the analysis element is present in sufficiently high concentration. Still others may require only the addition of an ionization suppressor or releasing agent (to overcome interferences in the atomization process) prior to nebulization.

Nonaqueous solutions can sometimes be nebulized directly provided standards of similar composition can be prepared and the properties of the solvent are compatible with spraying and burning. Aromatic solvents tend to produce unsteady flames with high background emission. Chlorinated hydrocarbons produce HCl and possibly phosgene when burned, placing high demands on the exhaust system. Oxygenated hydrocarbons such as methyl isobutyl ketone (4-methyl-2-pentanone) and amyl acetate are generally preferred. Such samples as engine oils, paints, and vegetable oils are frequently nebulized directly following dilution with a solvent such as methyl isobutyl ketone.

Organic solids generally must be digested before carrying out atomic absorption measurements. Both wet and dry ashing procedures have been used. For some samples the elements of interest can be obtained in solution by an acid extraction procedure in which the sample is shaken several hours in contact with a hydrochloric or nitric acid solution. Such a procedure has been used, for example, in the determination of nickel in fats.

Methods for the dissolution of metals, rocks, ores, cements, etc., have been described in the reference works appropriate to the field. Where final solutions contain more than about 0.5% total dissolved solids, standards should also contain the major constituents to match viscosity, surface tension, and particle-size properties and avoid "matrix" interference effects.

Sometimes it proves desirable to separate the element of interest from its matrix prior to measurement. There are two chief reasons for such a step:

(a) the concentration of the element of interest is so low that preconcentration is needed, and
(b) other dissolved solids will interfere in the determination.

The separation methods used for atomic absorption spectroscopy do not differ greatly from those used for any other purpose. One distinction is worth noting: the atomic absorption method is quite selective and hence the separation method need not be very selective. In fact, a method suitable for separating a whole group of elements is generally preferable to a single element separation procedure.

Solvent extraction techniques have received the greatest attention as separation and concentration methods for atomic absorption spectroscopy. The choice of complexing agent for the extraction is not limited, as in colorimetric procedures, by the need for strong color development or specificity. Single-stage batch extractions are generally used with reagents capable of extracting a number of elements. The systems based on ammonium pyrrolidine dithiocarbamate as the complexing agent are typical. This reagent is said to be useful for extraction of 30 elements with proper choice of pH range.

F. Calibration and Measuring Techniques

Analytical atomic absorption spectroscopy is not an absolute method but, in common with most instrumental methods, must be calibrated. There are two general approaches for calibration: the analytical curve method and the standard additions method.

For the analytical curve method, a series of standards is prepared to contain known concentrations of the element of interest. These standards should have overall compositions as nearly like the sample as possible, and the range of concentrations should include the concentrations of analyte element in all the samples. Absorption measurements are then carried out for all the samples and standards. The results for the standards are used to construct the analytical curve. The fraction absorbed can be plotted versus concentration if only a narrow range of concentrations is of interest or if the fraction absorbed does not exceed 0.1. For a wider range of concentrations absorbance should be plotted since it will provide the most nearly linear analytical curve. The concentrations of the samples can then be read directly from the analytical curve. Because of the large number of variables which affect the absorption reading (see, e.g., Fig. 9), frequent checks of the calibration should be made.

The standard additions method requires only one or two standard solutions. It may be implemented in various ways, but one procedure suitable for atomic absorption measurements is as follows. Take four separate, equal aliquots of the sample solution. To one of these add an exactly known amount of the analysis element approximately equal to the amount initially present. To a second add approximately twice as much, and to a third add approximately three times as much of the analysis element. To all four add whatever additional reagents are required (e.g., ionization suppression reagent), dilute all to the same volume, and read the absorbance of each. Plot the absorbance versus the added amount of the element of interest as shown in Fig. 32. The straight line drawn through the four points intercepts the negative concentration axis at a value corresponding to the amount of the analysis element in the aliquot with no added element. If the standard additions method is to give an accurate result, the analytical curve for the element must be linear and pass through zero.

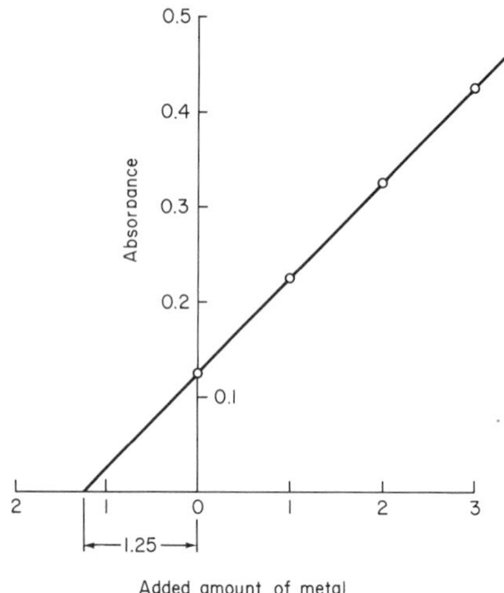

Fig. 32 Graphical evaluation using the standard additions method.

Two situations suggest the use of the standard additions method. Often the sample matrix is not completely known and cannot be sufficiently well duplicated to allow use of the analytical curve method. If the volume of the standard added to the sample in the additions method is small, then the matrix is undisturbed, and any matrix effect will be the same for measurements with the original sample and samples with additions. Secondly, the standard additions method requires preparation of only a single standard solution of which different volumes can be used to prepare the samples with additions. Therefore, if only a few samples are to be determined, the standard additions method requires less labor than the analytical curve method.

The preceding descriptions of the analytical curve and standard additions methods speak of plotting the data. With modern data handling equipment such "plots" may reside in the memory of a computer or calculator only as the coordinates of a best fit line. When properly applied such an approach should improve the accuracy of atomic absorption measurements by reducing the uncertainties associated with any graphical approach.

In preparing standard solutions it is customary to prepare a fairly concentrated stock solution (usually 100 or 1000 μg/ml) and prepare standards as required by serial dilution. Dilute standards (1 μg/ml or less) should not be stored because they often decrease rapidly in concentration, apparently due to adsorption of metal ions on the walls of the storage container. Dean and Rains (1971) have provided detailed directions for the preparation of standard stock solutions for most of the elements determinable by atomic absorption spectroscopy.

G. *Sensitivity, Limit of Detection, Precision, and Accuracy*

A commission of the International Union of Pure and Applied Chemistry (1972) has attempted to clarify the meaning of these terms as they apply to spectrochemical procedures. The commission has, in particular, noted ambiguity in the use of the term sensitivity in various disciplines and recommends that sensitivity be defined as "the differential quotient of the 'characteristic response' function, or—what is the same—the slope of the 'characteristic' curve." The "characteristic" curve for atomic absorption spectroscopy is, of course, what we have referred to as the analytical curve. This definition corresponds to the ordinary perception of sensitivity. We say a procedure is sensitive when a small variation in concentration produces a large change in the quantity observed (e.g., in the absorbance). Note that sensitivity is a concentration dependent quantity unless the analytical curve is linear, and it has units of response per unit concentration interval. Unfortunately, in atomic absorption, sensitivity has been used in a way not consistent with the IUPAC recommendations. It has been defined as that concentration which causes 1% absorption.

By either definition of sensitivity, it can be seen from Eq. (34) that the sensitivity depends on the efficiency with which the element in the sample is converted to atoms in the atomizer, on the absorption path length, and on the oscillator strength of the transition. Thus sensitivity is primarily dependent on the properties of the element and the atomizer and largely independent of the rest of the instrumental system.

A fraction absorbed of 1% (0.0044 absorbance) represents a convenient index point, and it is interesting to attempt to estimate the solution concentration required to produce this absorption for typical properties of the element and flame atomizer. Inserting the values $f_{lu} = 1.00$, $l = 10$ cm, $\Delta\nu_D = 7.0 \times 10^9\ \text{sec}^{-1}$, into Eq. (34), the atomic concentration required is approximately 3×10^8 atoms/cm^3. With typical flame atomizer properties and assuming complete conversion of the element to the atomic form, approximately 10^{14} atoms/cm^3 are produced for a 1 *M* solution. Thus a 3×10^{-6} *M* solution should produce approximately 1% absorption. This corresponds to a concentration, in the usual units of atomic spectroscopy, in the range 0.1–1 μg/ml.

With most instruments, less than 1% absorption can be measured with good precision. The lowest value which can be read is ultimately determined by fluctuations in the signal (Winefordner and Vickers, 1964). Such fluctuations may arise from fluctuations in the intensity of the source, fluctuations in the atomization process, noise in the detector, and noise in the amplifier and readout system. This lowest value, however specified, is thus a figure of merit for the complete analytical procedure. The limit of detection is such a figure of merit. It is the quantitative answer to the question: What is the least concentration (or smallest weight) of a substance whose presence can be reported with a chosen degree of certainty by a complete analytical procedure? The definition is usually stated in terms of a signal-to-noise ratio, where "noise" is understood to include all uncontrolled fluctuations in the system.

The noise is defined in statistical terms. It is the scatter of measurements of the blank value as described by the standard deviation. It is found experimentally by making a sufficiently large number (at least 20) of blank measurements and calculating the standard deviation in the usual fashion. The limit of detection is then defined to be that concentration which produces a background corrected signal equal to three times the standard deviation of the blank measures. The choice of the factor of 3 leads to a confidence level of about 90% in establishing that the measured signal differs from random fluctuations in the blank.

Detection limits may be relative or absolute. Relative detection limits are given in concentration units. Absolute detection limits are given in weight units. For flame atomizers, relative detection limits are appropriate. Absolute detection limits are usually reported for electrothermal atomizers since with these devices the signal depends not on the concentration of the solution but on the total amount of material placed in the atomizer.

The precision of an analytical procedure is most conveniently expressed in terms of the relative standard deviation. It is defined as the ratio of the standard deviation to the mean, both in the same units of concentration or amount. It should be noted that when the analytical curve is nonlinear, the relative standard deviation of the signal is different from the relative standard deviation of the analytical result (concentration or weight).

Accuracy is the proximity of the analytical result to the true analyte concentration in the sample. It is, therefore, much more difficult to assess than precision. The best way to assess the accuracy of a procedure is to carry out an analysis on a certified reference material and compare the result with the certificate value. If no standard reference material is available, then it may be possible to compare the results with the results by other reliable methods on the same sample. If neither of these procedures is possible, as is often the case for trace analysis, then recovery and dilution tests provide some evidence of the validity of the result.

VII. Practical Atomic Fluorescence Spectrometry

A. Typical Equipment

The basic equipment for atomic fluorescence spectrometry is shown in Fig. 33. There are no commercially available systems designed specifically for atomic fluorescence measurements, and hence the burden of instrument design falls on the individual experimenter. There are many similarities between the equipment for atomic fluorescence and that for atomic absorption, shown in Fig. 27, but the differing requirements of the two kinds of measurements must be recognized if an atomic absorption spectrophotometer is to be adapted for fluorescence measurements. In absorption measurements the monochromator-readout system is viewing a rather intense light source. As a consequence, flame background emission is relatively unimportant, and a rather inefficient light gathering system with a small solid

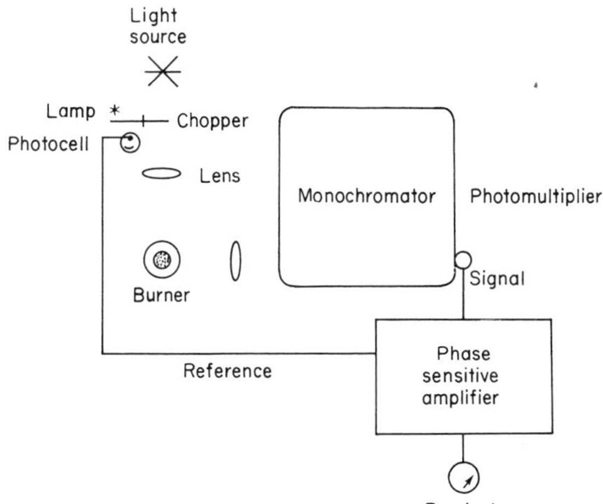

Fig. 33 Basic components of an atomic fluorescence spectrometer.

angle of acceptance is adequate. In fluorescence measurements the radiation signal is weak, flame background emission can be a very important contributor to noise in the measurement, and an efficient light gathering system with a large solid angle of acceptance is desirable.

The radiation source has most often been an electrodeless discharge lamp. Pulsed hollow cathode lamps and tunable lasers seem likely to grow in importance.

Premixed flames, square or circular in cross section, with provision for flame separation, have proven most useful for fluorescence measurements. Nitrous oxide–acetylene and air–acetylene flames seem to have the greatest analytical utility, but many atomic fluorescence measurements have been made with hydrogen flames, including relatively cool flames, such as hydrogen–argon–entrained air, because of the low background emission of these flames.

The entrance optics play a more significant role in atomic fluorescence measurements than in atomic absorption measurements, because of the need for efficient transfer of exciting radiation from the source to the atomizer

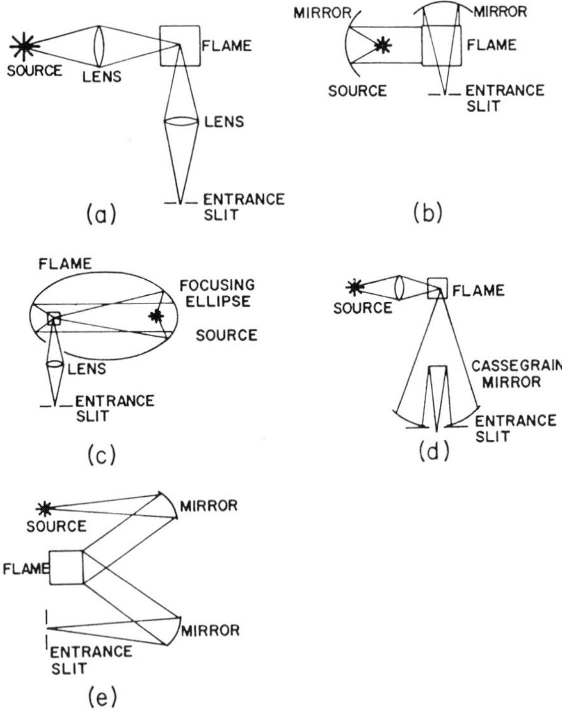

Fig. 34 Arrangements of entrance optics used for atomic fluorescence spectrometry. [Reprinted with permission (Winefordner, J. D., and Elser, R. C. (1971). *Anal. Chem.* **43**, 34A). Copyright © by the American Chemical Society.]

and of fluorescence radiation from the atomizer to the monochromator–detector system. Some of the arrangements which have been used are shown in Fig. 34.

The discussion of monochromators in the section on atomic absorption is pertinent here. Dispersion is less important and aperture more important for atomic fluorescence measurements than for atomic absorption measurements. The fluorescence spectrum of an element is always less complex than the emission spectrum of the source lamp since radiation will be absorbed, exciting fluorescence, at only a few wavelengths for a given element. Thus less demand is placed on the dispersing power of the monochromator for fluorescence measurements. If a single element lamp is used for excitation, only the fluorescence spectrum of that element will be excited in most cases. Since the fluorescence intensities are additive, it is not necessary, or even desirable, that any single line be isolated from the other lines of the fluorescence spectrum. The chief role of the monochromator is to limit the background emission reaching the detector by isolating a relatively narrow wavelength range.

Detectors and amplifiers are basically the same for both absorption and fluorescence measurements.

B. Choice of Analytical Line and Possible Spectral Interferences

Direct overlap of spectral lines can give rise to a spectral interference in atomic fluorescence just as in atomic absorption. Atomic fluorescence offers the possibility of avoiding such interferences by the use of nonresonance fluorescence. When overlap occurs in the resonance transition it is quite unlikely that overlap will also occur in the nonresonance lines of the fluorescence spectrum.

Deliberate use has occasionally been made of spectral overlap for exciting atomic fluorescence. For example, the fluorescence of bismuth has been excited with an iodine electrodeless discharge lamp. Iodine has a strong nonresonance line at 2061.6 Å which overlaps the bismuth absorption line at 2061.7 Å sufficiently well to excite resonance fluorescence (at the same wavelength) and direct line fluorescence at 2696.7 and 3024.6 Å.

When a multielement or continuous excitation source is used for atomic fluorescence the likelihood of a spectral interference is greatly increased, since the fluorescence lines of elements other than the analysis element may fall within the bandwidth of the monochromator. Such interferences can sometimes be eliminated by decreasing the slit width, and hence the spectral bandwidth, of the monochromator. Alternately, a more selective excitation source can be used or the offending exciting line can be filtered out.

A more serious problem, which can not be eliminated by improving spectral resolution, is scattering of the source radiation by particles in the atomizer. Light scattering depends on the size and number of the particles per unit volume. In general, therefore, greater scattering is observed with total consumption than with premixed burners, and scatter decreases with increasing height in the flame or increasing flame temperature or when more volatile solvents are used. Thus light scattering is particularly troublesome in the "cool" flames which might otherwise offer desirable features for atomic fluorescence measurements.

The most elegant approach to avoiding spectral interference due to scatter is to measure a line not present in the exciting radiation. This can be accomplished by measuring nonresonance fluorescence where the corresponding line has been filtered out of the excitation source. Excitation of bismuth fluorescence with an iodine lamp offers a slightly different approach for achieving the same end. The fluorescence is excited at 2061.7 Å and measured at 2696.7 or 3024.6 Å. The lamp emits the iodine line at 2061.6 Å but no radiation corresponding to the two longer wavelength lines.

Unfortunately, there are relatively few instances where the approach described above is practical. Often the nonresonance fluorescence is not sufficiently sensitive or the source line corresponding to the nonresonance fluorescence is so close to the wavelength of the exciting line that it is not possible to find a filter which will remove one while transmitting the other and still provide sufficient intensity for reasonable sensitivity.

Another way of dealing with scattered radiation is to estimate its contribution and subtract it from the measured signal. If a continuous source is used for excitation, this is a simple matter of scanning across the line to measure the background contribution on both sides of the line. If a line source is used for excitation, scattering is usually less important but more difficult to correct for. The scattering contribution may be measured by observing the apparent fluorescence with a control sample with a matrix identical to the original sample but free of the analysis element. Where this proves difficult it may be possible to measure the scatter by using a nearby line which can not produce fluorescence. Such a procedure is more difficult for fluorescence measurements than for absorption measurements since the fluorescence and scatter signals are, unlike absorption signals, dependent on the source intensity and detector sensitivity. Thus to correct for scatter by this procedure it is necessary to first establish equal detector response for the correcting and analysis line. An example of this type of procedure has been described by Sychra and Matousek (1970). In determining Ni in oils, the scatter signal was measured with an Au lamp at 2428 Å and subtracted from the scatter plus fluorescence signal measured with an Ni lamp at 2320 Å. The current of the Au hollow cathode lamp was set to give the same detector signal as the Ni lamp when pure xylene was sprayed with the flame turned off.

Rains *et al.* (1974) have described an automatic system to correct for light scatter in atomic fluorescence measurements. Radiation from an electrodeless discharge lamp and a 150-W xenon lamp continuous source is alternately passed through the flame atomizer. The signal produced with the electrodeless discharge lamp is due to fluorescence plus scatter but the signal produced with the xenon lamp is due almost entirely to scatter. The detector output is amplified with a phase sensitive amplifier. The phase of the reference signal derived from the rotating mirror chopper is such that the signal due to the line source is positive and that due to the continuous source is negative so that the amplifier output is the fluorescence signal corrected for scatter. The signals due to the two sources were initially made equal by spraying a strongly scattering solution (no fluorescence) and adjusting the intensity of the continuous source with an iris diaphragm until the amplifier output was zero.

From Eq. (35) it can be seen that the intensity of the fluorescence emission depends on the intensity of radiation absorbed and on the quantum efficiency. Thus in choosing a fluorescence line we must consider the strength of both the absorption and emission transitions.

Figure 35 shows an energy level diagram for Mg. The other alkaline earth elements and Zn, Cd, and Hg have similar diagrams. With a diagram such as this it is easy to anticipate a simple fluorescence spectrum with the

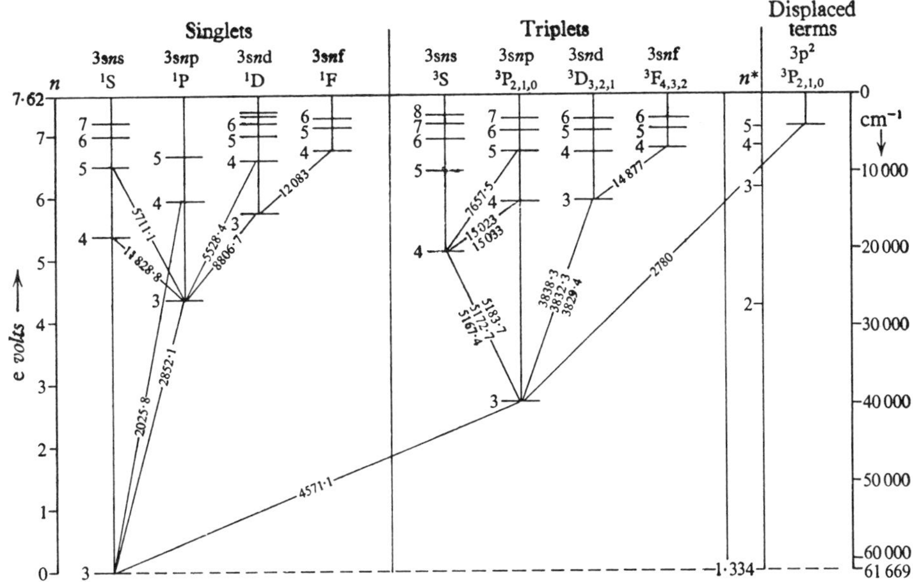

Fig. 35 An energy level diagram for Mg. [Kuhn, H. G. (1969). "Atomic Spectra," 2nd ed., p. 186. Academic Press, New York.]

strongest line due to resonance fluorescence. For Mg the strongest line is at 2852 Å. The 4571 Å line is formally forbidden because of violation of the selection rule governing multiplicity. For Hg the corresponding lines occur at 1850 and 2537 Å. The 1850 Å line is strongly absorbed by oxygen and the strongest line normally observed in the fluorescence spectrum is the line at 2537 Å arising from the "forbidden" triplet to singlet transition.

Lead provides an example of an element with a somewhat more complicated fluorescence spectrum. A partial energy level diagram for Pb is shown in Fig. 36. Only the $^3P_0$ ground term is appreciably populated at flame temperatures, and the 2833 Å line is the strongest absorption line. Unlike the Mg case, however, there are four routes for radiational deactivation of the excited state, and, neglecting reabsorption, the four fluorescence lines will, from Eq. (35), have radiant fluxes in the ratio

$$\phi_{f1}:\phi_{f2}:\phi_{f3}:\phi_{f4} = Y_{p1}:Y_{p2}:Y_{p3}:Y_{p4} \tag{53}$$

From the definition of Y_p [Eqs. (36) and (37)], it can be shown that

$$\phi_{f1}:\phi_{f2}:\phi_{f3}:\phi_{f4} = A_1:A_2(v_2/v_1):A_3(v_3/v_1):A_4(v_4/v_1) \tag{54}$$

where the subscripts indicate the transitions as labeled in Fig. 36. The pertinent data are shown in Table III. The transition probability for the 4057.8 Å line is by far the largest and thus this line will be the strongest in the Pb fluorescence spectrum.

In choosing a measurement wavelength the quantity of concern is not just the inherent intensity of the line but the detector signal due to the line. The signal depends on the transmission properties of the optical system and

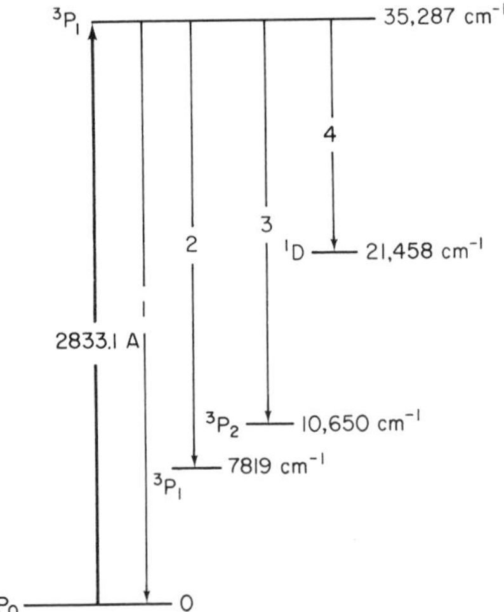

Fig. 36 A partial energy level diagram for Pb.

TABLE III

Data for Pb Transitions of Fig. 35

Line	Wavelength (Å)	A_{ul} (sec^{-1})
1	2833.1	0.6×10^8
2	3639.6	0.4×10^8
3	4057.8	3.1×10^8
4	7229.0	0.02×10^8

the spectral response characteristics of the detector in addition to the line intensity. Experimentally the wavelength providing the strongest fluorescence signal is easily established by scanning the spectrum.

C. *Nondispersive Atomic Fluorescence*

As noted in the discussion of monochromators in Section VII.A, the primary role of the monochromator in atomic fluorescence measurements with a single-element source is to limit the amount of atomizer background emission reaching the detector. If the background emission is sufficiently low, considerable benefit may be derived by eliminating the monochromator and expanding the spectral bandwidth of the measurement system:

(a) greater energy throughput can be achieved if the system is no longer limited by the narrow slits and relatively small solid angle of acceptance of most monochromators, and

(b) for elements with a multiline fluorescence spectrum, perhaps several lines can contribute to the detector signal.

Several nondispersive systems have been developed to obtain these advantages (Vickers *et al.*, 1972; Muscat *et al.*, 1975). Some systems use bandpass filters to limit the half-intensity spectral bandwidth to approximately 100 Å. With others the spectral bandwidth is limited only by the spectral response characteristics of a "solar-blind" detector, thus providing an instrument which can respond to any fluorescence in the range from 3200 Å to the ultraviolet cutoff of the atmosphere.

The background emission spectra of several flame atomizers are shown in Fig. 37. These spectra were recorded using a "solar-blind" photomultiplier and thus indicate the background signal which would be present with a pure "solar-blind" nondispersive system. (Specifically, the background signal for each atomizer would be proportional to the area under the spectrum shown.) The strong band emission at the long wavelength end of the spectra is due to OH and obviously would be a major contributor to background

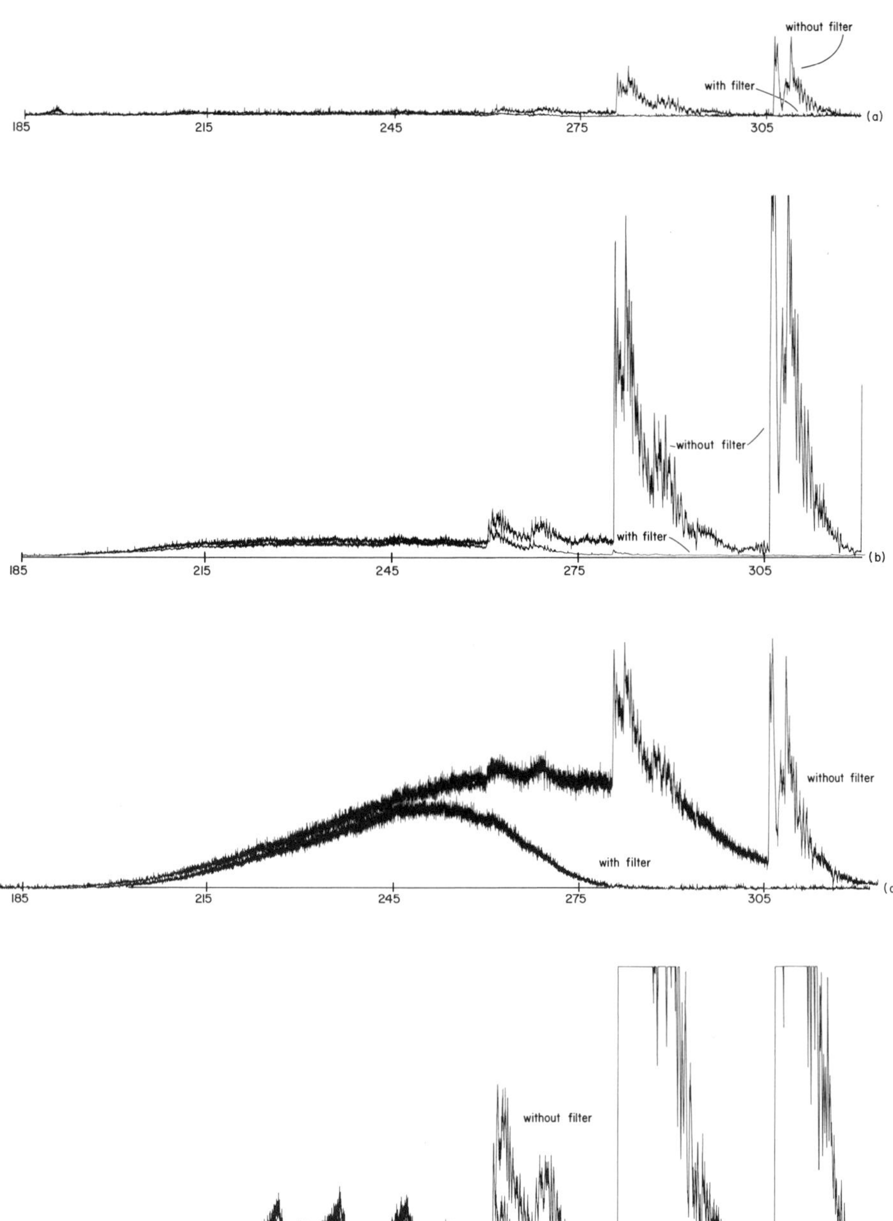
without filter
with filter
(a)
185
215
245
275
305
without filter
with filter
(b)
185
215
245
275
305
without filter
with filter
(c)
185
215
245
275
305
without filter
with filter
(d)
185
215
245
275
305

signal with a nondispersive system. The OH band emission can be greatly reduced by flame separation or by use of a cutoff filter with the transmission characteristics shown in Fig. 38. This filter consists of an approximately 50-mm thickness of chlorine gas at 1-atm pressure contained in a quartz tube. The effect of the filter on the background signal is shown in Fig. 37.

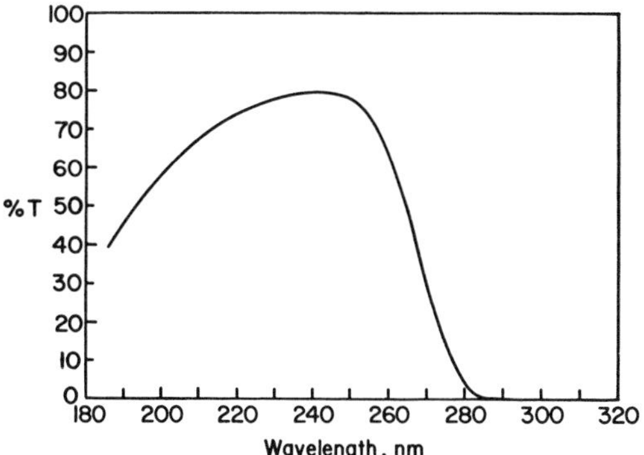

Fig. 38 Transmission characteristics of chlorine filter. [Reprinted with permission (Vickers, T. J., Slevin, P. J., Muscat, V. I., and Farias, L. T. (1972). *Anal. Chem.* **44**, 932). Copyright © by the American Chemical Society.]

Several examples may serve to demonstrate the advantages and limitation of nondispersive atomic fluorescence measurements.

The fluorescence spectrum of Hg consists of the single line at 2537 Å. Thus the greater spectral bandwidth of a nondispersive system offers no advantage. Compared to a dispersive system, Muscat *et al.* (1975) found the energy throughput advantage provided enhancement of the fluorescence signal by a factor of about 50 with a "solar-blind" system. When a bandpass filter centered on the Hg line was used, the enhancement factor was about 10. The bandpass filter had a peak transmittance of about 0.2 and a half-intensity spectral bandwidth of 80 Å.

Iron has a complex spectrum as shown in Fig. 39. Spectrum A shows the emission from a hollow cathode lamp in the 2450 to 2550 Å range. The number of lines suggests that considerable possibility for spectral interference exists and if the sample matrix were complex, scattering of the source radiation could be a severe problem. Spectrum B shows the fluorescence spectrum in the 2450–2550 Å range. It is evident that for this sample,

---

Fig. 37 Background emission from several flame atomizers recorded in the 1850 to 3200 Å range with and without chlorine filter. (a) H_2–air, (b) H_2–O_2–Ar, (c) C_2H_2–air, (d) H_2–N_2O. [Reprinted with permission (Vickers, T. J., Slevin, P. J., Muscat, V. I., and Farias, L. T. (1972). *Anal. Chem.* **44**, 930). Copyright © by the American Chemical Society.]

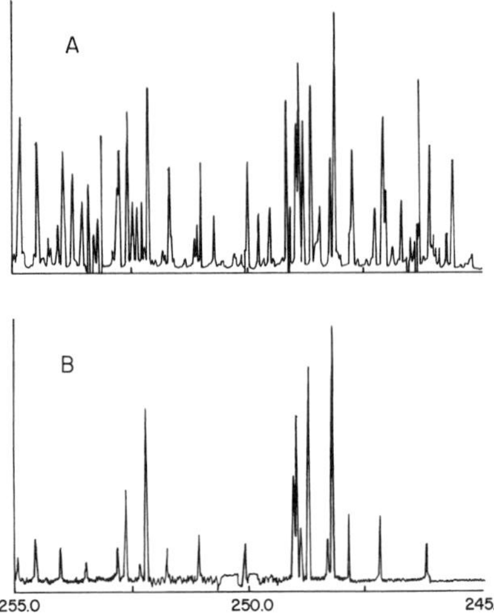

Fig. 39 Iron spectrum in the 2450–2550-Å range. (a) Fe hollow cathode lamp, (b) Fe atomic fluorescence with a separated C_2H_2–air flame atomizer. [Muscat, V. I., Vickers, T. J., Rippetoe, W. E., and Johnson, E. R. (1975). *Appl. Spectrosc.* **29**, 52.]

at least, scattering is not a problem since many lines prominent in the source spectrum are entirely missing from the fluorescence spectrum. All the lines shown can contribute to the fluorescence signal with either a filter or a "solar-blind" nondispersive system. With a dispersive system, usually only the lines at 2483 and 2488 Å would make an appreciable contribution to the signal. Muscat *et al.* (1975) observed signal enchancement factors of approximately 180 and 75 for the "solar-blind" and filter systems, respectively.

Despite the signal enhancement, the detection limit will generally be poorer with a nondispersive system than with a reasonably well designed dispersive system. The detection limit, of course, depends on the noise as well as the signal, and Muscat *et al.* (1975) found that for most flame atomizers the noise due to the background emission of the atomizer increased even more rapidly than the signal for both the filter and "solar-blind" nondispersive systems.

D. Multielement Atomic Fluorescence

Nondispersive atomic fluorescence provides a convenient basis for the development of a system for multielement analysis on a single sample. The instrument shown in Fig. 40 is a good example (Mitchell and Johansson, 1970, 1971; Palermo *et al.*, 1974). This instrument uses pulsed single-element hollow cathode lamps, a rotating interference filter wheel, and logic circuitry to select the proper lamp and measure the fluorescence of the selected element when the proper interference filter is in place.

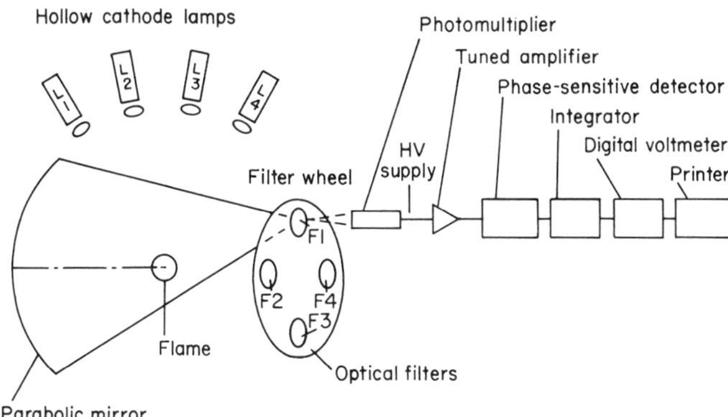

Fig. 40 Schematic diagram of a four-channel atomic fluorescence spectrometer. [Mitchell, D. G., and Johansson, A. (1971). *Spectrochim. Acta* **26B**, 678.]

The system shown in Fig. 41 represents a quite different approach to multielement atomic fluorescence spectroscopy (Johnson *et al.*, 1975). A xenon-arc continuous-excitation source is used with a computer-controlled rapid scan spectrometer. The wavelength at which fluorescence is to be measured is entered by the keyboard for each element. The spectrometer then slews rapidly to a desired wavelength, the fluorescence is measured until the desired signal-to-noise ratio is obtained, and then the spectrometer moves on to the next wavelength. The speed and flexibility of this system has been demonstrated by analysis for five elements (Fe, Mg, Cu, Ag, and Cr) in jet engine lubricating oils. About 1 min was required for analysis for each sample.

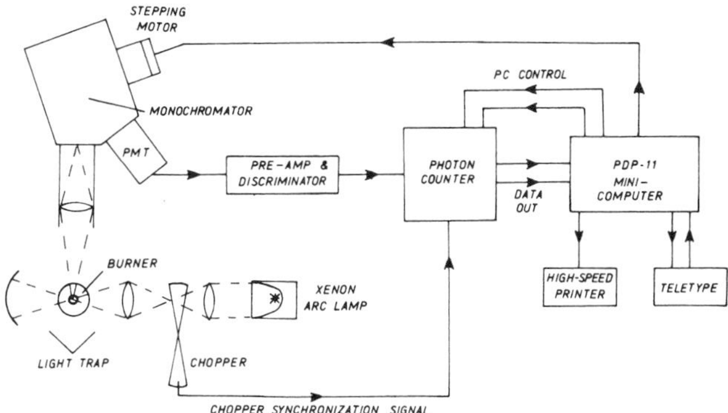

Fig. 41 Schematic diagram of a rapid scan computer-controlled atomic fluorescence spectrometer. [Reprinted with permission (Johnson, D. J., Plankey, F. W., and Winefordner, J. D. (1975). *Anal. Chem.* **47**, 1740). Copyright © by the American Chemical Society.]

E. *Other Considerations*

The discussions of sample preparation, calibration and measuring techniques, sensitivity, limits of detection, precision, and accuracy in connection with atomic absorption measurements apply equally well to atomic fluorescence measurements.

TABLE IV

Comparison of Experimental Limits of Detection (in μg/ml) in Atomic Flame Spectrometry[a]

Element	Absorption	Fluorescence	Emission
Ag	0.0005	0.0001	0.02
Al	0.04	0.1	0.005
As	0.1	0.1	50
Au	0.01	0.005	4
Be	0.002	0.01	0.1
Bi	0.05	0.005	2
Ca	0.0005	0.02	0.0001
Cd	0.0006	0.000001	2
Co	0.005	0.005	0.05
Cr	0.005	0.05	0.005
Cu	0.003	0.001	0.01
Fe	0.005	0.008	0.05
Ga	0.07	0.01	0.01
Ge	0.1	0.1	0.5
Hg	0.2	0.0002	40
In	0.05	0.1	0.005
Mg	0.0003	0.001	0.005
Mn	0.002	0.006	0.005
Mo	0.03	0.5	0.1
Ni	0.005	0.003	0.6
Pb	0.01	0.01	0.2
Pd	0.02	0.04	0.05
Rh	0.03	3	0.3
Sb	0.07	0.05	20
Si	0.1	0.6	5
Se	0.1	0.04	—
Sn	0.03	0.05	0.3
Sr	0.004	0.03	0.0002
Te	0.1	0.005	200
Tl	0.02	0.008	0.01
V	0.02	0.07	0.01
Zn	0.002	0.00002	50

[a] Reprinted with permission [Winefordner, J. D., and Elser, R. C. (1971). *Anal. Chem.* **43** (4), 24A]. Copyright © by the American Chemical Society.

VIII. Applications

Tables IV and V list detection limits for some of the elements determined with flame and electrothermal atomizers (Winefordner and Elser, 1971). Flame atomic emission results are included in Table IV for comparison purposes. In general, atomic absorption and atomic fluorescence provide better detection limits than atomic emission flame spectroscopy for elements with resonance lines below about 3500 Å.

TABLE V

Absolute Limits of Detection (in ng) for Atomic Absorption and Atomic Fluorescence Spectroscopy with Graphite Electrothermal Atomizers[a]

Element	Absorption	Fluorescence
Ag	0.0005	0.005
As	0.2	0.5
Au	0.07	0.01
Be	0.003	0.03
Bi	0.02	0.01
Ca	0.003	0.0001
Cd	0.00006	0.00003
Cu	0.006	0.005
Fe	0.02	0.01
Ga	0.02	0.05
Hg	0.4	—
Mg	0.003	0.0000001
Pb	0.02	0.01
Tl	0.002	0.05
Zn	0.001	0.00005

[a] Reprinted with permission [Winefordner, J. D., and Elser, R. C. (1971). *Anal. Chem.* **43** (4), 24A]. Copyright © by the American Chemical Society.

The three techniques suffer in approximately equal measure from interferences. All interferences that pertain to the atomization process affect all three techniques in identical fashion. Spectral interferences have been discussed in some detail for both absorption and fluorescence measurements.

The precision of measurements should also be about the same for all three methods. The precision achieved depends on the design of the measurement system and on the concentration level but, in general, a relative standard deviation of the order of 1 to 2% should be achievable at concentrations greater than ten times the detection limit. The accuracy of measurement depends on the proper choice of standards, elimination of interferences, and careful sample preparation rather than on the measurement technique.

Instrumentation for atomic absorption and atomic fluorescence is rather similar and relatively inexpensive. Analytical atomic absorption developed earlier than atomic fluorescence. Perhaps as a consequence, commercial equipment is available for absorption measurements but not for fluorescence measurements. For this reason it seems likely that most trace metal determinations will continue to be done by atomic absorption spectrophotometry. We have noted in describing atomic fluorescence measurements the suitability of this technique for nondispersive and multielement measurements. One can anticipate a role for atomic fluorescence in instruments designed to take advantage of these capabilities for special purposes (see, e.g., Muscat *et al.*, 1972).

The range of materials analyzed by atomic absorption and atomic fluorescence is impressively large. Trace elements may have profound effects on the health of plants and animals and on the physical and chemical properties of materials. Thus atomic spectroscopy is routinely applied to the determination of trace elements in biological materials, such as blood, serum, urine, tissue, plant materials, soils and soil extracts, fertilizers and foodstuffs, and in such other materials as metals, ores, glass, plating solutions, petroleum products, and cement.

References

A chapter of this type cannot be comprehensive. The reader is referred to the various monographs on atomic absorption and atomic fluorescence listed as General References for a fuller presentation of the historical development of the fields and a more detailed presentation of such topics as sample treatment and applications. Most of the books include an element by element discussion of applications.

This chapter is not exhaustive in literature citation. The literature cited has been selected to provide examples or because it would provide a convenient entry point for the reader to discover other references on a particular topic. Two sources may be quite useful in obtaining a complete literature survey of the field or in adding to the material of this chapter as the field continues to grow. Biennial reviews of analytical atomic spectroscopy are published in *Analytical Chemistry* (Winefordner and Vickers, 1970, 1972, 1974; Hieftje *et al.*, 1976), and the Chemical Society (London) publishes the very helpful *Annual Reports on Analytical Atomic Spectroscopy* (Hubbard, 1972, 1973; Woodward, 1974, 1975).

Chuang, F. S., and Winefordner, J. D. (1975). *Appl. Spectrosc.* **29**, 412.
Cordos, E., and Malmstadt, H. V. (1972a). *Anal. Chem.* **44**, 2277.
Cordos, E., and Malmstadt, H. V. (1972b). *Anal. Chem.* **44**, 2407.
Cordos, E., and Malmstadt, H. V. (1973). *Anal. Chem.* **45**, 27.
Corliss, C. H., and Bozman, W. R. (1962). "Experimental Transition Probabilities for Spectral Lines of Seventy Elements" (NBS Mono. 53). U.S. Govt. Printing Office, Washington, D.C.

Dean, J. A., and Rains, T. C. (eds.) (1971). "Flame Emission and Atomic Absorption Spectrometry," Vol. 2, p. 327. Dekker, New York.

Delves, H. T. (1970). *Analyst (London)* **95**, 431.

Fassel, V. A., Mossotti, V. G., Grossman, W. E. L., and Kniseley, R. N. (1966). *Spectrochim. Acta* **22**, 347.

Fraser, L. M., and Winefordner, J. D. (1971). *Anal. Chem.* **43**, 1963.

Fraser, L. M., and Winefordner, J. D. (1972). *Anal. Chem.* **44**, 1666.

Grime, J. K., and Vickers, T. J. (1974). *Anal. Chem.* **46**, 1810.

Grime, J. K., and Vickers, T. J. (1975). *Anal. Chem.* **47**, 432.

Hicks, J. M., Gutierrez, A. N., and Worthy, B. E. (1973). *Clin. Chem.* **19**, 322.

Hieftje, G. M., Copeland, T. R., and deOlivares, D. R. (1976). *Anal. Chem.* **48**, 142R.

Hooymayers, H. P. (1968). *Spectrochim. Acta* **23B**, 567.

Hubbard, D. P. (1972). "Annual Reports on Analytical Atomic Spectroscopy, 1971," Vol. 1. Society for Analytical Chemistry, London.

Hubbard, D. P. (1973). "Annual Reports on Analytical Atomic Spectroscopy, 1972," Vol. 2. Society for Analytical Chemistry, London.

International Union of Pure and Applied Chemistry (1972). "Appendices on Tentative Nomenclature, Symbols, Units and Standards—Number 26." IUPAC Secretariat, Oxford.

Johnson, D. J., Plankey, J. W., and Winefordner, J. D. (1975). *Anal. Chem.* **47**, 1739.

Kahn, H. L., Peterson, D. E., and Schallis, J. E. (1968). *At. Absorpt. Newsletter* **7**, 35.

L'vov, B. V. (1961). *Spectrochim. Acta* **17**, 761.

L'vov, B. V. (1969). *Spectrochim. Acta* **24B**, 53.

L'vov, B. V. (1970). "Atomic Absorption Spectrochemical Analysis" (English transl. by J. H. Dixon). Hilger, London.

Manning, D. C., and Fernandez, F. (1970). *At. Absorpt. Newsletter* **9**, 65.

Massmann, H. (1968). *Spectrochim. Acta* **23B**, 215.

McCullough, M. R., and Vickers, T. J. (1976). *Anal. Chem.* **48**, 1006.

Mitchell, D. G., and Johansson, A. (1970). *Spectrochim. Acta* **25B**, 175.

Mitchell, D. G., and Johansson, A. (1971). *Spectrochim. Acta* **26B**, 677.

Muscat, V. I., Vickers, T. J., and Andren, A. (1972). *Anal. Chem.* **44**, 218.

Muscat, V. I., Vickers, T. J., Rippetoe, W. E., and Johnson, E. R. (1975). *Appl. Spectrosc.* **29**, 52.

Olsen, E. D., and Jatlow, P. I. (1972). *Clin. Chem.* **18**, 1312.

Omenetto, N., Fraser, L. M., and Winefordner, J. D. (1973a). *Appl. Spectrosc. Rev.* **7**, 147.

Omenetto, N., Hatch, N. N., Fraser, L. M., and Winefordner, J. D. (1973b). *Anal. Chem.* **45**, 195.

Omenetto, N., Hatch, N. N., Fraser, L. M., and Winefordner, J. D. (1973c). *Spectrochim. Acta* **28B**, 65.

Palermo, E. I., Montaser, A., and Crouch, S. R. (1974). *Anal. Chem.* **46**, 2154.

Parsons, M. L., McCarthy, W. J., and Winefordner, J. D. (1966). *Appl. Spectrosc.* **20**, 223.

Parsons, M. L., Smith, B. W., and Bentley, G. E. (1975). "Handbook of Flame Spectroscopy." Plenum Press, New York.

Poesner, D. W. (1959). *Aust. J. Phys.* **12**, 184.

Rains, T. C., Epstein, M. S., and Menis, O. (1974). *Anal. Chem.* **46**, 207.

Rann, C. S., and Hambly, A. N. (1965). *Anal. Chem.* **37**, 879.

Soffer, B. F., and McFarland, B. B. (1967). *Appl. Phys. Lett.* **10**, 266.

Svoboda, V., Browner, R. J., and Winefordner, J. D. (1972). *Appl. Spectrosc.* **26**, 505.

Sychra, V., and Kolihova, D. (1973). *Int. Congr. At. Abs., At. Fl. Spectrom. Pap., 3rd, 1971* **1**, 265.

Sychra, V., and Matousek, J. (1970). *Anal. Chim. Acta* **52**, 376.

Veillon, C. (1972). "Handbook of Commercial Scientific Instrumentation, Atomic Absorption," Vol. 1. Dekker, New York.

Vickers, T. J., Slevin, P. J., Muscat, V. I., and Farias, L. T. (1972). *Anal. Chem.* **44**, 930.

West, T. S., and Cresser, M. S. (1973). *Appl. Spectrosc. Rev.* **7**, 79.

West, T. S., and Williams, X. K. (1969). *Anal. Chim. Acta* **45**, 27.

Winefordner, J. D., and Vickers, T. J. (1964). *Anal. Chem.* **36**, 1947.

Winefordner, J. D., and Vickers, T. J. (1970). *Anal. Chem.* **42**, 206R.

Winefordner, J. D., and Elser, R. C. (1971). *Anal. Chem.* **43(4)**, 24A.

Winefordner, J. D., and Vickers, T. J. (1972). *Anal. Chem.* **44**, 150R.

Winefordner, J. D., and Vickers, T. J. (1974). *Anal. Chem.* **46**, 192R.

Woodward, C. (1974). "Annual Reports on Analytical Atomic Spectroscopy, 1973," Vol. 3. Society for Analytical Chemistry, London.

Woodward, C. (1975). "Annual Reports on Analytical Atomic Spectroscopy, 1974," Vol. 4. Chemical Society, London.

General References

Christian, G. D., and Feldman, F. J. (1970). "Atomic Absorption Spectroscopy. Applications in Agriculture, Biology, and Medicine." Wiley, New York.

Dean, J. A., and Rains, T. C. (1969). "Flame Emission and Atomic Absorption Spectrometry," Vol. 1. Dekker, New York.

Dean, J. A., and Rains, T. C. (1971). "Flame Emission and Atomic Absorption Spectrometry," Vol. 2. Dekker, New York.

Elwell, W. T., and Gidley, J. A. F. (1966). "Atomic Absorption Spectrophotometry," 2nd ed. Pergamon, Oxford.

Kirkbright, G. I., and Sargent, M. (1974). "Atomic Absorption and Fluorescence Spectroscopy." Academic Press, New York.

Mavrodineanu, R. (1970). "Analytical Flame Spectroscopy." Macmillan, New York.

Parker, C. R. (1972). "Water Analysis by Atomic Absorption Spectroscopy." Springvale, Australia.

Pinta, M. (1975). "Atomic Absorption Spectrometry." Halsted, New York.

Price, W. J. (1972). "Analytical Atomic Absorption Spectrometry." Heyden and Sons, London.

Ramirez-Munoz, J. (1968). "Atomic Absorption Spectroscopy and Analysis by Atomic Absorption Flame Photometry." American Elsevier, New York.

Reynolds, R. J., and Aldous, K. (1970). "Atomic Absorption Spectroscopy. A Practical Guide." Griffin, London.

Robinson, J. W. (1975). "Atomic Absorption Spectroscopy," 2nd ed. Dekker, New York.

Rubeska, I., and Moldan, B. (1969). "Atomic Absorption Spectrophotometry" (English transl. by P. T. Woods). CRC Press, Cleveland, Ohio.

Slavin, W. (1968). "Atomic Absorption Spectroscopy." Wiley (Interscience), New York.

Sychra, V., Svoboda, V., and Rubeska, I. (1975). "Atomic Fluorescence Spectroscopy." Van Nostrand–Reinhold, Princeton, New Jersey.

Welz, B. (1972). "Atomic Absorption Spectroscopy," Verlag Chemie, Weinheim, West Germany.

Winefordner, J. D. (1970). "Spectrochemical Methods of Analysis." Wiley, New York.

Flame and Plasma Emission Analysis

Peter N. Keliher

Chemistry Department
Villanova University
Villanova, Pennsylvania

I. General Considerations–Definitions

This section considers the two related techniques of flame emission spectrometry (FES) and plasma emission spectrometry (PES) with regard to their suitability for qualitative and quantitative trace analysis, most particularly for metals, but in certain cases for metalloids and nonmetallic species.

A "flame" may be defined as a self-sustaining burning gas or vapor (fuel) undergoing combustion, usually with oxygen in some form. Although many fuels have been used in analytical FES, the most common fuels are acetylene,

ISBN 0-12-430801-5

hydrogen, and propane. The oxygen may be provided as pure oxygen, air (diluted oxygen), or nitrous oxide, depending upon the analytical requirement and the burner system used.

A "plasma" may be defined as a self-sustaining discharge of an inert gas, usually argon or helium. Various types of plasma systems are currently popular, e.g., DC plasmas, microwave plasmas, and inductively coupled plasmas, and these various types will be considered later in detail.

It is appropriate here to comment on the relationship between FES/PES and atomic absorption spectrometry (AAS) and atomic fluorescence spectrometry (AFS). These techniques are closely related to the emission technique except that in AAS the signal is dependent upon the population of atoms in the ground state whereas in emission one is concerned with increasing the population of atoms in the excited state. AFS may be considered as an absorption of an atom followed by subsequent emission of that atom. AAS and AFS are considered in detail in the chapter by Vickers.

In emission spectrometry, it is the outermost electrons of atoms that are excited and the emitted radiation normally consists of sharp, well-defined lines in the ultraviolet, visible, or near-infrared region of the spectrum. By monitoring an appropriate spectral region, one can accomplish *qualitative* analysis whereas the measurement of relative line intensities allows *quantitative* analysis.

II. Flame Emission Spectrometry

A. Historical Background

In his "Opticks, or a Treatise of the Reflections, Refractions, Inflections, and Colours of Light," Sir Issac Newton (1704) stated, "Do not all fix'd Bodies, when heated beyond a certain degree, emit light and shine; and is not this Emission perform'd by the vibrating motions of their parts?" Newton used a glass prism to study the spectrum of the sun and observed a spectrum of seven colors—red, orange, yellow, green, blue, indigo, and violet. In 1826, Talbot performed some interesting simple experiments on flames. By impregnating the cotton wick of a spirit lamp with table salt, Talbot observed an intense yellow light. He noted (Talbot, 1826)

> I have found that the same effect takes place whether the wick of the lamp is steeped in the muriate, sulphate, or carbonate of *soda*, while the nitrate, chlorate, sulphate, and carbonate of *potash*, agree in giving a blueish tinge to the flame. Hence, the yellow rays may indicate the presence of *soda*, but they, nevertheless, frequently appear where no soda can be supposed to be present.

Talbot also observed that candles, and platinum touched by the hand or rubbed with soap, also gave a yellow coloration. Table salt sprinkled on platinum showed the yellow coloration for some time and the effect could be renewed again by wetting the platinum. This latter phenomenon led Talbot to postulate that the light might be due to water of crystallization rather than to sodium. But then there arose the problem of explaining why potassium salts should not produce a yellow color. Talbot finally concluded that water could not produce the yellow color since the color was also produced by sulfur in the absence of water.

By this time, it was possible to use flame analysis to *visually* distinguish between the salts of several elements including sodium, potassium, calcium, strontium, and copper. Using a visual prism, further refinements became possible. In 1834, Talbot observed that while ordinarily it was difficult to distinguish red lithium and red strontium flames, it was markedly easy using a prism. He wrote (Talbot, 1834)

> The strontia flame exhibits a great number of redy rays well separated from each other by dark intervals, not to mention an orange and a very definite bright blue ray. The lithia exhibits one single red ray. Hence, I hesitate not to say that optical analysis can distinguish the minutest portions of these two substances from each other with as much certainty, if not more, than any other known method.

An important advance in emission spectrometry occured about 1855 when Bunsen developed a burner that gave a colorless and fairly hot flame. Using this burner, Bunsen and Kirchhoff (1860, 1861a) studied chlorides, bromides, iodides, hydrated oxides, sulfates, and carbonates of potassium, sodium, lithium, strontium, calcium, and barium. They concluded that

> the different bodies with which the metals employed were combined, the variety of the chemical processes occuring in several, and the wide differences of temperature which these flames exhibit, produce no effect upon the position of the bright lines in the spectrum which are characteristic of each metal.

In other words, they realized that the emission from each salt did not depend upon the nonmetallic component of the salt but was dependent solely on the metallic component of the salt. It should be noted here that the spectra that these workers obtained were quite simple since their source only allowed excitation of the principal characteristics of a spectrum. Also, Bunsen and Kirchhoff compared their flame spectra (for chlorides) with "electric" spectra obtained by discharge of a coil through a glass tube containing two platinum electrodes to which small quantities of the appropriate metal were fastened. Line spectra were observed to be identical to the flame spectra.

In 1862, Roscoe and Clifton noted that for some elements, particularly barium and strontium, broad bands that could be seen in the low temperatures of flames disappeared in high temperature spark discharges and were replaced by intense lines *not coincident* with the band spectra. Furthermore, the alkali elements, sodium, potassium, and lithium did not exhibit this effect. Roscoe and Clifton (1862) suggested a possible explanation of their observations. They considered that at the lower temperatures of a flame or weak spark the spectrum is produced by the glowing vapor of some compound, probably the oxide of the metal. This effect would only be expected for "difficultly reducible" metals. This was in line with their observations. At the relatively high temperatures of intense spark discharges, the compounds would be decomposed and the true metallic spectrum would be observed.

At about the same time, and unaware of the work of Roscoe and Clifton, Mitscherlich (1862) observed two bright green bands associated with a substance containing barytes, which seemed to suggest the presence of a new metal. On closer examination, however, Mitscherlich observed that the two green bands were seen, sometimes alone but frequently together with barium lines, by a solution in sal ammoniac. With specially designed apparatus he could produce either the two green bands or the barium spectrum. He did this as follows: He had a solution containing one part of acetate of barytes and ten parts of ammonium acetate to give an intense barium spectrum when introduced into a flame. When a solution of hydrochloric acid was also introduced into the flame at the same time, the barium spectrum disappeared and was replaced by the green band spectrum. Mitscherlich was also able to observe a similar effect with calcium chloride and with strontium chloride.

Using the qualitative knowledge then available on the identification of characteristic spectra for different metals, Bunsen and Kirchhoff (1861b) used spectral analysis and were able to identify two new metals that had not previously been known, i.e., cesium and rubidium. The discovery of several other elements in the 19th century was also based on spectroscopic techniques. For example, following on Kirchhoff and Bunsen's work, Crookes (1861) discovered thallium and, some years later, Ramsey (1895) showed that helium was a terrestial element.

This new utilization of spectroscopy for the identification of new elements led to a much greater interest in the technique. Up until this time, the usual experimental arrangement for the introduction of samples into a flame was the primitive "injection" technique using a platinum loop wire. This is the same technique that is sometimes used today in undergraduate general chemistry laboratories. Mitscherlich (1862), however, used a capillary system in conjunction with fine platinum wires and was able to obtain an intense and stable light signal of about two hours duration. This represented, of course, a great improvement over the transient-type signals obtained with the platinum loop system. Mitscherlich also investigated various copper halides and found characteristic band spectra.

In 1864, Mitscherlich reported spectral characteristics of all metals available, and also many of their compounds, and from the results concluded that "every compound of the first order which is not decomposed, and which is heated to a temperature adequate for the production of light, exhibits a spectrum peculiar to this compound, and independent of other circumstances" (Mitscherlich, 1864). The alkali metals, cadmium, magnesium, silver, and zinc were cited as examples of metals whose compounds easily decomposed giving only spectra of the metal, whereas alkaline earth metals showed "broad luminosities with narrow lines which recur at definite intervals" (Mitscherlich, 1864). Mitscherlich's 1864 publication is the first case of clear distinction between line and band spectra.

Quantitative flame emission methods were pioneered by Janssen (1870) and by Champion *et al.* (1873). They determined sodium over the range of 0.05–1% by visual comparison of two flames. An accuracy of 2–5% was claimed for the determination of sodium in plant ashes.

It was about this time that Salet reported on the emission spectra of phosphorus (Salet, 1869) and sulfur (Salet, 1871) containing flames. Sulfur was observed to give a blue-violet coloration while phosphorus showed a green color. In a later publication, Salet (1873) reported some investigations on the spectra of the metalloids.

In 1879, Gouy developed a reliable and efficient pneumatic nebulizer that charged the air supply to the burner with droplets of the sample solution that could be continuously fed through the aspirator. The introduction of dry photographic plates occured at about the same time and this led to the development of practical spectrographs (Gouy, 1879) but the various analytical flame techniques proposed all had considerable limitations and were restricted to the alkali and alkaline earth elements.

Other developments during the 19th century that eventually led to the acceptance of FES (and ultimately PES, AAS, and AFS) included improvements in photography, prism and grating development, improvements in wavelength measurements, and cataloging of spectra. For a detailed discussion of these advances, the reader should consult the excellent book, "Nineteenth Century Spectroscopy" by McGucken (1969).

In the early part of the 20th century, there was very little in the way of further development of FES for analytical purposes. There are two principal reasons for this. First, spectroscopic systems were considered to be of primary value in the teaching of optical principles and were used largely by physicists.

Second, the use of low temperature flames restricted the number of elements that could be determined. For some of these elements, particularly sodium of course, there was too much sensitivity.

The next major development in FES occurred in 1928 when Lundegardh developed a relatively sophisticated system involving nebulizers, burners, and gas control devices (Lundegardh, 1928). A microphotometer was used to measure spectral line density so that analyses could be accomplished quite rapidly. An air–acetylene flame was used in conjunction with ultraviolet spectrography (photographic readout) and it was possible to analyze about 40 elements.

Lundegardh (1930) also developed the first photoelectric flame instrument using a vacuum photocell. It was difficult, however, to stabilize due to the unsophisticated electronic amplifiers then available. However, the basic concepts and systems developed by Lundegardh are still used in systems today and, following on the work of Lundegardh, many workers developed more sophisticated systems, gradually improving burner design, replacing filters with monochromators, and developing better detection systems.

Commercial instrumentation became available in Germany during the 1930s, and after the war rapid developments continued with American manufacturers introducing commercial systems from about 1950.

B. Instrumentation

1. *Burner Systems*

There are two basic burner systems used in analytical flame spectrometry; these are the direct injection burner, shown in Fig. 1, and the premixed burner, shown in Fig. 2. The direct injection burner is often called, somewhat incorrectly, a "total consumption" burner. With the direct injection burner, fuel and support gas do not come into contact until they reach the burning area. The flame produced by this burner is turbulent and the support gas also serves to nebulize liquid sample into the flame. In those cases where the fuel serves this purpose, the term reversed direct injection burner is recommended (IUPAC, 1977).

The most common direct injection burner is the Beckman burner, which was developed as an accessory to the Beckman DU spectrophotometer to allow the determination of alkali and alkaline earth elements. Despite its apparent simplicity in combining the nebulization and burner into one system, the direct injection burner suffers from several drawbacks.

1. Much of the sample is actually carried through the flame without being dissociated, hence sample is not totally consumed.

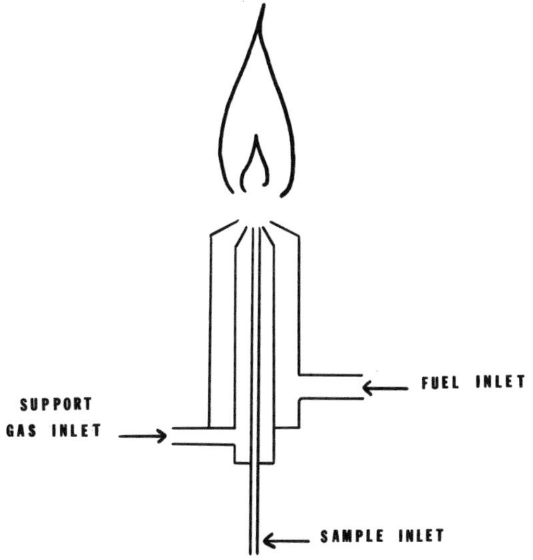

Fig. 1 Direct injection burner.

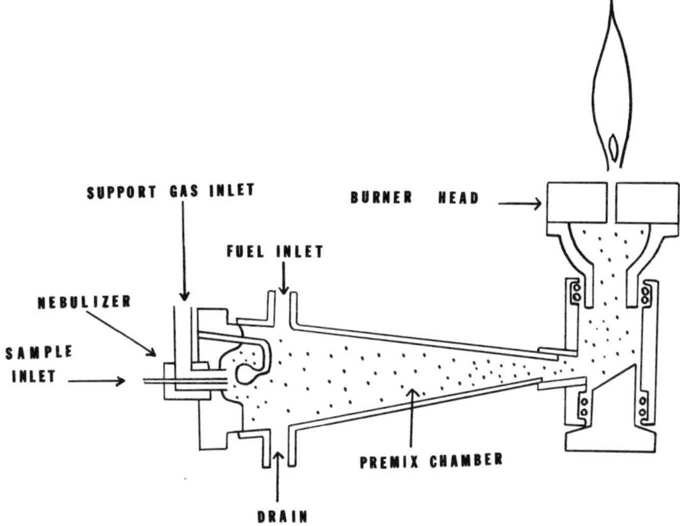

Fig. 2 Premixed burner (laminar flow burner).

2. There is a limit in physical size to the flame and furthermore, there is flame flicker fluctuation that leads to high noise levels.
3. The flame itself is quite audible, generating a high pitched, whining type of sound.

An important advantage of the direct injection burner, however, is its ability to be used with pure oxygen as a support gas. Oxygen cannot be used with conventional premixed burner systems since its burning velocity with fuels is so high that a flashback (explosion of the flame back into the premixed chamber) would occur. Flashbacks cannot occur with the direct injection burner since, as pointed out previously, fuel and support gas emerge from separate flow streams directly into the burning area. Although much of the past work in FES has been accomplished with direct injection burners, they have gradually been replaced, particularly in very recent years, by premixed burner systems.

In the conventional premixed burner system, the support gas and fuel are intimately mixed in a spray chamber. The support gas is also used to aspirate the liquid sample and it is usual to have some sort of bead, usually glass, placed just behind the nebulizer to aid in dispersing the drops into a fine mist. Sample, support gas, and fuel, having been intimately mixed, then pass into the burner portion of the system. Premixed burners themselves are referred to as Bunsen, Meker, or slot burners according to whether they have one large hole, a number of small holes, or a slot as outlet port(s) for the gas mixture (IUPAC, 1977). When several parallel slots are present, they are referred to as multislot burners. A common burner of this sort is the three-slot Boling burner. The small diameter of the holes in the Meker burner or the narrowness of the slot(s) in the slot burner prevents the unwanted flashback of the flame into the burner housing. Different burner heads are chosen for the particular fuel-support gas and analytical conditions.

The inner zone of the flame is generally referred to as the primary reaction zone, while the edge of the flame where secondary burning via contact with surrounding air occurs is called the secondary combustion zone. The region of the flame confined by the inner and outer zones, where in many instances the conditions for flame analysis are optimum, is called the interzonal region or, where the combustion zones have the form of a cone, the interconal zone (IUPAC, 1977).

It should be noted that not all of the sample reaches the flame since there is waste of solution in the spray chamber because of the deposition of droplets on the walls. In fact, with a conventional premixed system, only about 5–10% of the sample is atomized; the rest of the sample is discarded through a drain tube at the bottom of the spray chamber. The term efficiency of nebulization refers to the ratio of the amount of analyte entering the flame to the amount of analyte aspirated (IUPAC, 1977).

Although this relatively low efficiency of nebulization may seem a waste of sample, and it is, the fact remains that the premixed burner is superior to the direct injection burner in two important respects.

1. There is less noise in the system when operated properly and the system does not emit an audible noise.
2. As noted, it is possible to use different burner heads for different purposes and this increases the versatility of the premixed system for different systems.

An important point with respect to the operation of premixed burners is that the flow velocity of the fuel and support gas must be greater than the burning velocity, otherwise flashback will occur in the spray chamber. In practice, this is not a problem if two important factors are considered. First, the flow rates of the gases must be high enough to ensure that the flow velocity is greater than the burning velocity; instrument manufacturers generally indicate recommended safe flow rates with a good margin for safety. Second, the correct burner head must be used for the particular choice of gases. Manufacturers generally provide warnings, sometimes printed onto the burner head itself, so that a 10-cm slot burner designed to be used with air–acetylene would have the following warning: *Not to be used with nitrous oxide or oxygen.* A shorter length, narrower diameter, slot burner would be used with nitrous oxide–acetylene. Oxygen, as indicated previously, would not be used at all in conventional premixed burner systems.

It is appropriate to comment here on the nebulization and atomic processes which occur when a sample is transported into a flame. If we consider an aqueous sample of metal plus anion MX, then we can represent the atomization processes as

$$[M^+X^-]_{sol} \underset{\text{precipitation}}{\overset{\text{nebulization}}{\rightleftharpoons}} [M^+X^-]_{mist}$$

$$[M^+X^-]_{mist} \xrightarrow{\text{evaporation}} [M^+X^-]_{solid}$$

$$[M^+X^-]_{solid} \xrightarrow{\text{fusion}} [M^+X^-]_{liquid}$$

$$[M^+X^-]_{liquid} \xrightarrow{\text{vaporization}} [M^+X^-]_{gas}$$

$$[M^+X^-]_{gas} \xrightarrow{\text{dissociation}} M^o_{gas} + X^o_{gas}$$

The atomic processes which the metallic atom might undergo could be represented as follows:

Absorption $M^o + h\nu \rightarrow M^{o*}$

Ionization $M^o + \Delta \rightarrow M^+ + e^-$

Oxidation $M^o + O^o \rightarrow MO$

Emission $M^{o*} \rightarrow M^o + h\nu$

Combination $M^o + Y^o \rightarrow MY$

where O and Y represent, respectively, oxygen and any other foreign atom.

Clearly, ionization, oxidation, and combination are undesirable phenomena since they reduce or even eliminate the atomic population of the element of interest. If the temperature of the flame is relatively high, e.g., nitrous oxide–acetylene, there will be an increase in ionization for elements that are easily ionized. For this reason, a relatively low temperature flame, e.g., air–propane, is recommended for the alkali elements. When a high temperature flame is used, it is a common practice to add an "ionization buffer" that will be preferentially ionized and thus prevent ionization of the element to be determined. The most common ionization buffers are potassium and sodium.

Oxidation and combination occur most commonly when low temperature flames are used for refractory oxides such as aluminum, vanadium, titanium, and the rare earths. For this reason, the premixed nitrous oxide–acetylene flame provides the most satisfactory "atom reservoir" for these elements. Its combination of high temperature and reducing environment prevents, or at least minimizes, oxidation and combination that will occur in lower temperature flames. The air–acetylene flame cannot be used for the analytical determination of refractory oxides. Although other nitrous oxide-supported flames, e.g., nitrous oxide–hydrogen, have been studied (Cresser *et al.*, 1970), they are not nearly as satisfactory for the determination of refractory oxides.

2. *Wavelength Selection*

The simplest instruments for FES generally use filters for wavelength selection and are restricted to the determination of the alkali metals and sometimes calcium. They are popularly known as "flame photometers" although that term is discouraged by the International Union of Pure and Applied Chemistry (IUPAC, 1977). These simple instruments find their principal use in clinical analysis. More sophisticated instruments have used either prisms or gratings, and quite often systems are designed primarily for atomic absorption with flame emission as an "accessory" item. The theories of prisms (Faust, 1971) and gratings (Barnes and Jarrell, 1971) are discussed in extensive detail in a text devoted exclusively to analytical emission spectrometry (Grove, 1971).

In recent years, gratings have largely replaced prisms as wavelength isolation devices. Grating instruments generally have better dispersion than prism instruments, particularly at longer wavelengths. They have a further advantage in that wavelength is presented on a linear scale. The dispersion of prisms decreases rapidly with increasing wavelength and thus a nonlinear wavelength scale is obtained. As pointed out previously, most of the atomic absorption–flame emission systems that are available commercially were designed primarily as atomic absorption units with emission capabilities of secondary importance.

One of the limitations of FES is the lack of specificity that can occur when a line or band from a contaminant element appears within the spectral bandwidth of the monochromator. A typical example of this is the interference of the CaOH band emission on the analytical determination of barium at 535.5 nm. Various approaches may be taken to increase selectivity in FES. Some years ago, Southern Analytical Ltd. (Camberly, Surrey, United Kingdom) introduced a dual-photomultiplier system whereby background radiation was focused onto a second (background) photomultiplier while the radiation from the line of interest went directly to the primary (line) photomultiplier. The system is shown schematically in Fig. 3. The signal from the background photomultiplier was subtracted from the line photomultiplier in order to get a corrected signal. Unfortunately, this system is no longer available commercially. The author, however, has found great success with the system, and two units are still in use in his laboratory.

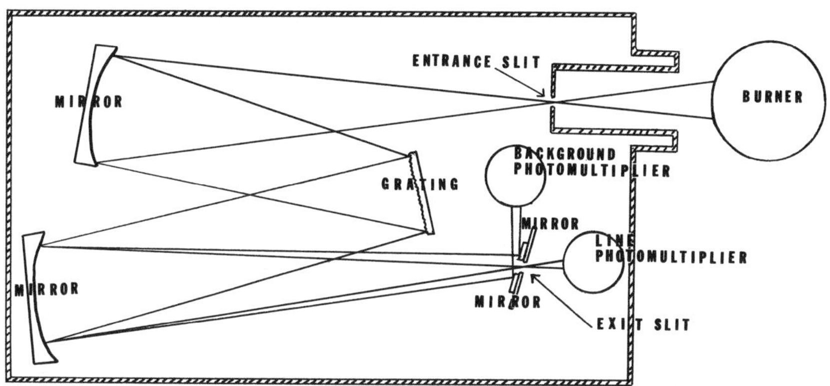

Fig. 3 Optical system for Southern Analytical Ltd. Model 1740-A flame emission spectrophotometer.

An obvious approach to increasing specificity is to increase the resolving power of a grating system so that the spectral bandpass of the monochromator approximates the line width (about 0.01 Å) of the element of interest. Unfortunately, however, this is not practical with conventional systems since the resultant system would not have sufficient light throughput and would require very long, inconvenient, focal lengths. An alternative spectral system, which has been used successfully, involves an "echelle" grating as part of a spectroscopic system. The term *echelle* comes from the French, meaning *ladder* or *scale*. An echelle monochromator differs from a conventional monochromator in that high orders are used rather than the first (conventional) order used with other monochromators. Also, in an echelle monochromator the resulting spectrum is presented as a two-dimensional pattern with the orders observed in a vertical direction and the wavelengths within a particular order in a horizontal direction. Thus, a photograph of the spectrum emitted by a continuum source would appear as a series of bands (the orders) running across a plate as seen in Fig. 4. Other echelle spectrographs are shown elsewhere (Keliher and Wohlers, 1976; Keliher, 1976).

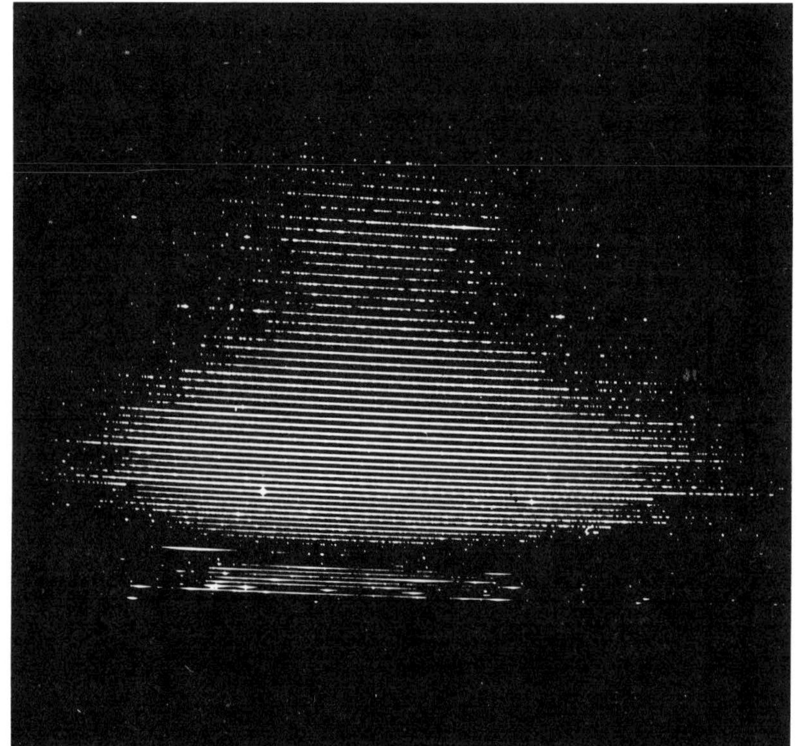

Fig. 4 Echelle spectrogram of continuum source. Note distinct orders.

An important point regarding the use of an echelle monochromator in FES or, for that matter, PES is that resolution and dispersion are approximately an order of magnitude greater than with conventional monochromators of the same focal length (Keliher and Wohlers, 1976; Keliher, 1976; Cresser *et al.*, 1973a). Using an echelle monochromator, a prototype Spectraspan

(SMI, Inc., Andover, Massachusetts), Cresser *et al.* (1973a) were able to eliminate spectral interferences in FES such as the interference of the CaOH band emission on the barium 553.5-nm line emission and the gallium 403.30-nm line on the manganese 403.08-nm line in premixed air–acetylene flames, and the interferences of cobalt at 341.23 and 341.26 nm on nickel at 341.48 nm, in a premixed nitrous oxide–acetylene flame. Comparison was made with a conventional Varian Techtron AA-4 atomic absorption spectrophotometer operating in the emission mode. One of the more severe of the many mutual spectral interferences of the rare earths in FES—namely that of samarium at 492.41 nm on neodymium at 492.45 nm—was also studied in the nitrous oxide–acetylene flame. The signal-to-background ratio was optimized for the Varian Techtron AA-4 by adjusting flame conditions for the 200-μm slits, and these conditions were used for all measurements on that instrument. For the Spectraspan echelle instrument, progressively more fuel-rich conditions were used for each of the two slit widths. Results are shown in Fig. 5. The flame conditions for the Spectraspan 100-μm slits were as fuel-rich as could be obtained without rapid clogging of the burner with carbon deposits. The background at this wavelength was appreciable and increased with the fuel flow. Neodymium requires an extremely fuel-rich flame for optimum sensitivity intensity (far more so than samarium) because of the high stability of its oxide. As the background decreases with spectral slit width, it was possible to use very fuel-rich conditions with the echelle instrumentation. Therefore, the apparent spectral interference from samarium decreases not only because of the small spectral slit width of the echelle instrument, but also because a more fuel-rich flame becomes analytically useful. It would seem that, in the future, instrumentation designed around the echelle grating would play an ever increasing role in both FES and PES.

3. *Gas Control Devices*

In order to have a stable flame, it is necessary to be able to properly regulate the flow of fuel and support gases to the burner. Two stage regulators are generally used in conjunction with needle valves to adjust the gas flow. It is conventional to set the support gas flow at a certain optimized setting, ensuring a constant flow rate to keep a constant amount of sample uptake to the burner, and to adjust the fuel flow rate for optimum signal-to-noise.

Most gas control devices incorporate a switchover system to rapidly change from one support gas to another. The reason for this is as follows. Because of the danger of flashback, a nitrous oxide–acetylene flame is not usually ignited directly. Instead, an air–acetylene flame is lit and when this flame is made very fuel-rich by increasing the acetylene flow, the switchover

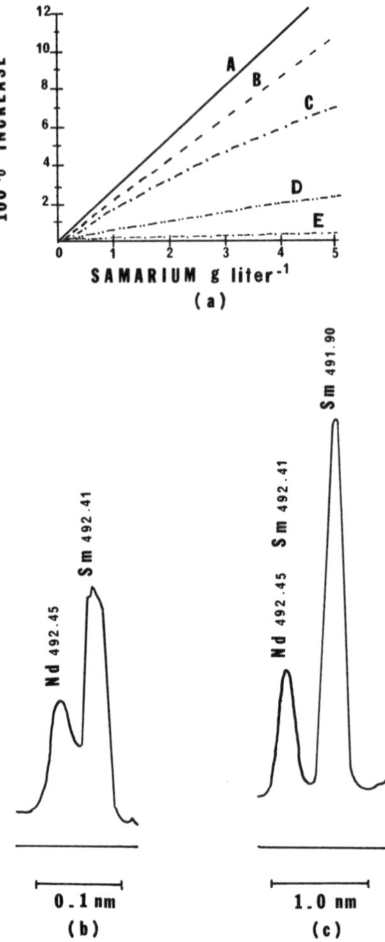

Fig. 5 (a) Percent interference of added samarium on neodymium (250 ppm neodymium constant) at 492.5 nm. A, Techtron 200 μm slits; B, Techtron 100 μm slits; C, Techtron 25 μm slits; D, Spectraspan 300 μm slits; E, Spectraspan 100 μm slits. (b,c) Wavelength scans of neodymium 492.5 nm region with 250 ppm neodymium and 5000 ppm samarium. Wavelength increases from right to left. (b) Spectraspan 300 μm slits; (c) Techtron 50 μm slits. [Reprinted with permission (Cresser *et al.* (1973a). *Anal. Chem.* **45**, 111). Copyright © by the American Chemical Society.]

valve is then used to replace the air with nitrous oxide. A nitrous oxide–acetylene flame is not normally extinguished directly. Instead, the reverse procedure is followed, the switchover valve is used to replace the nitrous oxide with air, the acetylene flow to the air–acetylene flame is then reduced, and finally the flame is extinguished by stopping the acetylene flow.

4. *Radiation Detectors*

Although some inexpensive filter instruments use simple barrier-layer cells, most modern systems use conventional side-on photomultiplier tubes (PMTs). These tubes consist of a photocathode, having a thin photoemissive surface, a series of dynodes with electron emissive surfaces where secondary electrons are produced, and finally an anode that serves as the collector electrode.

PMTs are available from several companies although the most popular in AAS and/or FES instrumentation is probably the RCA 1P28. This tube has a 9-dynode stage system and a relatively broad response from about 200 to 650 nm. Another common PMT is the Hamamatsu TV Company Ltd. R446, which has a spectral response very similar to that of the 1P28. Of course, many other PMTs, both side-on and end-on, have been used in FES.

Very recently, Morrison and co-workers have used a commercially available silicon diode vidicon (TV-type) tube (SSR Instruments; present manufacturer, Princeton Applied Research Company, Princeton, New Jersey) as a detector in multielement FES (Busch *et al.*, 1974a). A conventional monochromator and premixed nitrous oxide–acetylene burner system were used in conjunction with the vidicon detector. With the system, a spectral window of 20 nm is monitored simultaneously and atomic lines 1.4 Å may be resolved. Some detection limits obtained under compromise flame conditions are shown in Table I and compared with single-element detection limits, also obtained from a premixed nitrous oxide–acetylene flame, taken from a review article by Christian and Feldman (1971).

In another publication, Morrison and co-workers have used their vidicon flame spectrometer for "spectral stripping," i.e., removal of molecular band

TABLE I

Comparison of Single-Element and Multielement Detection Limits

Element	Line[a]	Multielement[b]	Single-Element[c]
Aluminum	396.1	0.14	0.05
Manganese	403.1	0.29	0.008
Molybdenum	390.3	0.31	0.2
Titanium	399.9	1.33	0.2
Tungsten	400.8	25.9	0.6

[a] In nanometers.
[b] Busch *et al.* (1974a).
[c] Christian and Feldman (1971).

or undesired concomitant interferences from analytical lines of interest with resultant increase in specificity (Busch *et al.*, 1974b). Three cases were considered:

1. Spectral interference of a flame band with an analytical line; e.g., the interference of the OH band emitted by the nitrous oxide–acetylene flame from the bismuth resonance line at 306.7 nm.
2. Spectral interference of molecular bands emitted by sample concomitants on an analytical line; e.g., the CaOH band emission on barium resonance emission at 553.5 nm.
3. Spectral interference of an analytical line with a line emitted by a sample concomitant; e.g., strontium 553.481-nm nonresonance emission with barium 553.480 resonance emission.

Using the vidicon flame spectrometer, it was possible in the three specific examples cited above to eliminate the interference and, therefore, to increase the specificity. The authors noted, however, that case 3 represented the most difficult problem for their system.

In another publication, Morrison and co-workers used the vidicon flame spectrometer as an efficient means of performing "internal standardization" analyses (Howell *et al.*, 1976). In the technique of internal standardization, a known amount of an element (not present in the sample) is added and the emission of that element is monitored to compensate for changes in the system caused by variables such as fuel-to-oxidant ratio shifts, aspiration rate variations, sample viscosity, and surface tension differences. Morrison and his colleagues performed flame emission analyses of sodium, potassium, and calcium simultaneously using lithium as the internal standard, and reported results for serum samples. They also analyzed magnesium using manganese as the internal standard for both serum and bovine liver samples. The time required for spectral stripping analysis using the vidicon flame spectrometer was reported to be less than one minute per sample.

5. *Readout Systems*

The actual presentation of data is usually presented on a meter on a linear scale and, most often, the electronic system is set up so that scale expansion may be easily used. Most instruments have provision to take the reading onto a suitable strip chart recorder. Some commercial systems also allow an "integration" time of 5, 10, 20, or more seconds, and this can be particularly useful with noisy systems or at relatively low concentration levels. When atomic absorption units are used in the flame emission mode, a small mechanical chopper in phase with the AC electronics of the amplifier may be used to provide the in-phase AC signal.

C. Contemporary FES—A New Look at an Old Method

In 1969, Pickett and Koirtyohann published an excellent review article entitled, Emission Flame Photometry—a New Look at an Old Method. In their review, Pickett and Koirtyohann make several statements regarding the place and future of FES in comparison with AAS and AFS. For example, they state that total consumption burners with converted spectrophotometers (Beckman Instruments) no longer represent the "state of the art" and that a very desirable system is a modern grating spectrometer equipped with suitable burners, especially the nitrous oxide–acetylene slot burner, and a good recorder. This system is stated to be well suited for AAS and FES. The authors comment

> It is no accident that most of these parameters are the same in FES and AAS. Several manufacturers of AA instruments now advertise their products as being usable in both ways. It is likely that many prospective purchasers are unaware of the significance of this fact and will fail to make the most of the FES capabilities of the more recent (AA) models.*

The authors also comment on reported detection limits for the three related techniques. First, they report that reporting of detection limits is a "hazardous business" (this author concurs) since these limits are not accurately reproducible and change rapidly with new developments. Also, it should be noted that different groups of workers have defined the term "detection limit" in different ways, making any comparison difficult. Bearing that in mind, however, Pickett and Koirtyohann (1969) further consider that, *as a guide to the analyst concerned with method selection*, detection limits can be quite useful. Pickett and Koirtyohann present an extensive list of detection limits in their review. Their detection limit is defined as the concentration of metal, in μg/ml, giving a signal twice as great as the rms noise level in the background signal. Data are presented only for aqueous solutions with ordinary premixed air–acetylene or nitrous oxide–acetylene flames. With respect to FES–AAS relative detection limits, 24 elements are reported better (i.e., *lower* detection limits) by FES, 17 are stated to be about the same, whereas 21 are stated better by AAS. As a generality, elements having wavelengths above 320 nm are usually better determined by FES, whereas lower wavelength elements are better determined by AAS.

Pickett and Koirtyohann's excellent review article is highly recommended for anyone contemplating the use of FES as an analytical technique.

* Reprinted with permission [Pickett and Koirtyohann (1969). *Anal. Chem.* **41**, 28A]. Copyright © by the American Chemical Society.

D. Molecular Emission Spectrometry in Cool Flames

Nonmetals cannot be determined by direct atomic emission techniques in flames since their principal atomic resonance lines are in the vacuum region (below 200 nm) of the ultraviolet. However, as noted in Section II.A, in the 19th century Salet had observed emission spectra for phosphorus (Salet, 1869) and sulfur (Salet, 1871) compounds when an air–hydrogen flame was allowed to impinge on a vertical glass surface that was water cooled. This emission has recently been utilized for analytical purposes by several groups of workers. The origin of these emissions is chemiluminescent in nature. For sulfur it is caused by S_2 molecular emission, whereas for phosphorus it is caused by HPO molecular emission. These spectra may be observed in a fuel-rich air–hydrogen flame but are better seen in a premixed entrained air–hydrogen flame where the "support gas" is an inert gas (nonburning) such as argon or nitrogen. In effect, this is a diluted hydrogen diffusion flame.

Sulfur emission appears as a series of 16 bands extending from 320 to 460 nm with particularly strong bands at 370, 383, 394, and 405 nm (Syty and Dean, 1968; Dagnall *et al.*, 1967), whereas phosphorus emission consists of three bands extending from 480 to 560 nm with the major band at 528 nm (Syty and Dean, 1968; Dagnall *et al.*, 1968). Syty and Dean (1968) and also Dagnall *et al.* (1967, 1968) have reported on the analytical applications and limitations of this chemiluminescent phenomenon. Both groups of workers commented on the remarkable sensitivity for the elements and also observed a good linearity with concentration. However, they also observed that the emission intensity was very dependent upon the form in which the element was introduced. This effect was also noted by Kerber (1970). An example of this is shown in Table II, taken from Dagnall *et al.* (1968).

Another observation made by both groups of workers was that the emission was easily depressed by either organic solvents or metallic ions. The depressive effect was observed to increase with increasing concentrations and eventually there is quenching of the emission. In order to overcome the depressive effect, Dagnall and co-workers (1967, 1968) suggested a preliminary ion-exchange separation.

Syty (1971) and Elliott and Mostyn (1971) used cool flame molecular emission spectrometry to determine phosphorus in detergents; in both cases cation-exchange resin was used to remove metallic interferences. Elliott and Mostyn (1971) reported results on detergent samples containing up to 20% phosphates (expressed as P_2O_5) and gave a precision of about 2–4% for the method. In another publication, Syty (1973) determined phosphorus in phosphate rock using the same technique, again with a cation-exchange resin to remove metals. Results compared favorably with gravimetric procedures.

TABLE II

Emission Intensities of Various Phosphorus Compounds in the Nitrogen–Hydrogen Diffusion Flame[a]

Compound, 2×10^{-3} M	Emission Reading, Peak-to-Trough Height
Orthophorphoric acid	48
Sodium dihydrogen orthophosphate	15
Disodium hydrogen orthophosphate	6
Disodium hydrogen phosphite	5.5
Sodium pyrophosphate	5
Calcium hydrogen orthophosphate	12

[a] From Dagnall *et al.* (1968).

All of the above results were obtained using instrumentation with monochromators. Since the band emission is very broad, wide slits were used in all cases in order to obtain as intense a signal as possible. Aldous *et al.* (1970), however, constructed a specially designed filter photometer with the end-on PMT placed directly behind the filter. Limits of detection for aqueous solutions (with conventional nebulizing) of sulfur and phosphorus were, respectively, 0.2 and 0.01 μg/ml. Using a heated nebulizer chamber, they were able to reduce detection limits for sulfur to 0.08 μg/ml and phosphorus to 0.007 μg/ml. They were able to use their system for the determination of sulfur in solid samples, using an oxygen flask. The oxygen flask was used to convert the sulfur in the sample into sulfate in aqueous solution followed by measurement with their filter photometer. Crider (1965) has used molecular flame spectrometry to detect sulfur dioxide and sulfuric acid in airborne droplets. S_2 and HPO emission are also used in gas chromatography in the fairly specific "flame photometric" detector (Natusch and Thorpe, 1973).

Molecular emission spectrometry has also been used to determine tin in the range 3–3000 μg/ml (Dagnall *et al.*, 1969a). SnH molecular emission in a nitrogen–hydrogen-entrained air diffusion flame shows a sharp band at 610 nm and the reported detection limit was given as 1.5 μg/ml.

Dagnall and co-workers (1969b) also used molecular emission spectrometry for the determination of chloride, bromide, and iodide, by measurement of the intense InCl, InBr, and InI band emission signals, which were found to occur in the cool nitrogen–hydrogen-entrained air flame. Emission from InCl was observed at 360 nm, InBr was observed at 376 nm, and the InI emission was observed at 410 nm. The main emission peaks for the three halide species were well separated, and it was possible to determine chloride in the presence of a large excess of bromide and iodide, bromide in the presence of a large excess of chloride and iodide, and iodide in the presence of a large excess of chloride. Gallium halides were also studied but the observed emission spectra were too weak to be of any analytical utility.

E. *Organic Flame Spectra*

uv–visible spectra produced in flames by organic molecules have also been studied from a potential analytical point of view. The spectra produced are from diatomic fragments of the parent molecule and, as for all diatomic molecules, exhibit characteristic band spectra. The emitting species of primary interest are shown in Table III, which also shows the most intense wavelength for each band. A more complete description of these spectra and their theoretical basis can be found in the books by Gaydon (1957) and Mavrodineanu and Boiteaux (1965).

TABLE III

Emitting Species of Primary Interest in Organic Analysis

Species	Wavelength[a]
C_2	516.5
	468
CH	431.5
CN	388.3
	358.4
NH	336.0
PO	324.0
	230–253 (4 bands)
OH	306.4–309.0
	281.1–294.5
CS	257.6
SH	328
CCl	258, 277

[a] In nanometers.

Because of the high energy of this spectral region and the destructive nature of the flame, one would expect to find only small molecules of the kind noted in Table III, and no evidence of whole molecules. Since it is generally not possible to observe whole molecules, the utility of this technique for structural determination would appear to be limited. However, the relative intensity of observed emission peaks has been shown to be dependent on structure, particularly in a homologous series (Kroeten *et al.*, 1970).

A fundamental study of organic flame spectra in analytical flame spectrometry was undertaken by Robinson and Smith (Robinson and Smith, 1966; Smith and Robinson, 1969, 1970a,b). They first studied the effect of various experimental conditions on the organic spectra. A direct injection oxy-hydrogen burner was used because of the larger primary reaction zone as compared with the premixed burner. Hydrogen was used as a fuel since any organic fuel (e.g., acetylene) would obviously obscure any C_2 or CH bands. The oxy-hydrogen, or air–hydrogen, flame does, of course produce intense OH bands at about 310 nm and this limits their analytical usefullness. It was found, however, that alcohols aspirated into the flame greatly increased the OH intensity (Robinson and Smith, 1966). It was also observed that the flame composition had a significant effect on band intensity, with the fuel-rich, reducing flame being preferred. The most intense emission was generally found in the primary reaction zone of the flame with the OH band progressively obscuring all else as one went higher in the flame (Robinson and Smith, 1966; Smith and Robinson, 1969). The various organic solvents examined showed different aspiration rates, which could introduce changes in band intensity *not due* to structure. This could be overcome by using a motor-driven syringe to deliver a constant supply of liquid to the burner, as described by Parsons and co-workers (Kroeten *et al.*, 1970).

Robinson and Smith also reported on the absorption of organic flames by a continuum (Smith and Robinson, 1970a) and by hollow cathode lamps (Smith and Robinson, 1970b). They found that absorption was intense at most of the wavelengths already mentioned for emission, and that absorption should be useful for the determination of organic molecules, particularly those containing sulfur or phosphorus.

Dagnall and co-workers (1969c) examined the emission spectra of organic molecules using a nitrogen–hydrogen-entrained air flame and also a premixed air–hydrogen flame. They found that the emission intensities from the premixed air–hydrogen flame were generally an order of magnitude higher than for the nitrogen–hydrogen-entrained air flame. The very low background of the entrained air flame made it more useful, however, in certain circumstances. No comparisons to other flames used by other workers were given. In addition to the bands noted in Table III, Dagnall and co-workers also found a series of NO bands for all nitrogen-containing compounds extending from 235 to 270 nm, with measurements being taken at 259 nm. They found that the CN to NO band intensity ratio increases in amines as one goes from primary to secondary to tertiary, and that the CN to NO ratio increases as RNO_2–RNH_2–R_2NH–R_3N–RCH–pyridine.

Dagnall *et al.* (1970) also examined the emission from organic molecules in a high voltage AC spark in an argon or nitrogen atmosphere. Nitrogen was preferred as a carrier gas due to less background and greater stability. The CN band at 389 nm was generally the most intense and was observed for all organic compounds. In addition, tungsten atomic lines (from the electrodes) were observed when halogenated compounds were introduced, and atomic iodine emission at 206.2 nm was observed for ethyl iodide.

Braman (1966) studied flame emission of organics as a detector for gas chromatography. He found that the response for each compound studied was linear with the amount injected into the chromatograph, and also that response at each wavelength studied varied with the compound injected for a chlorinated methane series and for a substituted benzene series. McCrea and Light (1967) studied the response of various hydrocarbons in methanol solution. They found linear response with molality of solution and increasing emission intensity with the series hexane–cyclohexane–benzene–toluene–*o*–xylene.

Parsons and co-workers (Kroeten *et al.*, 1970) examined the relationship of structure with emission intensity at various wavelengths for a homologous series. The ratio of CH to C_2 emission intensity for two series of normal and branched alcohols as solutions in methanol was studied. It was found that the data for the two series was statistically different.

Parsons (1969) also examined the CH, CN, and C_2 emission intensity for various series of alcohols and amines as solutions in methanol. It was found that the intensity of the NH band was relatively constant for all compounds studied, so this band could not be used. The CN emission was moderately insensitive to carbon number, while the C_2 and CH intensities were observed to increase with increasing carbon number.

The utility of FES for structure determination is obviously limited, primarily because of the lack of information obtainable from each molecule. One generally obtains only the relative intensity of a few peaks, which is certainly not enough to completely categorize an organic compound. The main use of organic flame spectra, therefore, would seem to be in distinguishing among the members of a series whose members differ from each other in a simple way such as carbon number or geometrical isomers. A system such as this might prove useful as a detector in gas chromatography, as shown by Braman (1966). A flame emission detector using simple interference filters in conjunction with a detector not sensitive to structure could be used to characterize peaks in gas chromatography. The presence of CN, NH, MO, PO, SH, etc. bands can be used as a simple *qualitative* test for these elements. It should be possible to test qualitatively for almost any element in an organic compound using FES. However, much more work is needed to exploit flame spectra for organic structure determination.

F. Molecular Emission Cavity Analysis

MECA is an acronym for Molecular Emission Cavity Analysis, a technique pioneered, and to a large extent developed, by Belcher and his research associates at the University of Birmingham, England.* This technique, somewhat related to cool flame molecular spectrometry, employs a small cavity at the end of a rod into which samples are deposited; the cavity is introduced into a nitrogen–hydrogen-entrained air flame in line with a detector. A MECA system is shown in Fig. 6. The cavity is introduced into the flame at an angle of 7° downward to optimize contact with the flame gases. The flame serves the dual function of heating the cavity to vaporize the sample and of providing the radicals that maintain the molecular emission. It is desirable to adjust the flame gas flow and the position of the cavity in order to restrict the emission to the cavity (Belcher *et al.*, 1973). Emission intensity is measured with a conventional spectrophotometer equipped with a chart recorder or other readout device. The response must be calibrated using known concentrations of the particular compound under investigation. It is, of course, necessary to place a known amount of sample (either solid or liquid) into the cavity.

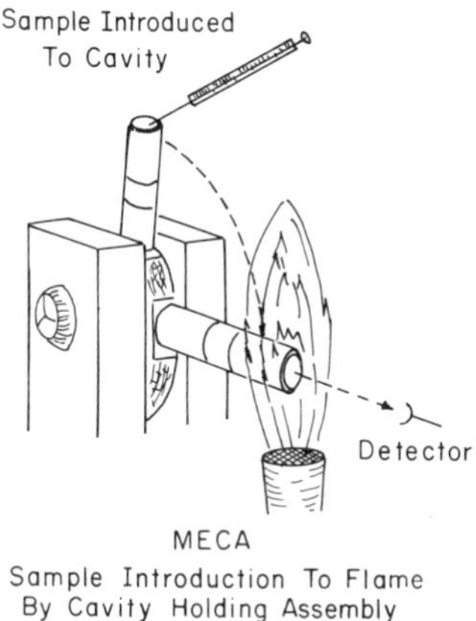

Fig. 6 MECA system.

With MECA techniques, it is sometimes possible to differentiate between two species on the basis of thermal stability and relative volatility of the compound. Sulfur compounds show the characteristic S_2 chemiluminescent emission discussed earlier. Organic sulfur compounds, however, emit almost

* See Belcher *et al.* (1973, 1974, 1975), Bogdanski (1973), and Anacon (no date).

immediately after the sample is placed in the flame, but inorganic compounds such as manganese sulfate and sodium sulfate require a longer dwell time in the flame. This effect is shown in Fig. 7. It arises because of the time required for the cavity to be heated to the necessary breakdown temperature by the flame.

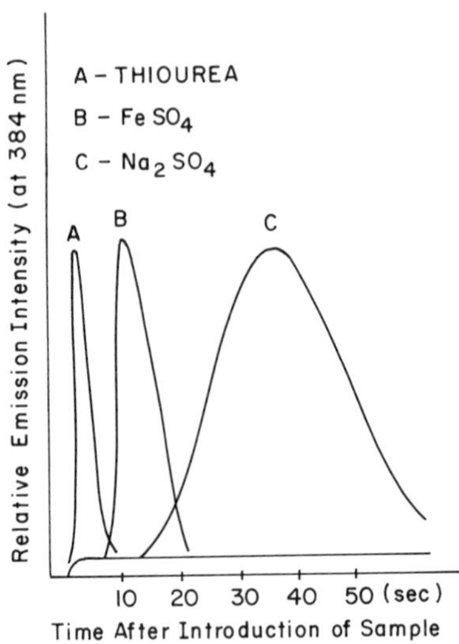

Fig. 7 MECA S_2 emission-time response profiles.

The cavity also exhibits an interesting effect not observed in aspiration systems. It has been reported (Belcher *et al.*, 1973) that when air is introduced into the nitrogen–hydrogen-entrained air diffusion flame sulfur emission by MECA is destroyed. This is not a surprising result since this effect also occurs with conventional aspiration. However, when the air concentration in the flame is further increased, emission from the MECA cavity reappears and becomes more sensitive than in the absence of added air.

The MECA technique is reported to be equally applicable to organic and inorganic sulfur compounds as either solids or liquids. Reported detection limits for thiourea and dimethyl sulfoxide are respectively, 2.5 ng and 30 pg. As with conventional molecular emission spectrometry for S_2, the presence of organic solvents can quench the emission and the authors (Belcher *et al.*, 1973, 1974; Bogdanski, 1973; Anacon, no date) recommend solvent evaporation before MECA measurement.

It is possible to use MECA for the analytical determination of phosphorus using the HPO green band emission. The experimental setup is similar to that for sulfur, and nanogram amounts of phosphorus may be determined.

An interesting application of MECA is for the analysis of selenium and tellurium (Belcher *et al.*, 1974). Unlike sulfur and phosphorus, selenium compounds do not emit chemiluminescence when aqueous solutions are aspirated into nitrogen–hydrogen-entrained air flames. They do, however, show a clear blue-white emission spectrum when measured by MECA techniques and the authors ascribe this emission to Se_2 chemiluminescence. The emission is measured at 411 nm and it is possible to determine as little as 50 ng of selenium. It is also possible to determine selenium in organic or inorganic compounds. Certain materials may interfere, e.g., large amounts of arsenic, antimony, sulfur, and many metal ions, but these may be eliminated by reduction of the selenium to the element. The red selenium is filtered onto a fine porosity glass-fiber pad, and the latter is placed in the cavity where MECA emission is then observed. Belcher and co-workers (1974) have used MECA for the determination of selenium in sulfuric acid. Alternatively, organic compounds can be burned in an oxygen flask and selenium determined in the aqueous solution obtained. This relatively rapid technique has been used successfully to determine very small quantities of selenium in shampoo formulations and also to determine percent amounts of selenium in organo-selenium compounds.

Tellurium compounds give a very faint blue emission when aspirated into a nitrogen–hydrogen-entrained air flame. With MECA, however, using some oxygen in the flame, a green emission centered at about 500 nm is concentrated in the cavity with a pale blue emission in the flame above the cavity. As the cavity heats up, two emission peaks appear in succession, corresponding to temperatures of 500 and 780°C (Belcher *et al.*, 1974). Using the second peak, it is possible to detect tellurium at concentrations as low as 0.5 ng.

Several other applications of MECA have also been considered. For example, boron, arsenic, antimony, and silicon form oxides in flames which are responsible for characteristic emissions. Of these, only boron emits when nebulized into a hydrogen diffusion flame but all give intense MECA emission (Anacon, no date). MECA may also be used for halide analysis via indium halide molecular emission, either by adding indium(III) ions to the test solution, or by using an indium plated cavity. Nanogram amounts of the halides can be easily determined at 360 nm for chloride, 376 nm for bromide, and 410 nm for iodide. It has been reported (Belcher *et al.*, 1975) that interhalogen interferences are not important at low concentrations.

G. *Separated Flames in FES*

In the 19th century, Teclu (1891) and independently Smithells and Ingle (1892) demonstrated the existence of two separate zones with premixed air–hydrocarbon flames. In order to achieve this, they surrounded the flames on Bunsen-type burners with glass tubes. It was possible to raise the outer combustion zone to the top of the glass tube. This is shown in Fig. 8. This outer flame corresponds to a diffusion flame in which the combustible gases generated in the primary reaction zone are burned (Kirkbright and West, 1968; Cresser and Keliher, 1970b; Cresser *et al.*, 1973b).

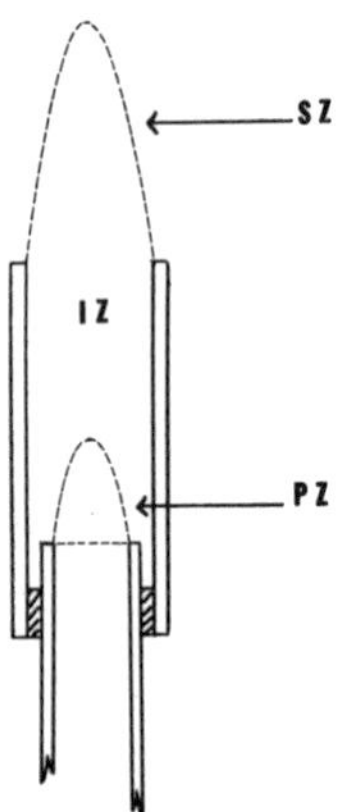

Fig. 8 The Smithells separator: PZ, primary reaction zone; IZ, interconal zone; SZ, secondary reaction zone.

Kirkbright and co-workers (Kirkbright and West, 1968; Kirkbright *et al.*, 1967) first exploited the Smithells and Ingle separator for analytical purposes. They used a separated air–acetylene flame and noted a decrease of about two orders of magnitude in the OH and CO emission from the interconal zone. Although the emission intensity of most of the elements investigated decreased to 10–30% of the values in the unseparated flame, substantial improvements in the signal-to-background ratios were noted. The decrease in signal intensity was caused by a flame temperature decrease of about 30°C. However, of much more importance, molecular background was significantly lowered, which led to lower, improved detection limits. Emission spectra of normal and separated air–acetylene flames are shown in Fig. 9. In a later publication, Kirkbright *et al.* (1968) reported on the analytical usefulness of a mechanically separated nitrous oxide–acetylene flame. Detection limits were reported for aluminum, beryllium, and molybdenum. In the most favorable case, molybdenum 319.4 nm, a detection limit of 20 μg/ml was improved to 0.5 μg/ml with the separator.

Although mechanical flame separators can lead to improved detection limits in FES, as shown by the work of Kirkbright and co-workers, they are somewhat inconvenient to use and although other designs have been pro-

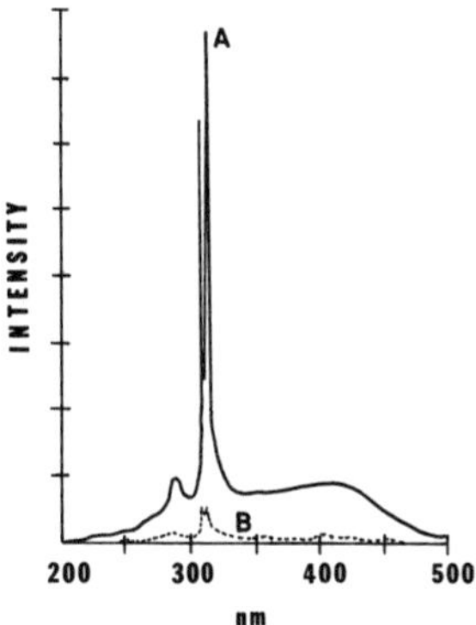

Fig. 9 Emission spectrum of (A) normal, and (B) separated air–acetylene flame.

posed (Ure and Berrow, 1970; Hingle *et al.*, 1968, 1969), they are not in general use today. An alternative and much better means for flame separation is to surround the flame with a "wall" of an inert gas, such as argon or nitrogen, as shown in Fig. 10. The stiff wall of the inert gas lifts the secondary diffusion zone above the observation height, i.e., the area viewed by the detector. The principal advantages of inert gas flame separation, as noted

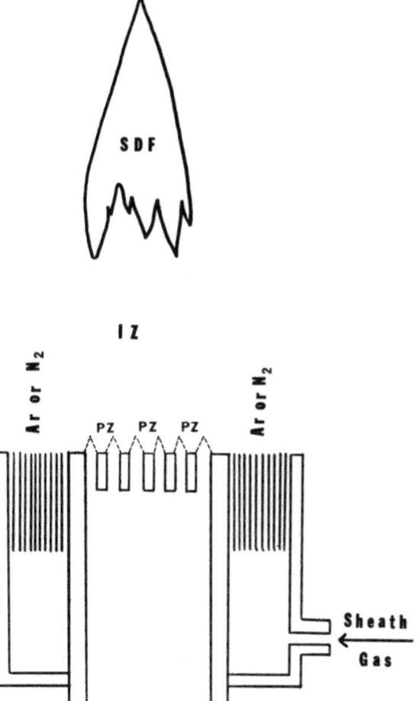

Fig. 10 Inert gas separator for Meker-type burner head: PZ, primary reaction zones; IZ, interconal zone; SDF, secondary diffusion flame.

in a recent review (Cresser *et al.*, 1973b), are the greater range of fuel–oxidant mixture strengths, over which stable separated flames can be supported, and the ability to nebulize a wide range of organic solvents into the flame without inducing flame instability or carbon deposition. Reflection losses that occur at the walls of the silica tube are also eliminated.

Kirkbright and co-workers (Hobbs *et al.*, 1968; Hingle *et al.*, 1970) used an inert gas separated air–acetylene flame for the determination of bismuth at 306.8 nm. This line occurs in the region of very strong OH molecular band emission and flame separation gave a much improved bismuth detection limit. As recently pointed out (Cresser *et al.*, 1973b), however, improvements in detection limits and specificity with separators in FES will depend not only on the burner–separator combination (Meker-type or long path burner) but also on the optical system and monochromator aperture. Meker-type separators should be used for FES when a monochromator with a narrow field of view (most AAS instruments operated in the emission mode) is used (Cresser *et al.*, 1973b).

Inert gas separators have also been used in conjunction with the nitrous oxide–acetylene flame (Kirkbright *et al.*, 1969). As with air–acetylene, the advantages of inert gas separation (over mechanical separation) are maintenance of a stable flame without carbon buildup, when organic solvents are nebulized into the flame, and stability over an extended range of fuel–oxidant ratios. Separation of the nitrous oxide–acetylene flame causes an elongation of the highly reducing interconal zone (Kirkbright *et al.*, 1969; Kirkbright and Vetter, 1972) accompanied by a corresponding increase in emission intensity from the CN molecular band emission from this zone. Kirkbright and Vetter (1972) have studied the influence of acetylene flow rate on the emission intensity observed for elements that form refractory oxides in conventional and inert gas-separated flames; for the nine elements investigated, an appreciable reduction in fuel flow rate was invariably required *on separation* so as to obtain maximal atomic line emission intensity. Kirkbright and Vetter (1972) reported detection limits of an order of magnitude lower with flame separation in conjunction with an $f/5$ aperture monochromator. Amos and co-workers (1970), however, stated that separation did not significantly improve detection limits, provided that a monochromator having a resolving power better than 0.1 nm was used. The apparent discrepancy between the two groups of workers was probably inadequate optimization of experimental parameters, rather than monochromator resolution differences (Cresser *et al.*, 1973b).

An important future use of the inert gas-separated nitrous oxide–acetylene flame might be for multielement FES where some compromise set of conditions (flame stoichiometry, observation height, etc.) would have to be used for the different elements to be determined. Boumans and DeBoer (1972) studied an argon-separated nitrous oxide–acetylene flame with respect to its suitability for multielement FES. *They considered this separated flame to be the most promising flame for multielement FES.*

Inert gas-separated nitrous oxide–hydrogen and nitrous oxide–MAPP Gas* (methyl acetylene–propadiene) flames have also been studied (Cresser *et al.*, 1970; Cresser and Keliher, 1970b). Detection limits, however, are much higher than for the separated nitrous oxide–acetylene flame, particularly for elements forming refractory oxides, such as aluminum and vanadium. Minor improvements for some elements (over the corresponding unseparated flames) were noted by Cresser *et al.* (1970).

H. Wavelength Modulation in FES

The process of modulation, which is the transposition of a DC signal to an AC waveform, is a technique used for signal-to-noise enhancement in spectrometric systems. Ideally, only that portion of the total experimental signal that contains the desired information is modulated while all other extraneous noise sources are unmodulated (O'Haver, 1972; Hieftje and Sydor, 1972; Epstein, 1976; Epstein and O'Haver, 1975). As discussed by Epstein (1976), there are several ways for a signal to be modulated, e.g., electronic modulation, light intensity modulation, sample modulation, solution modulation, and wavelength modulation. When a modulation technique is used in FES, it is generally light intensity modulation. Many commercial instruments (particularly AAS units operated in the emission accessory mode) use AC detection systems that are designed for this type of modulation. Light intensity modulation, however, will not reduce excitation source noise components, since they will be modulated at the same frequency as the analytical signal.

Wavelength modulation is the rapid, repetitive scanning of a small wavelength interval ($\Delta\lambda$) in a spectrometric system and the demodulation of the AC component of the detector waveform at the frequency of modulation or its higher harmonics (Epstein, 1976). As shown in Fig. 11, the AC waveform is generated by intensity differences across the wavelength modulation interval. If the radiation intensity over the wavelength modulation interval is increasing or decreasing monotonically with wavelength, an AC photodetector signal will be generated at the frequency of modulation ($1f$), the amplitude of which is proportional to the difference in intensity at the extremes of the modulation interval. Alternatively, if the intensity distribution has a single maximum or minimum within the wavelength modulation

* MAPP Gas is a registered trademark of the Air Reduction Company.

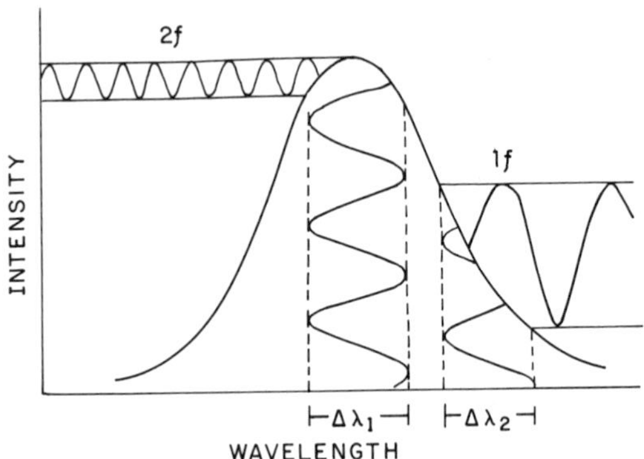

Fig. 11 The generation of AC waveforms by wavelength modulation over a spectral intensity distribution.

interval, an AC signal will be generated at twice the frequency of modulation (2*f*).

As noted by Epstein (1976), detection at twice the frequency of modulation, the second harmonic mode, is more generally applicable to quantitative analysis by FES. In the first harmonic mode, a signal is generated by modulation over a linearly increasing intensity distribution, whereas no signal is generated in the second harmonic mode unless curvature exists in the intensity distribution. Most background radiation will exhibit a linear or linearly sloping intensity distribution over the wavelength modulation signal, whereas an atomic emission signal will show a curvature in the intensity distribution. Therefore, wavelength modulation provides a continuous background correction.

The first application of wavelength modulation to analytical emission spectrometry was reported by Snelleman and co-workers (1970). These workers showed that the technique was useful for the elimination of broadband spectral interferences due to sample matrix components. A reduction in analysis time and sample required was noted. Lichte and Skogerboe (1973) used wavelength modulation for the analysis of solution samples. Maines *et al.* (1972) used wavelength modulation for qualitative and quantitative analysis with a nitrous oxide–acetylene flame and a repetitive scan superimposed over a continuous scan. Rains and Menis (1974) used wavelength modulation for the study of flame emission profiles of aluminum in the nitrous oxide–acetylene flame and also for the reduction of the interfering CH band system on the quantitative determination of aluminum by FES. Other interference studies in the nitrous oxide–acetylene flame have been reported by Sydor and Hieftje (1972) and Epstein and O'Haver (1975). Epstein and O'Haver considered the wavelength modulated flame emission system to be superior to conventional FES in several respects, notably with respect to optimization of flame parameters being simplified. Changes in flame background are not seen, since there is usually no signal generated by the background radiation. It is possible to optimize burner position, fuel–oxidant ratio, and sample aspiration rate without going back to the blank to check for background. Random errors due to fluctuations in the DC level of the background are reduced, improving precision. Analysis time and sample consumption are reduced, since background correction off of the analytical line or scanning of the line is unnecessary. Epstein and O'Haver (1975) stated that the wavelength modulation system could easily be added to a spectrometric system and that, furthermore, a low or medium resolution monochromator could be used with wavelength modulation for FES because the wavelength modulation system would discriminate against background flicker noise (Epstein, 1976; Epstein and O'Haver, 1975).

III. Plasma Emission Spectrometry

A. Plasma Classification

As indicated in Section II, flame spectrometry in its three complementary branches of FES, AAS, and AFS, has become widely accepted for trace qualitative and quantitative analysis. However, although chemical combustion flames are a very efficient means for the generation of atoms, they are unfortunately inadequate for many elements. As noted previously, it is not possible to utilize conventional atomic flame emission spectrometry for elements such as the halogens, sulfur, phosphorus, and carbon. Many metals require the use of the high temperature nitrous oxide–acetylene flame and, furthermore, despite advances with sophisticated techniques such as flame separation and wavelength modulation, detection limits via flame techniques are often not as low as would be desired.

For these reasons, there has been for many years a demand for newer and more refined "atom reservoirs." It is this need that has led to the development of various types of "flamelike" plasmas as alternative atom reservoirs, most often for emission spectrometry but in some cases for absorption and fluorescence spectrometry.

Fassel (1973) has defined a plasma as "any luminous gas in which a significant fraction (more than one percent) of its atoms or molecules are ionized." Although this definition would include chemical combustion flames, it is conventional to restrict the term "plasma" to those supported by electrical means. Furthermore, popular usage of the word "plasma" further restricts the term to flamelike plasmas, referred to by Fassel (1973) as "electrical flames," so that nonflamelike plasmas such as the DC arc, AC arc, and AC spark would not be included. Although flamelike plasmas resemble chemical combustion flames in general appearance, they have many important differences, particularly with respect to chemical environment and temperature. Plasmas are generally thousands of degrees hotter than chemical combustion flames and generally exhibit the spectrum of the inert gas discharge.

Flamelike plasmas may be divided into two general categories: static or low frequency plasmas, and high frequency plasmas. High frequency plasmas, in turn, may be further divided into single-electrode high frequency plasmas and electrodeless high frequency plasmas. Two types of electrodeless high frequency plasmas have been studied in great detail; these are the inductively coupled radiofrequency plasma (ICP) and the microwave induced plasma (MIP) in various configurations.

B. DC Plasmas

DC plasmas are generated by a DC arc discharge whereby a primary arc is struck in a chamber between the anode and cathode electrodes. The resultant plasma is a well-defined, temperature gradient column. A variety of electrode geometries have been used and the electrodes themselves may be graphite, tungsten, or doped tungsten, usually thoriated. DC plasmas are sometimes called plasma jets, plasmatrons, or transferred plasmas because the arc column is transferred away from the arc column by the vortex coolant inert gas flow. The inert gas used is usually argon, although helium, argon–helium mixtures, and nitrogen have also been used.

The flamelike DC plasma was developed by Weiss (1954). Margoshes and Scribner (1959) were the first to utilize the potential of such an analytical technique and their efforts were immediately reinforced by the independent work of Korolev and Vainshstein (1959) in the Soviet Union. The "plasma jet" developed by Margoshes and Scribner (1959) used graphite disk electrodes for both the anode (lower electrode) and cathode (upper electrode). A direct injection nebulizer (see Section II, B, 1) aspirated 0.5–1.0 ml/min of sample into the plasma. Argon was used as the spraying gas and helium was introduced tangentially into the interelectrode space. A standard DC arc power supply was used with plasma currents of about 15 to 20 amps. Margoshes and Scribner (1959) observed that at lower currents ($\sim$10 A) the plasma tended to be unstable and further observed that flow rates in excess of 2–3 ml/min. (common FES flow rates) tended to extinguish the discharge. The authors noted that "the plasma has an intensely bright blue color when no sample is introduced, but takes on a red tint due to the hydrogen spectrum when water is atomized (i.e., nebulized) into the discharge." When no sample was introduced, spectroscopic measurements showed helium and argon lines, as well as CN bands. This was ascribed to nitrogen in the atmosphere, or possibly nitrogen as an impurity in the helium. When water was sprayed into the discharge, the hydrogen spectrum and OH and NH bands appeared. All metals introduced into the plasma jet gave sparklike spectra with many ion lines, indicating (see Section II, B, 1) a very high temperature in the discharge.

Margoshes and Scribner (1959) applied their plasma to the spectrographic analysis of stainless steel for iron, chromium, and nickel. Relative standard deviations of about 2% were obtained. Other elements studied included vanadium, copper, cadmium, calcium, manganese, and zinc.

The plasma system set up by Korolev and Vainshstein (1959) was quite similar to the Margoshes and Scribner (1959) arrangement except that nitrogen was used with a chamber-type nebulizer.

In 1961, Owen showed that a considerable improvement in plasma stability could be obtained by using a third (external) electrode maintained at the same potential as the cathode. Various cathode configurations were investigated by Owen (1960) and both graphite and tungsten external electrodes were used. Graphite electrodes were observed to exhibit little electrical erosion and did not contribute to the spectrum even when grossly contaminated. The life of the tungsten electrode was observed to be a function of current and time. Several inches/hour were consumed at 25 A. The tungsten electrode did not contribute to the spectra in the discharge region between it and the graphite anode electrode.

Following Owen's work, Scribner and Margoshes (1962) described an improved version of their plasma, incorporating some of the changes suggested by Owen. Owen's radial cathode was replaced by an axial one and the top graphite ring was left floating electrically. Serin and Ashton (1964) studied the effects of tangential gas flow, arc current, and water-miscible solvents on the sensitivity of their plasma jet. Webb and Wildy (1963) used a plasma jet for the determination of calcium in biological materials.

Improvements with DC plasma devices continued in the 1960s with modifications reported by such workers as Yamamoto (1962, 1963), Kranz (1964, 1968), Goto and Atsuya (1967), and Jahn (1961). Yamamoto (1962, 1963) used argon in his device but observed that when argon was replaced by helium, the plasma showed a considerable increase in electrode erosion. This rate of erosion could be reduced by using an argon–helium (30–70) mixture; this mixture reduced the intensity of the background continuum. The maximum solution aspiration rate was about 2 ml/min. The electrodes used by Yamamoto were copper (anode) and thoriated tungsten (cathode). Flame temperatures were reported to be as high as 11,500 K when the arc was operated at 300 A with 7 kW power into the discharge. This was a much higher current than used by subsequent workers.

Goto and Atsuya (1967) studied the effects of acids and organic solvents on their plasma. Acids eroded the copper anode used, affecting stability and precision, whereas organic solvents lowered metal emission intensity and eroded electrodes. The authors were not able to use satisfactorily organic solvents with their plasma.

In the mid 1960s, Spex Industries (Metuchen, New Jersey) introduced a commercially available DC plasma, similar in design to the modified Scribner and Margoshes (1962) system. This is shown in Fig. 12. Helium was normally used as the swirl (tangential) gas and argon served as the nebulizing (atomizing) gas. Later, Collins (1967) used a modified Scribner–Margoshes-type plasma to spectrographically determine barium, boron, iron, manganese, and strontium in oil field waters; the method was reported to be rapid, sensitive, and precise. Szivek *et al.* (1968) used a modified Spex Industries plasma for the determination of magnesium in blood plasma. They had found that the unmodified Spex plasma accumulated residue in the nebulizer annulus. This problem was circumvented by the use of a simple Zeiss flame (direct injection) burner, which could be easily dismantled and cleaned. Chapman and co-workers (1973) also modified a Spex plasma by replacing the direct injection nebulizer with a premixed chamber nebulizer. Improved stability and sensitivity was noted. Heemstra (1970) described a controlled atmosphere plasma arc, i.e., a device to shield the plasma from the atmosphere, and this was used to analyze nonmetals such as nitrogen and sulfur in crude oils.

Up until 1970, most DC plasma systems used very high inert gas flow rates with their attendant high operating costs. Then, however, Valente and Schrenk (1970) described a plasma device that could be operated with a relatively low argon flow rate, ~ 2 liter/min. The plasma was operated from a conventional DC arc power supply with a current of approximately

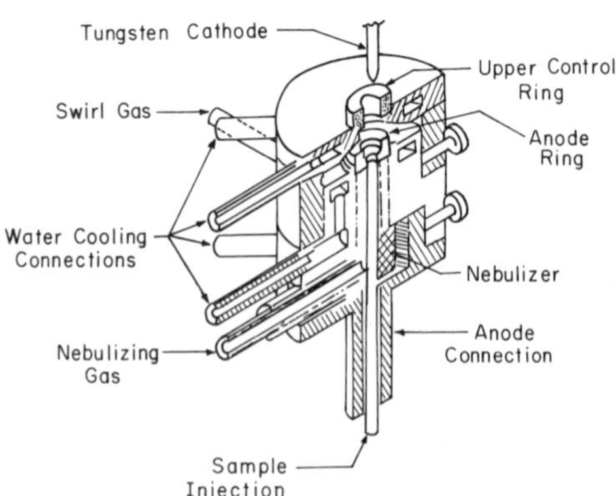

Fig. 12 Spex Industries' DC plasma.

10 A. The Valente and Schrenk device used commercially available spectrographic electrodes and it was possible to operate the plasma continuously for up to several hours before electrode erosion caused instability. Instead of direct nebulization, a heated chamber condensation system (Veillon and Margoshes, 1968b) was used to remove water vapor to provide a dry aerosol for the plasma. Detection limits were reported for 12 elements and were, generally, in the low parts-per-billion (ng/ml) range. Response was found linear over several orders of magnitude. Unfortunately, Valente and Schrenk (1970) noted memory effects under certain circumstances. This was attributed to two different sources. Sample accumulated on the wall of the heated spray chamber could reenter the injection system gas stream as a continuous supply of sample. This problem could be minimized by periodic dismantling and cleaning of the heated chamber. Another memory effect was observed when sample accumulated on the cylindrical wall of the control orifice insert slowly and continuously reentered the discharge. This was found to be a problem only for boron, probably due to the formation of the relatively volatile oxide, B_2O_3, which has a boiling point of only 1260°C.

In 1971, Marinkovic and Vickers (1971) described a long-path stabilized DC arc plasma and reported radial atomic distributions and detection limits for several refractory elements. The arc used a graphite anode, a carbon cathode, and argon as the flow gas. The minimum argon flow rate for a stable discharge was about 1.8 liter/min.

In the same year, Elliott (1971) described a right-angle plasma arrangement, which he called a "Spectrajet." This was introduced commercially by Spectrametrics Inc. (Andover, Massachusetts) at about the same time and is shown in Fig. 13. This plasma device consists of a Teflon-lined stainless steel cylindrical outer housing that fits onto a ceramic circular base. A stainless steel disk top is recessed into the cylinder, and a graphite ring is recessed in the disk. A water-cooled anode holder is located at the center of the ceramic disk base. This holder provides a coaxial argon flow surrounding a $\frac{1}{8}$-in.-diameter thoriated tungsten rod electrode. The argon flow surrounds this electrode and is guided by a replaceable alumina cylinder. The cathode is mounted externally in an adjustable water-cooled holder. A 0.040-in.-diameter thoriated (2%) tungsten electrode is clamped in a collet at the end of a tube that supports the electrode and also provides coaxial "focused flow" argon feed. An alumina cylinder guides the argon flow at the plasma end of the electrode. The tip of the electrode is positioned above the periphery of the graphite thermal pinch ring, away from the central axis of the plasma column. The thermal pinch results from the tangential introduction of argon around the DC discharge. The electrode geometry produces the bend in the plasma. Pure argon is introduced coaxially with the anode and cathode, and the argon carrying the sample is introduced

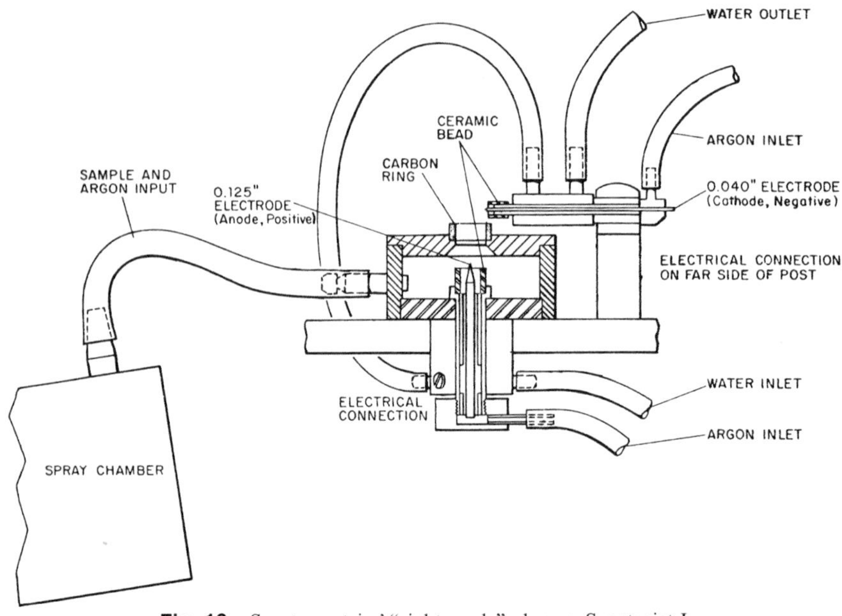

Fig. 13 Spectrametrics' "right-angle" plasma, Spectrajet I.

into the cylindrical housing. According to Elliott (1971), turbulent mixing of the sample and plasma column occurs in the region at the base of the graphite thermal pinch ring. Sample uptake is about 1 ml/min. For best results, Elliott recommended a heated spray chamber.

The appearance of the plasma is as an intense blue-white central right-angle bend core with a flamelike plume above the bent plasma. Using the line ratio method (Pearce, 1961), Elliott calculated a core plasma temperature of about 10,000 K with temperatures in the plume ranging from about 4000 to 8000 K. It should be noted that Elliott's right-angle plasma was developed specifically to be used with a high resolution echelle grating spectrometer (Keliher and Wohlers, 1976; Keliher, 1976; Cresser *et al.*, 1973a) and, because of the optical geometry of the echelle monochromator employed, only a small portion of the plasma is observed by the spectrometer system at any one time. By easy adjustment of an X–Y positioner, it is possible to optimize signal-to-background for a particular element. In general, measurements were best made in the region just above the plasma core. Karicki and Corcoran (1973) used the Spectrametrics system for coal ash analysis while Corcoran *et al.* (1972) described a method for the rapid determination of calcium and phosphorus in phosphate rock. Results compared favorably with classical methods. In another study, Corcoran and Elliott (1972) used the "Spectrajet" for the direct determination of chlorine using lines at 725.6 and 837.5 nm. A chlorine gas generation technique was developed employing a mixture of 35% (by volume) phosphoric acid containing about 3% (by weight) potassium permanganate. This mixture was found to be relatively efficient in the generation of chlorine gas from various inorganic and organic materials, and the generated chlorine gas was introduced into the plasma via the heated spray chamber.

In 1974, Merchant and Veillon described a right-angle DC plasma jet similar to that of Elliott's; their observed temperatures in the plasma core were in general agreement with Elliott. Unlike Elliott, however, they noted a very rapid drop in apparent temperature immediately above the cathode region of the plasma. They also observed that, with their device, much of the sample aerosol did not enter the high temperature plasma core region. In addition to emission techniques, Merchant and Veillon (1974) also used their plasma for some AAS measurements.

In the same year, Murdick and Piepmeir (1974) described a stable DC plasma similar to that previously constructed by Marinkovic and Vickers (1971). Solutions were sprayed into the main argon stream using a chamber-type nebulizer constructed from a Beckman direct injection (total consumption) burner. Sample uptake was about 1.9 ml/min. Detection limits were reported for hafnium (1.5 ppm), copper (0.06 ppm), and calcium (0.01 ppm).

Also in that year, Elliott (1974) described a new version of a highly original plasma system (Elliott, 1971). This was subsequently introduced by Spectrametrics, Inc. (Andover, Massachusetts) and referred to as a "Spectrajet-II." The Spectrajet-II is shown in Fig. 14. This configuration utilizes two of the "focused flow" (the cathode from the previous design–Spectrajet I) electrodes; one as the anode and the other as the cathode. The flow of argon from the anode and cathode intersect to form a continuous plasma shaped somewhat like a hairpin. Sample is introduced by flowing aerosol vertically upward from the intersection. There is, therefore, no possibility for premixing of the argon-carrying aerosol with the argon forming the plasma discharge. Also, a region of intense excitation exists in this configuration. With this device, a peristaltic pump is recommended (Elliott, 1974) to introduce sample into the heated spray chamber. Elliott observed that conventional nebulization would quench the discharge. In effect, the pump serves to retard (i.e., reduce) sample flow to the plasma. As with the previous Spectrajet, this device was designed to be used with a high resolution echelle monochromator (Keliher and Wohlers, 1976; Keliher, 1976; Cresser *et al.*, 1973a) and the optimum region of observation was just below the current-carrying arc column.

In 1975, Rippetoe and co-workers (1975) described a device similar to the stablized plasmas of Marinkovic and Vickers (1971) and Murdick and Piepmeier (1974) but measurements were made in the plasma plume rather than in the plasma core. In a later publication, Rippetoe and Vickers (1975)

Fig. 14 Spectrametrics' "hairpin" plasma, Spectrajet II.

described a rotating arc plasma jet where the arc column rotated at speeds up to 600 Hz around a graphite disk electrode.

At this point in time, despite considerable development by many workers from Margoshes and Scribner (1959) and Scribner and Margoshes (1962) to the present day, the DC plasma must still be considered in late infancy with respect to widespread use in routine analyses. More effort is needed with respect to optimum electrode geometries, sample introduction, interelement studies, etc. Unfortunately, detection limits obtainable with DC devices are not as low as presently obtainable with other types (particularly inductively coupled) of plasmas (Fassel, 1973; Keirs and Vickers, 1977).

As noted by Keirs and Vickers (1977) in a recent review of DC plasmas,

> the key to the future for DC plasma devices is sample introduction. If the sample introduction problem can be solved, then for elemental analysis by emission spectroscopy, the DC plasma arc should offer exactly the same advantages as any other high temperature plasma and may offer some additional advantages in simplicity and cost of equipment and operator acceptance.

It is to be hoped that further research along these lines will lead to a wider acceptance of DC plasmas in the near future. Very recently, Spectrametrics has introduced a three-electrode plasma with an improved spray chamber introduction system (Spectrajet III), and preliminary results (Keliher, 1978) are most promising.

C. Inductively Coupled Plasmas

An inductively coupled plasma (ICP)* is a special type of plasma that derives its sustaining power by induction from high frequency magnetic fields (Fassel, 1973; Fassel and Kniseley, 1974a,b; Greenfield *et al.*, 1976a; Sharp, 1976). In the early 1940s, Babat (1942, 1947) observed an ICP discharge in air. Babat used extremely high power levels (100 kW) and made no effort to flow gases through the discharge in order to achieve a plasma torch configuration. This was subsequently accomplished, however, by Reed (1961a,b, 1962, 1963) who successfully generated a stable argon plasma at 1 atm pressure using a commercial radio frequency heating unit operated at a frequency of 4 MHz with a maximum power output of 10 kW. Reed's torch consisted of a quartz tube with a brass base having a tangential gas entry, placed within the work coil of the generator. Reed also operated his plasma with nitrogen (Reed, 1961b) in order to grow refractory crystals. The discharge produced from Reed's ICP was capable of melting many refractory materials due to the intense heat of association of the dissociated N_2 molecules. Although Reed recognized that ICPs had a great potential for exciting spectra (Fassel and Kniseley, 1974a), he did not use his plasma for analytical purposes. His pioneering efforts, however, in the design of ICPs led to their use for analytical purposes by Greenfield and co-workers (Greenfield, 1965; Greenfield *et al.*, 1964, 1968) in the United Kingdom and, independently, by Fassel and co-workers (Wendt and Fassel, 1965, 1966; Barnett and Fassel, 1968; Fassel and Dickinson, 1968; Dickinson and

* Also referred to as ICAPs (inductively coupled argon plasma) and ICPTs (inductively coupled plasma torch).

Fassel, 1969) in the United States. Both groups of workers used ICPs as excitation sources in conjunction with conventional spectrometer (monochromator, lenses, PMT, etc.) systems.

Greenfield and his colleagues (1964) used a 36-MHz, 2.5-kW generator to produce an annular plasma, i.e., a discharge with a cooler central region through which the central stream and sample flow. Samples were introduced by solution nebulization. It was stated, however, that direct injection of liquids, slurries, and solid powders would also be possible. Wendt and Fassel's (1965) original ICP system used a frequency of 3.4 MHz with a power of 5 kW. Ultrasonic nebulization (Dunken *et al.*, 1963; Issaf and Morgenthaler, 1975a,b) was used to introduce liquid samples into the plasma. The lower frequency of Wendt and Fassel's plasma produced a "tear drop" shaped plasma without a cool axial channel. Following the initial publications (Greenfield *et al.*, 1964; Wendt and Fassel, 1965), several other groups of workers studied radiofrequency plasmas for analytical purposes.*

In 1968, Veillon and Margoshes (1968a) evaluated a 4.8-MHz ICP. They observed significant interelement effects and found their plasma torch considerably less convenient to operate than most chemical combustion flames. They stated,

> The sensitivity of the discharge to molecular gases, particularly water vapor, makes it necessary to employ a more elaborate sample system. Except for a few refractory elements the plasma torch does not appear to be a suitable replacement for the chemical flame.

Despite this 1968 observation, ICPs have turned out to be very desirable "atom reservoirs" in emission spectrometry. Veillon and Margoshes (1968a) had also noted, "It is possible that the interferences are absent at some other frequencies or else that there is some as yet unidentified difference in operating conditions that accounts for the interferences being present or absent." Hoare and Mostyn (1967) had studied a 36-MHz ICP but these workers did not observe interelement effects. In the determination of boron (249.7 nm) and zirconium (349.6 nm), the presence of 10,000 ppm of other material, principally nickel, chromium, cobalt, and iron, did not influence the analysis. In a separate experiment, the addition of 1000 ppm sodium did not affect the boron emission signal. Greenfield *et al.* (1976a) have recently commented that Veillon and Margoshes's observed matrix effects (Veillon and Margoshes, 1968a) may have been due to modification of the electrical properties of the discharge by sample aerosol flowing around it. Greenfield *et al.* (1976a) also feel that it is possible that the temperatures encountered

* For example, Hoare and Mostyn (1967), Dunken and Pforr (1966), Britske *et al.* (1967), Kirkbright *et al.* (1972, 1973), Kirkbright and Ward (1974), Morrison and Talmi (1970), Veillon and Margoshes (1968a), Boumans and DeBoer (1975a,b, 1976), and Boumans *et al.* (1976).

in the Veillon–Margoshes plasma are insufficient to dissociate refractory species.

In the late 1960s and continuing in the 1970s, Fassel and co-workers and Greenfield and co-workers continued their extensive in-depth studies of ICPs. The acceptance of ICPs today must, to a very large extent, be ascribed to these studies. In the early 1970s, Fassel and Kniseley (1974a,b), Kniseley *et al.* (1973), and Scott *et al.* (1974) described an ICP system operated at a frequency of 27.12 MHz with a power of 2 kW. A relatively low argon consumption rate, ~10 liter/min was stated (Scott *et al.*, 1974) to result in an operating cost comparable to that of a premixed nitrous oxide–acetylene flame. A diagram of the plasma tube (torch) configuration used by Fassel and co-workers (Kniseley *et al.*, 1973; Scott *et al.*, 1974) is shown in Fig. 15. The torch is fabricated of fused quartz tubing and consists of concentric outer (coolant) and inner (plasma) tubes and a removeable aerosol injection tube. Argon is introduced tangentially through a side tube. The argon flow velocity is caused to increase toward the top of the plasma tube by the constriction resulting in an increase in both the cooling velocity and the degree of vortex stabilization. The induction coil surrounds the quartz tube and, of course, no electrodes are used. The induction coil is connected to the high frequency generator. To form the plasma, it is necessary to "plant" a "seed" of electrons in the induction coil space after first flushing the sample chamber and torch with argon. A Tesla coil is normally used to initiate the argon ionization. The energy in the induction coil inductively couples with the "seed" of electrons introduced by the Tesla coil so that there is almost instantaneous formation of a stable plasma. Proper adjustment of the radio frequency power is necessary for optimal operation.

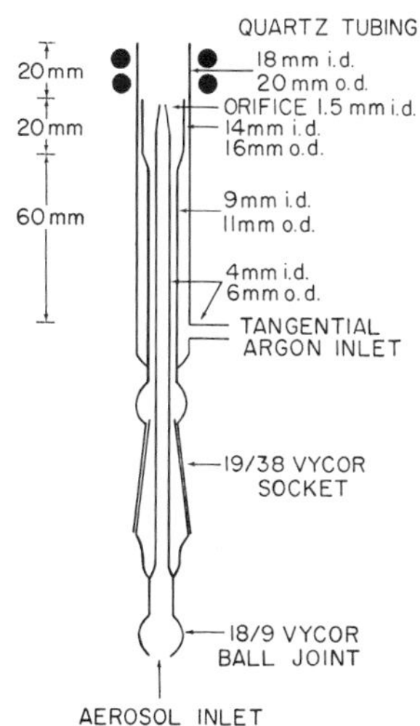

Fig. 15 Diagram of plasma tube configuration. [Reprinted with permission (Scott *et al.* (1974). *Anal. Chem.* **46**, 75). Copyright © by the American Chemical Society.]

Because of the high temperatures (9,000–10,000 K) of the plasma, it is necessary to thermally isolate the plasma from the quartz tube walls. Otherwise, melting of the walls would occur. The tangential argon flow serves three purposes: it cools the walls, keeps the plasma away from the walls, and also creates a low pressure zone in the axial channel of the tube. In addition to the tangential stabilizing flow, a second inlet carries sample aerosol into the plasma. Argon flow rate is about 1–2 liter/min. Because the aerosol is injected into what Fassel has described as the "doughnut hole" of the plasma and since most of the radiofrequency power is absorbed within the skin depth (the "doughnut"), the aerosol is principally heated by conduction and radiation from the surrounding argon. According to Fassel and co-workers (Fassel and Kniseley, 1974a,b; Kniseley *et al.*, 1973; Scott *et al.*, 1974), this indirect heating process is favorable because major changes in sample composition would not be expected to change the absorbed power significantly.

Although Fassel's earlier ICP systems had used ultrasonic nebulization, the 1974 plasma used pneumatic (conventional) nebulization with a solution uptake rate of about 3.0 ml/minute. A dual-tube aerosol chamber, shown in Fig. 16, was used in order to reduce random fluctuations in signal intensity. The purpose of the central tube is to separate the forward and reverse aerosol flows, the latter being produced as a result of the low pressure in the region of high aerosol velocity. Kniseley *et al.* (1973) and Scott *et al.* (1974) stated that the turbulence within the chamber was reduced with less condensation occuring on the nebulizer, resulting in less signal noise in the ICP. The aerosol is fed directly to the plasma without any external desolvation. The rate at which the aerosol enters the plasma was reported to be 0.1 ml/min for an argon flowrate of 1.4 liter/min, indicating a nebulizer efficiency of about 3%. Detection limits were reported for 12 elements and were generally in the ppb range, 0.1–10 ng/ml.

Although this ICP system was used with a conventional monochromator

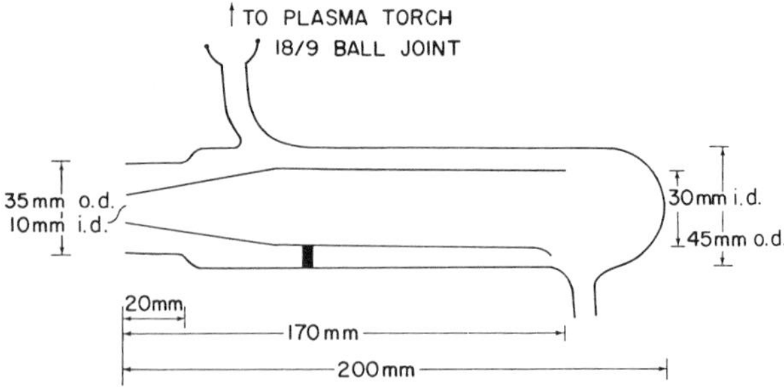

Fig. 16 Diagram of dual-tube sample chamber. [Reprinted with permission (Scott *et al.* (1974). *Anal. Chem.* **46**, 75). Copyright © by the American Chemical Society.]

readout system, it was stated that it would be possible to easily use the ICP for multielement analysis (Scott *et al.*, 1974). In another report (Fassel and Kniseley, 1974a), it was suggested that the ICP could easily be used with a polychromator, positioning PMTs at appropriate wavelengths.

Fassel and co-workers also studied interelement effects in their ICP system (Larson *et al.*, 1975). A slightly simpler spray chamber [compared to that of Kniseley *et al.* (1973) and Scott *et al.* (1974)] was used in this investigation. The authors were able to show clearly that two solute vaporization interferences such as Ca–PO_4 and Ca–Al interferences often observed in chemical combustion flames were eliminated or reduced to negligible proportions in their plasma. They also showed that increasing concentrations of an easily ionizable element such as sodium (up to concentrations as high as 6900 ppm) exerted a very low influence on the observed emission intensities of three selected elements (calcium, chromium, and cadmium) of widely differing degrees of ionization. It was also noted (Larson *et al.*, 1975) that a variety of matrices did not affect the emission intensity of molybdenum to any extent. It was indicated, however, that interelement effects could be easily changed as a function of ICP variables such as the power dissipated in the plasma, the flow velocity of the aerosol carrier gas, the height of observation, and the viewing field of the optical system. The ICP designed by Fassel and co-workers has been applied to the determination of trace amounts of metals in microliter volumes of biological fluids (whole blood, serum, and plasma) using a novel sample introduction system (Kniseley *et al.*, 1973).

Fassel *et al.* (1976) have also used an ICP for the simultaneous determination of 15 different wear metals in lubricating oil. Detection limits ranged from 0.0004 ppm (calcium) to 0.3 ppm (lead) for the elements studied. Results were in good agreement with AAS values. In another study, Fassel and co-workers (Butler *et al.*, 1975) used their ICP for the determination of residual impurities and alloying constituents in solutions of high and low alloy steels of widely varying composition. The presence of an iron matrix did not influence the detection limits for the 12 elements studied. The analytical curves exhibited linearity over three to four orders of magnitude.

In a very recent publication, Fassel and co-workers (Olsen *et al.*, 1977) used a polychromator for the simultaneous analysis of 14 elements of bioenvironmental interest. Conventional pneumatic and ultrasonic nebulizers were compared in this study. The ultrasonic nebulizer, when combined with a conventional aerosol desolvation apparatus (Veillon and Margoshes, 1968b), provided an order of magnitude or more improvement in simultaneous multielement detection limits. Direct sample introduction, convenient sample change, and rapid cleanout, were stated to be other important characteristics of the ultrasonic system making it an attractive alternative to conventional pneumatic nebulization.

Greenfield and Smith (1972) have used a 7-MHz ICP in conjunction with a cross flow nebulizer in a heated chamber for the spectrographic analysis of 1–25 μl samples of fuel oil, organophosphorus compounds, and blood samples. Detection limits for various elements (e.g., aluminum, copper, iron, lead, magnesium, phosphorus, silicon, and silver in human blood) were in the range 10^{-9}–10^{-10} g. At the 10^{-9}-g level, precision was reported to be about 5%. Greenfield and co-workers (1976b) have also studied the effect of high acid concentrations on signal intensity in the ICP. With mineral acids the intensity is reduced by a factor that correlates well with the expected reduction in sample uptake caused by increased viscosity. With organic acids such an expected reduction is outweighed by other enhancements that can be represented as a function of the viscosity, surface tension, and density of the solution. In all cases, acid interferences were ascribed to the aspiration and nebulization systems rather than to the actual ICP. Greenfield and co-workers (1976a) have also used their ICP plasma for multi-element analysis.

Fassel and co-workers (Nixon *et al.*, 1974) have interfaced a tantalum filament vaporization apparatus (of the type normally used in flameless AAS and AFS) to an ICP. Due to the increased concentration of the analyte (already desolvated and vaporized by the filament) in the plasma, detection limits are about 1–2 orders of magnitude lower than conventional nebulization–ICP results (Scott *et al.*, 1974).

Although most work accomplished to date with ICPs has involved their use with solutions (either direct nebulization via conventional or ultrasonic techniques, or transferred via microliter sampling devices), there has been much interest in using ICP devices for the direct analysis of solid samples, something not possible via conventional FES techniques. As early as 1964, Greenfield *et al.* (1964, 1976a) used a pneumatic nebulizer to inject powders and slurries through an annular plasma and reported the "tail-flame" as being intense and the plasma stable. No quantitative results were reported.

Dagnall and co-workers (1971) used a fluidized-bed device to introduce alumina and magnesium oxide samples into an ICP and were able to determine boron and berrylium in the matrix. The system, however, produced severe segregation with the finer particles being removed first. As noted by Greenfield *et al.* (1976a) in a recent ICP review,

> First hand experience of the problem (direct solid introduction) leads to the impression that the partial success so far gained has been obtained on carefully selected matrices, and that a general, practical solution to the several problems of injecting powders into plasmas and performing quantitative analysis, has yet to be found.

Clearly, further research is necessary before direct solid introduction into plasmas can become a viable analytical technique.

TABLE IV

Inductively Coupled Plasma Detection Limits[a]

Element	1965[b]	1975[c]
Aluminum	3	2×10^{-4}
Arsenic	25	6×10^{-3}
Calcium	0.2	1×10^{-7}
Cadmium	20	2×10^{-4}
Chromium	0.3	1×10^{-4}
Iron	3	9×10^{-5}
Lanthanum	50	1×10^{-4}
Magnesium	2	3×10^{-6}
Manganese	1	2×10^{-5}
Nickel	1	2×10^{-4}
Phosphorus	10	0.02
Tin	50	3×10^{-3}
Strontium	0.09	3×10^{-6}
Tungsten	3	8×10^{-4}
Zinc	30	1×10^{-4}
Zirconium	15	6×10^{-5}

[a] Taken from and suggested by Keirs and Vickers (1977).
[b] Data from Wendt and Fassel (1965).
[c] Data from Boumans and DeBoer (1975a).

Although ICPs have been used primarily in emission spectrometry, they have also found some use as "atom reservoirs" in AAS (Greenfield *et al.*, 1968; Wendt and Fassel, 1966) and AFS (Montaser and Fassel, 1976). Their use in these techniques will, however, be restricted due to the excellent detection limits obtainable via emission.

As recently noted by Keirs and Vickers (1977) and as indicated in Table IV, there have been remarkable improvements in detection limits during the last decade. The ICP is becoming widely accepted for trace single-element and multielement analysis and it will certainly find many more uses in a variety of analytical fields, where rapid, extremely sensitive, trace element determinations are required.

D. *Microwave Induced Plasmas*

A microwave induced plasma (MIP) may be defined as a discharge operating at a frequency greater than 300 MHz. In practice, due to the commercial availability of medical diathermy units operated at 2450 MHz, that frequency has become, by far, the most commonly used frequency for analytical microwave studies. The development of MIPs for direct solution nebulization has been closely related to two ancillary uses for microwave radiation in analytical spectrometry, namely, the development of electrodeless discharge lamps (EDLs)* as spectral sources for AAS and AFS, and the development of MIPs *as element specific detectors* in gas chromatography (GC).

In the case of EDLs, the emission from the plasma discharge is not ordinarily used for sample analysis but rather the emission irradiates the atoms of interest allowing them to absorb and/or fluoresce. Several recent reviews discuss the preparation of these spectral sources (Cresser and Keliher, 1970a; Browner, 1974; Haarsma *et al*., 1974) and, therefore, this aspect will not be considered further here. An MIP as an "atom reservoir," however, might be visualized as a "continually flowing EDL" where the sample is, in some fashion, introduced into the discharge region. The MIP draws its energy from a standing electromagnetic wave resonating in a confined cavity (Sharp, 1976). The power to the cavity is supplied by a Magnetron tube and the microwave energy to the cavity is usually in the range 0–200 W. As with ICPs, the MIP discharge is normally initiated by a "seed" of electrons from a Tesla coil. Although various cavity designs have been used (Sharp, 1976), the most common are the $\frac{3}{4}$ wave "Broida" cavity and the $\frac{1}{4}$ wave "Evensen" cavity.

In 1958, Broida and Chapman (1958) analyzed nitrogen isotopes in a resonant microwave cavity. They noted that it was necessary to keep the sample tubes as free as possible from water, ammonia, organic solvents, and stopcock grease.

In 1965, Cooke and co-workers (McCormack *et al*., 1965) at Cornell University operated an atmospheric pressure argon plasma as an emission detector for compounds eluted from a gas chromatograph. Two types of cavities were used and the plasma discharge was formed within a quartz tube. Quartz was used (rather than glass) because of its higher melting point and superior transparancy to ultraviolet radiation. All subsequent workers have used quartz tubing. Species measured by Cooke and co-workers included CN, C_2, CH, I, CS, P, PO, PS, CCl, and F. The intensity of the emitted spectra were found to be greatly dependent on the microwave power supplied to the cavity. With some lines, the intensity increased with increasing power but with others a decrease was observed.

In the same year, Bache and Lisk (1965) (also at Cornell University) used a MIP in conjunction with a GC to determine organophosphorus residues in crops. Their system was essentially identical to that used by Cooke

* Also referred to as electrodeless discharge tubes (EDTs).

and co-workers (McCormack *et al.*, 1965). Argon flow rates were studied and varied from 20 to 115 ml/min. The phosphorus 235.6-nm line was monitored with conventional spectrometric instrumentation. Subsequent papers by these authors (Bache and Lisk, 1966, 1967, 1971) reported modifications to the original system and further applications. In 1967, Bache and Lisk operated a low pressure (5–10-torr) helium-MIP as a GC detector for the determination of organobromine, -chlorine, -iodine, -phosphorus, and -sulfur compounds (Bache and Lisk, 1967). A later publication reported on the utility of the GC-helium-MIP for the determination of organomercury compounds (Bache and Lisk, 1971). It was not possible to achieve an atmospheric pressure helium-MIP and maximum stability for the discharge was found to be in the range 5–10 Torr.

The early experiments at Cornell in the 1960s by Cooke and co-workers (McCormack *et al.*, 1965) and Bache and Lisk (1965, 1966, 1967, 1971) clearly showed the potential advantages of interfacing chromatography with MIPs. During the 1970s, many further reports have appeared.* Serravallo and Risby (1974) have used an oxygen-doped helium-MIP operated at low pressure (1 Torr) as a metal selective GC detector. The authors observe that the reduced pressure helium-MIP is better (as a GC detector) than the atmospheric pressure argon-MIP since it more completely dissociates molecular species and since it forms predominantly atomic (rather than ionic) species. The added oxygen (3%) maximizes atomic emission and simultaneously quenches molecular emission (Serravallo and Risby, 1975). In a subsequent paper, Serravallo and Risby (1976) determined vinyl chloride in air by direct injection into their plasma system. Unfortunately, the detection limit obtained, 390 ppm, was quite poor due to the presence of air, which quenches the ionic chlorine emission.

Talmi and co-workers (Talmi and Andrew, 1974; Talmi and Bostick, 1975; Talmi and Norvell, 1975) have also used MIPs as element specific detectors in GC. Arsenic and antimony have been determined (Talmi and Norvell, 1975) by cocrystallization of As^{+3} and Sb^{+3} with thionalid and reacting the precipitate with phenylmagnesium bromide (PMB). Following the decomposition of excess PMB, the triphenyl arsine and stilbine formed are extracted into ether, followed by GC–MIP detection. Samples analyzed via the technique included biological and plant tissues, coal and fly ash, and fresh and salt water. Talmi and Bostick (1975) have applied their GC–MIP to the determination of alkyl-arsenic acids in commercial pesticides and environmental samples.

* See, for example, Dagnall *et al.*, (1970a, 1972b), Braun (1971), McLean *et al.* (1973), Talmi and Andrew (1974), Talmi and Bostick (1975), Talmi and Norvell (1975), and Serravallo and Risby (1974, 1975, 1976).

Dagnall *et al.* (1972a) have described an unusual GC–MIP system. Rather than using spectral observation of the plasma for specific detection, the authors monitored the reflected microwave power as a function of species present. Although the detector operated in this fashion is completely nonselective, detection limits in the nanogram range were observed. West (1970) has compared a 2450-MHz MIP with a 30-MHz (noninductive) plasma for chromatography. The 30-MHz plasma was found to be less expensive, somewhat simpler to adjust and operate, and less likely to foul the quartz plasma tube. The MIP was found to have several advantages, however; it was more easily thermostated and perhaps more readily adapted to GC systems. Furthermore, its commercial availability was cited as an asset. In 1970, West made the following prophetic comment (West, 1970),

> Factors in favor of both plasmas as gas chromatographic detectors include great versatility, reasonable selectivity and stability and, in general, a large dynamic range. They require no make-up carrier and should be inexpensive to operate. The radio frequency (and probably also the microwave) torch can be made satisfactorily, simple, convenient, and reliable.*

Today, the GC–plasma combination (particularly the GC–MIP) has become widely accepted as a detector where selectivity based on elemental composition—e. g., phosphorus-, sulfur-, mercury-, or halogen-containing pesticides is desirable. The GC–MIP is also a versatile plasma system for research purposes.

Because of the competition of DC plasma and ICPs, there has been less attention paid to the use of MIPs as "atom reservoirs" for direct solution (or, for that matter, solid) analysis. However, in 1967, Runnels and Gibson (1967) used a low wattage atmospheric pressure argon-MIP for metal analysis. Solid samples were vaporized into the discharge by deposition onto a platinum filament, which was then heated to a high temperature. Microwave power levels of about 25 W gave maximum emission intensities when samples were introduced as a volatile metal chelate or volatile inorganic salt such as a halide. Aldous *et al.* (1971) have also used a platinum-loop sampling technique to determine metals in a MIP. Lichte and Skogerboe have used a cold vapor generation technique [more commonly used for flameless AAS (Hatch and Ott, 1968)] for the determination of mercury (Lichte and Skogerboe, 1972a) and arsenic (Lichte and Skogerboe, 1972b) with a MIP. Taylor *et al.* (1970a) have used an argon-MIP to determine trace contaminants (carbon, oxygen, nitrogen, and hydrogen) in the argon. Sulfur has also

* Reprinted with permission [West (1970). *Anal. Chem.* **42**, 811]. Copyright © by the American Chemical Society.

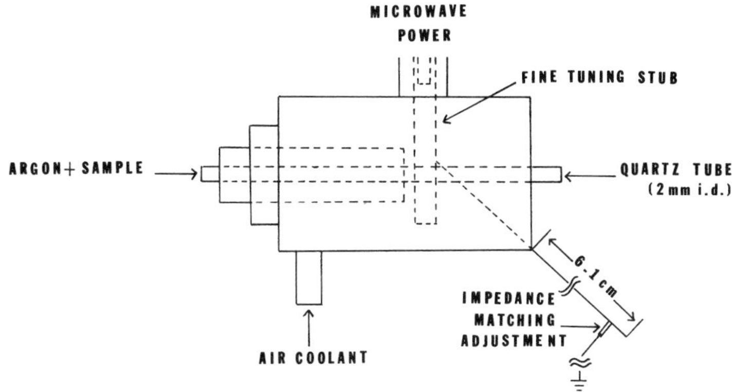

Fig. 17 Lichte and Skogerboe's modified Evensen cavity [Reprinted with permission (Lichte and Skogerboe (1973). *Anal. Chem.* **45**, 399). Copyright © by the American Chemical Society.]

been studied with an MIP (Taylor *et al.*, 1970b) using emission, absorption, and fluorescence.

Although MIPs are very easily quenched by water vapor (Sharp, 1976; Greenfield *et al.*, 1975; Busch and Vickers, 1973), several workers have devised continuous nebulization systems for these plasmas.* Lichte and Skogerboe (1973) modified an Evensen cavity, Fig. 17, and used an end-on configuration rather than the more usual transverse configuration. That change gave a more stable argon plasma, particularly when the argon was saturated with water vapor. A desolvation system (Veillon and Margoshes, 1968b) was used with this plasma. The system was stated to be convenient to use and relatively inexpensive. Skogerboe and Coleman (1976) have recently evaluated an MIP for multielement emission spectrometry. Detection limits for 12 elements studied were reported better than 0.01 μg/ml.

Unfortunately, chemical interferences in MIPs appear to be common (Sharp, 1976; Greenfield *et al.*, 1975). Classical Ca–PO_4 and Al–Ca interferences have been observed (Lichte and Skogerboe, 1973). Boumans *et al.* (1975) have compared a particular inductively coupled radio frequency plasma with a particular capacitively coupled microwave plasma. In all respects, the ICP was found superior to the microwave plasma. Although the microwave plasma was capacitively coupled (the authors did not have access to a MIP), the superiority of the ICP *for direct solution analysis* (with or without desolvation) is still clearly indicated over microwave plasma systems.

MIPs will probably find a much greater use in the future as specific elemental detectors in chromatographic systems.

* See Fallgatter *et al.* (1971), Kawaguchi *et al.* (1972), Lichte and Skogerboe (1973, 1974), Kawaguchi and Vallee (1975), Frick *et al.* (1975), and Skogerboe and Coleman (1976).

References

Aldous, K. M., Dagnall, R. M., and West, T. S. (1970). *Analyst* **95**, 417.
Aldous, K. M., Dagnall, R. M., Sharp, B. L., and West, T. S. (1971). *Anal. Chim. Acta* **54**, 233.
Amos, M. D., Bennett, P. A., and Brodie, K. G. (1970). *Resonance Lines* **2(1)**, 3.
Anacon (no date). "MECA Spectroscopy—A New Flame Analytical Technique." Anacon, Inc., Ashland, Massachusetts.
Babat, G. I., (1942). *Vestn. Elektroprom.* **2**, 1.
Babat, G. I., (1947). *J. Inst. Elec. Eng. (London)* **94**, 27.
Bache, C. A., and Lisk, D. J. (1965). *Anal. Chem.* **37**, 1477.
Bache, C. A., and Lisk, D. J. (1966). *Anal. Chem.* **38**, 1757.
Bache, C. A., and Lisk, D. J. (1967). *Anal. Chem.* **39**, 786.
Bache, C. A., and Lisk, D. J. (1971). *Anal. Chem.* **43**, 950.
Barnes, R. M., and Jarrell, R. F. (1971). Gratings and grating instruments, *in* "Analytical Emission Spectroscopy" (E. L. Grove, Ed.). Dekker, New York.
Barnett, W., and Fassel, V. A. (1968). *Spectrochim. Acta* **23B**, 643.
Belcher, R., Bogdanski, S. L., and Townshend, A. (1973). *Anal. Chim. Acta* **68**, 1.
Belcher, R., Kouimtzis, T., and Townshend, A. (1974). *Anal. Chim. Acta* **68**, 297.
Belcher, R., Bogdanski, S. L., Hendon, E., and Townshend, A. (1975). *Analyst* **100**, 522.
Bogdanski, S. L. (1973). Ph.D. Thesis, Univ. of Birmingham, England.
Boumans, P. W. J. M., and DeBoer, F. J. (1972). *Spectrochim. Acta* **27B**, 351.
Boumans, P. W. J. M., and DeBoer, F. J. (1975a). *Proc. Anal. Div. Chem. Soc.* **12**, 140.
Boumans, P. W. J. M., and DeBoer, F. J. (1975b). *Spectrochim. Acta* **30B**, 309.
Boumans, P. W. J. M., and DeBoer, F. J. (1976). *Spectrochim. Acta* **31B**, 355.
Boumans, P. W. J. M., DeBoer, F. J., Dahmen, F. J., Hoelzel, H., and Meier, A. (1975). *Spectrochim. Acta* **30B**, 449.
Boumans, P. W. J. M., VanGool, G. H., and Jansen, J. A. J. (1976). *Analyst* **101**, 585.
Braman, R. S. (1966). *Anal. Chem.* **38**, 734.
Braun, W., Peterson, N. C., Bass, A. M., and Kurylo, M. J. (1971). *J. Chromatogr.* **55**, 237.
Britske, M. E., Borisov, V. M., and Sukach, Y. S. (1967). *Zavod. Lab.* **33**, 252.
Broida, H. P., and Chapman, M. W. (1958). *Anal. Chem.* **30**, 2049.
Browner, R. F. (1974). *Analyst* **99**, 617.
Bunsen, R., and Kirchhoff, G. (1860). *Ann. Phys.* **110**, 160.
Bunsen, R., and Kirchhoff, G. (1861a). *Ann. Phys.* **113**, 337.
Bunsen, R., and Kirchhoff, G. (1861b). *Bull. Soc. Chim.* 70.
Busch, K. W., and Vickers, T. J. (1973). *Spectrochim. Acta* **28B**, 85.
Busch, K. W., Howell, N. G., and Morrison, G. H. (1974a). *Anal. Chem.* **46**, 575.
Busch, K. W., Howell, N. G., and Morrison, G. H. (1974b). *Anal. Chem.* **46**, 2074.
Butler, C. C., Kniseley, R. N., and Fassel, V. A. (1975). *Anal. Chem.* **47**, 825.
Champion, P., Pellet, H., and Grenier, M. (1873). *C. R. Acad. Sci. Paris* **76**, 707.
Chapman, J. F., Dale, L. S., and Whittem, R. N. (1973). *Analyst* **98**, 529.
Christian, G. D., and Feldman, F. J. (1971). *Appl. Spectrosc.* **25**, 660.
Collins, A. G. (1967). *Appl. Spectrosc.* **21**, 16.
Corcoran, F. L., Jr., and Elliott, W. G. (1972). Presented at the *Pittsburgh Conf. Anal. Chem. Appl. Spectrosc., 23rd, Cleveland, Ohio* March.
Corcoran, F. L., Jr., Keliher, P. N., and Wohlers, C. C. (1972). *Am. Lab.* **4(3)**, 51.
Cresser, M. S., and Keliher, P. N. (1970a). *Am. Lab.* **2(8)**, 8.
Cresser, M. S., and Keliher, P. N. (1970b). *Am. Lab.* **2(11)**, 21.
Cresser, M. S., Joshipura, P. B., and Keliher, P. N. (1970). *Spectrosc. Lett.* **3**, 267.
Cresser, M. S., Keliher, P. N., and Wohlers, C. C. (1973a). *Anal. Chem.* **45**, 111.

Cresser, M. S., Keliher, P. N., and Kirkbright, G. F. (1973b). *Sel. Ann. Rev. Anal. Sci.* **3**, 139.
Crider, W. J. (1965). *Anal. Chem.* **37**, 1770.
Crookes, W. (1861). *Chem. News* **3**, 193.
Dagnall, R. M., Thompson, K. C., and West, T. S. (1967). *Analyst* **92**, 506.
Dagnall, R. M., Thompson, K. C., and West, T. S. (1968). *Analyst* **93**, 72.
Dagnall, R. M., Thompson, K. C., and West, T. S. (1969a). *Analyst* **93**, 518.
Dagnall, R. M., Thompson, K. C., and West, T. S. (1969b). *Analyst* **93**, 643.
Dagnall, R. M., Smith, D. J., Thompson, K. C., and West, T. S. (1969c). *Analyst* **94**, 871.
Dagnall, R. M., Pratt, S. J., West, T. S., and Deans, D. R. (1970a). *Talanta* **17**, 1009.
Dagnall, R. M., Smith, D. J., and West, T. S. (1970b). *Anal. Lett.* **3**, 475.
Dagnall, R. M., Smith, D. J., West, J. S., and Greenfield, S. (1971). *Anal. Chem. Acta* **54**, 397.
Dagnall, R. M., Silvester, M. D., West, T. S., and Whitehead, P. (1972a). *Talanta* **19**, 1226.
Dagnall, R. M., West, T. S., and Whitehead, P. (1972b). *Anal. Chim. Acta* **60**, 25.
Dickinson, G. W., and Fassel, V. A. (1969). *Anal. Chem.* **41**, 1021.
Dunken, H., and Pforr, G. (1966). *Z. Chem.* **6**, 278.
Dunken, H., Pforr, G., and Mikkeleit, W. (1963). *Z. Chem.* **3**, 196.
Elliott, W. G. (1971). *Am. Lab.* **3(8)**, 45.
Elliott, W. G. (1974). Presented at the *Pittsburgh Conf. Anal. Chem. Appl. Spectrosc., 25th, Cleveland, Ohio March.*
Elliott, W. N., and Mostyn, R. A. (1971). *Analyst* **96**, 452.
Epstein, M. S. (1976). Ph.D. Thesis, Univ. of Maryland.
Epstein, M. S., and O'Haver, T. C. (1975). *Spectrochim. Acta* **30B**, 135.
Fallgatter, K., Svoboda, V. S., and Winefordner, J. D. (1971). *Appl. Spectrosc.* **25**, 347.
Fassel, V. A. (1973). Electrical plasma spectroscopy, *Colloq. Spectrosc. Int., 16th, Heidelberg, 1971.* Adam Hilger, London.
Fassel, V. A., and Dickinson, G. W. (1968). *Anal. Chem.* **40**, 247.
Fassel, V. A., and Kniseley, R. N. (1974a). *Anal. Chem.* **46**, 1110A.
Fassel, V. A., and Kniseley, R. N. (1974b). *Anal. Chem.* **46**, 1155A.
Fassel, V. A. Peterson, C. A., Abercrombie, F. N., and Kniseley, R. N. (1976). *Anal. Chem.* **48**, 516.
Faust, H. W. (1971). Prism systems, spectrographs, and spectrometers, *in* "Analytical Emission Spectroscopy" (E. L. Grove, Ed.). Dekker, New York.
Fricke, H. L., Rose, O., and Caruso, J. F. (1975). *Anal. Chem.* **47**, 2018.
Gaydon, A. J. (1957). "The Spectroscopy of Flames." Wiley, New York.
Goto, H., and Atsuya, I. (1967). *Z. Anal. Chem.* **225**, 121.
Gouy, A. (1879). *Ann. Chim. Phys.* **18**, 5.
Greenfield, S. (1965). *Proc. Soc. Anal. Chem.* **2**, 111.
Greenfield, S., and Smith, P. B. (1972). *Anal. Chim. Acta* **59**, 341.
Greenfield, S., Jones, I. L., and Berry, C. T. (1964). *Analyst* **89**, 713.
Greenfield, S., Smith, P. B., Breeze, A. E., and Chilton, N. M. D. (1968). *Anal. Chim. Acta* **41**, 385.
Greenfield, S., McGeachin, H. McD., and Smith, P. B. (1975). *Talanta* **22**, 553.
Greenfield, S., McGeachin, H. McD., and Smith, P. B. (1976a). *Talanta* **23**, 1.
Greenfield, S., McGeachin, H. McD., and Smith, P. B. (1976b). *Anal. Chim. Acta* **84**, 67.
Grove, E. L. (ed.) (1971). "Analytical Emmission Spectroscopy." Marcel Dekker, New York.
Haarsma, J. P. S., deJong, G. L., and Agterdenbos, J. (1974). *Spectrochim. Acta* **29B,** 1.
Hatch, W. R., and Ott, W. L. (1968). *Anal. Chem.* **40**, 2085.
Heemstra, R. J. (1970). *Appl. Spectrosc.* **24**, 568.
Hieftje, G. M., and Sydor, R. J. (1972). *Appl. Spectrosc.* **26**, 624.
Hingle, D. N., Kirkbright, G. F., and West, T. S. (1968). *Analyst* **93**, 522.

Hingle, D. N., Kirkbright, G. F., and West, T. S. (1969). *Analyst* **94**, 864.
Hingle, D. N., Kirkbright, G. F., Sargent, M., and West, T. S. (1970). *Lab. Pract.* 1069.
Hoare, H. C., and Mostyn, R. A. (1967). *Anal. Chem.* **39**, 1153.
Hobbs, R. S., Kirkbright, G. F., Sargent, M., and West, T. S. (1968). *Talanta* **15**, 997.
Howell, N. G., Ganjei, J. D., and Morrison, G. H. (1976). *Anal. Chem.* **48**, 319.
Issaq, H. J., and Morgenthaler, L. P. (1975a). *Anal. Chem.* **47**, 1661.
Issaq, H. J., and Morgenthaler, L. P. (1975b). *Anal. Chem.* **47**, 1668.
IUPAC (1977). Nomenclature, symbols, units, and their usage in spectrochemical analysis—III, International Union of Pure and Applied Chemistry, published in *Appl. Spectrosc.* **31**, 348.
Jahn, R. E. (1961). *Proc. Int. Conf. Ionization Phenomena, 5th, Munich.*
Janssen, M. J. (1870). *C. R. Acad. Sci. Paris* **71**, 626.
Karicki, S. S., and Corcoran, F. L., Jr. (1973). *Appl. Spectrosc.* **27**, 41.
Kawaguchi, H., and Vallee, B. L. (1975). *Anal. Chem.* **47**, 1029.
Kawaguchi, H., Hasegawa, M., and Mizuike, A. (1972). *Spectrochim. Acta* **27B**, 205.
Keirs, C. D., and Vickers, T. J. (1977). *Appl. Spectrosc.* **31**, 273.
Keliher, P. N. (1976). *Res./Dev.* **27(6)**, 26.
Keliher, P. N. (1978). Unpublished results.
Keliher, P. N., and Wohlers, C. C. (1976). *Anal. Chem.* **48**, 333A.
Kerber, J. D., Barnett, W. B., and Kahn, H. L. (1970). *At. Abstr. Newsletter* **9**, 39.
Kirkbright, G. F., and Vetter, S. (1972). *Spectrochim. Acta* **27B**, 351.
Kirkbright, G. F., and Ward, A. F. (1974). *Talanta* **21**, 1145.
Kirkbright, G. F., and West, T. S. (1968). *Appl. Opt.* **7**, 1305.
Kirkbright, G. F., Semb, A., and West, T. S. (1967). *Talanta* **14**, 1011.
Kirkbright, G. F., Semb, A., and West, T. S. (1968). *Spectrosc. Lett.* **1**, 7.
Kirkbright, G. F., Sargent, M., and West, T. S. (1969). *Talanta* **16**, 245.
Kirkbright, G. F., Ward, A. F., and West, T. S. (1972). *Anal. Chim. Acta* **62**, 241.
Kirkbright, G. F., Ward, A. F., and West, T. S. (1973). *Anal. Chim. Acta* **64**, 353.
Kniseley, R. N., Fassel, V. A., and Butler, C. C. (1973). *Clin. Chem.* **19**, 807.
Korolev, V. V., and Vainshstein, E. E. (1959). *J. Anal. Chem. (USSR)* **14**, 658.
Kranz, E. (1964). "Emissionsspektroskopie," p. 160. Akademi Verlag, Berlin.
Kranz, E. (1968). *Proc. Colloq. Spectrosc. Int., 14th*, p. 697. Adam Hilger, London.
Kroeten, J. J., Moody, H. W., and Parsons, M. L. (1970). *Anal. Chim. Acta* **52**, 101.
Larson, G. F., Fassel, V. A., Scott, R. H., and Kniseley, R. N. (1975). *Anal. Chem.* **47**, 238.
Lichte, F. E., and Skogerboe, R. K. (1972a). *Anal. Chem.* **44**, 1321.
Lichte, F. E., and Skogerboe, R. K. (1972b). *Anal. Chem.* **44**, 1480.
Lichte, F. E., and Skogerboe, R. K. (1973). *Anal. Chem.* **45**, 399.
Lichte, F. E., and Skogerboe, R. K. (1974). *Appl. Spectrosc.* **28**, 354.
Lundegardh, H. (1928). *Ark. Kemi. Mineral Geol.* **10A**, No. 1.
Lundegardh, H. (1930). *Z. Phys.* **66**, 109.
Maines, I. S., Mitchell, D. G., Rankin, J. M., and Bailey, B. W. (1972). *Spectrosc. Lett.* **5**, 251.
Margoshes, M., and Scribner, B. F. (1959). *Spectrochim. Acta* **15**, 138.
Marinkovic, M., and Vickers, T. J. (1971). *Appl. Spectrosc.* **25**, 319.
Mavrodineanu, R., and Boiteaux, H. (1965). "Flame Spectroscopy." Wiley, New York.
McLean, W. R., Stanton, D. L., and Penketh, G. E. (1973). *Analyst* **98**, 432.
McCormack, A. J., Tong, S. C., and Cooke, W. D. (1965). *Anal. Chem.* **37**, 1470.
McCrea, P. F., and Light, T. S. (1967). *Anal. Chem.* **39**, 1731.
McGucken, W. (1969). "Nineteenth-Century Spectroscopy." Johns Hopkins Press, Baltimore, Maryland.
Merchant, P., Jr., and Veillon, C. (1974). *Anal. Chim. Acta* **70**, 17.
Mitscherlich, A. (1862). *Ann. Phys.* **116**, 499.

Mitscherlich, A. (1864). *Ann. Phys.* **121**, 459.
Montaser, A., and Fassel, V. A. (1976). *Anal. Chem.* **48**, 1490.
Morrison, G. H., and Talmi, Y. (1970). *Anal. Chem.* **42**, 809.
Murdick, D. A., Jr., and Piepmeier, E. H. (1974). *Anal. Chem.* **46**, 678.
Natusch, D. F. S., and Thorpe, T. M. (1973). *Anal. Chem.* **45**, 1184A.
Newton, I. (1704). "Opticks, or a Treatise of the Reflections, Refractions, Inflections, and Colours of Light." Reissued by Dover Publications, New York, 1952.
Nixon, D. E., Fassel, V. A., and Kniseley, R. N. (1974). *Anal. Chem.* **46**, 210.
O'Haver, T. C. (1972). *J. Chem. Ed.* **49**, A-131.
Olsen, K. W., Haas, W. J., and Fassel, V. A. (1977). *Anal. Chem.* **49**, 632.
Owen, L. E. (1961). *Appl. Spectrosc.* **15**, 150.
Parsons, M. L. (1969). *Anal. Lett.* **2**, 229.
Pearce, W. J. (1961). Plasma-jet temperature measurements, *in* "Optical Spectrometric Measurement of High Temperatures" (P. J. Dickerman, ed.). Univ. of Chicago Press, Chicago, Illinois.
Pickett, E. E., and Koirtyohann, S. R. (1969). *Anal. Chem.* **41**, 28A.
Rains, T. C., and Menis, O. (1974). *Anal. Lett.* **7**, 715.
Ramsey, W. (1895). *J. Chem. Soc.* **67**, 1107.
Reed, T. B. (1961a). *J. Appl. Phys.* **32**, 821.
Reed, T. B. (1961b). *J. Appl. Phys.* **32**, 2534.
Reed, T. B. (1962). *Int. Sci. Technol.* **6**, 42.
Reed, T. B. (1963). *Proc. Nat. Electron Conf.* **19**, 654.
Rippetoe, W. E., and Vickers, T. J. (1975). *Anal. Chem.* **47**, 2082.
Rippetoe, W. E., Johnson, E. R., and Vickers, T. J. (1975). *Anal. Chem.* **47**, 436.
Robinson, J. W., and Smith, V. J. (1966). *Anal. Chim. Acta* **36**, 489.
Roscoe, H. E., and Clifton, R. B. (1862). *Proc. Lit. Phil. Soc., Manchester* **2**, 227.
Runnels, J. H., and Gibson, J. H. (1967). *Anal. Chem.* **39**, 1398.
Salet, G. (1869). *Bull. Soc. Chim. Fr.* **11**, 302.
Salet, G. (1871). *C. R. Acad. Sci. Paris.* **73**, 559.
Salet, G. (1873). *Ann. Chim. Phys.* **28**, 5.
Scott, R. H., Fassel, V. A., Kniseley, R. N., and Nixon, D. E. (1974). *Anal. Chem.* **46**, 75.
Scribner, B. F., and Margoshes, M. (1962). *Colloq. Spectrosc. Int., 9th* **II**, 309. GAMS, Paris.
Serin, P. A., and Ashton, K. H. (1964). *Appl. Spectrosc.* **18**, 166.
Serravallo, F. A., and Risby, T. H. (1974). *J. Chromatogr. Sci.* **12**, 585.
Serravallo, F. A., and Risby, T. H. (1975). *Anal. Chem.* **47**, 2141.
Serravallo, F. A., and Risby, T. H. (1976). *Anal. Chem.* **48**, 673.
Sharp, B. L. (1976). *Sel. Ann. Rev. Anal. Sci.* **4**, 37.
Skogerboe, R. K., and Coleman, G. N. (1976). *Appl. Spectrosc.* **30**, 504.
Smith, V. J., and Robinson, J. W. (1969). *Anal. Chim. Acta* **48**, 391.
Smith, V. J., and Robinson, J. W. (1970a). *Anal. Chim. Acta* **49**, 161.
Smith, V. J., and Robinson, J. W. (1970b). *Anal. Chim. Acta* **49**, 417.
Smithells, A., and Ingle, H. (1892). *Trans. Chem. Soc.* **61**, 204.
Snelleman, W., Rains, T. C., Yee, K., Cook, H., and Menis, O. (1970). *Anal. Chem.* **42**, 394.
Syty, A. (1971). *Anal. Lett.* **4**, 531.
Syty, A. (1973). *At. Abstr. Newsletter* **12**, 1.
Syty, A., and Dean, J. A. (1968). *Appl. Opt.* **7**, 1331.
Szivek, J., Jones, C., Paulson, E. J., and Valberg, L. E. (1968). *Appl. Spectrosc.* **22**, 195.
Talbot, W. H. F. (1826). *Edinburgh J. Sci.* **5**, 77.
Talbot, W. H. F. (1834). *Phil. Mag.* **4**, 112.
Talmi, Y., and Andrew, A. W. (1974). *Anal. Chem.* **46**, 2122.

Talmi, Y., and Bostick, D. T. (1975). *Anal. Chem.* **47**, 2145.
Talmi, Y., and Norvell, V. E. (1975). *Anal. Chem.* **47**, 1510.
Taylor, H. E., Gibson; J. H., and Skogerboe, R. K. (1970a). *Anal. Chem.* **42**, 876.
Taylor, H. E., Gibson, J. H., and Skogerboe, R. K. (1970b). *Anal. Chem.* **42**, 1569.
Teclu, N. J. (1891). *Prakt. Chem.* **44**, 246.
Ure, A. M., and Berrow, M. L. (1970). *Anal. Chim. Acta* **52**, 247.
Valente, S. E., and Schrenk, W. G. (1970). *Appl. Spectrosc.* **24**, 197.
Veillon, C., and Margoshes, M. (1968a). *Spectrochim. Acta* **23B**, 503.
Veillon, C., and Margoshes, M. (1968b). *Spectrochim. Acta* **23B**, 553.
Webb, M. S. W., and Wildy, P. C. (1963). *Nature* (*London*) **198**, 1218.
Weiss, R. (1954). *Z. Phys.* **138**, 170.
West, C. D. (1970). *Anal. Chem.* **42**, 811.
Wendt, R. H., and Fassel, V. A. (1965). *Anal. Chem.* **37**, 920.
Wendt, R. H., and Fassel, V. A. (1966). *Anal. Chem.* **38**, 337.
Yamamoto, Y. (1962). *Jpn. J. Appl. Phys.* **1**, 235.
Yamamoto, Y. (1963). *Jpn. J. Appl. Phys.* **2**, 62.

Index

G

J

K

N

Q

R

S